Springer Tracts in Modern Physics
Volume 234

Springer Tracts in Modern Physics

Springer Tracts in Modern Physics provides comprehensive and critical reviews of topics of current interest in physics. The following fields are emphasized: elementary particle physics, solid-state physics, complex systems, and fundamental astrophysics.
Suitable reviews of other fields can also be accepted. The editors encourage prospective authors to correspond with them in advance of submitting an article. For reviews of topics belonging to the above mentioned fields, they should address the responsible editor, otherwise the managing editor.
See also springer.com

Dedicated in gratitude to

Hildegard and Margaret.

Acknowledgements

It is our pleasure to express our gratitude to Prof. Dimitri A. Varshalovich for his interest and communications in the initial stage of our work. In a way, his impressive work provided the stimulus and touchstone for the subject of this book.

It is also obvious that, without the generous continuing support and infrastructure of our home institutions through many years this book would never have been compiled. One of us, E.B., wishes to express his appreciation of encouragement and friendship extended by Prof. Helmut Rauch as director of the Atominstitut Vienna, while S.W.L. is grateful to A.D. Taylor and N. Gidopoulos (ISIS Facility) and G. Materlik (Diamond Light Source Ltd.) for ongoing support.

Editorial assistance

We want to express our gratitude to the staff of Springer Verlag who, in many ways, have helped to create this book, and to acknowledge, in this respect, the support received from Adelheid Duhm and Jaqueline Lenz, who suggested technical adaptations of the final manuscript.

TeXnical assistance

In preparing this work we have benefited a great deal from the contributions of many people, too many to name them all. The foundation was laid by Donald E. Knuth who, driven by his quest for beauty in fonts, type-setting and book-printing, has presented the world of science with TeX.

This book, in particular, would have been very difficult to write without the PS-Tricks package of Timothy van Zandt which was used extensively to create the diagrams. We have used the TeXLive distribution and LaTeX2e, and the user interface was WinShell written and supported by Ingo de Boer.

Preface

Lectures on Quantum Mechanics will, in most cases, stop at Clebsch–Gordan coefficients and, perhaps, mention the Wigner–Eckart Theorem. Any student venturing beyond this limit into the theory of atomic states will, of course, be immediately confronted with further coupling coefficients of higher order.

Once familiar with these concepts and quantities, it is quite natural to seek and explore the conditions and options available. Drawing from our own experiences, we realized that a text like this introduction could encourage and facilitate getting deeper into the subject. However, an introductory chapter presents notation, relations between and symmetry properties of, more or less, familiar coupling coefficients as a bridge-head into unknown territory.

The graphical representations and methods for treating and manipulating higher-order coupling coefficients represent the main topic of this book. Although the graphical theory presented here has been around for some time, it has, perhaps, not received the attention it merits.

Starting from the simplest building blocks we gradually introduce more complex elements. Initially the emphasis is on the equivalence between algebraic and graphical representations, but later we shift to solving problems with the help of graphical methods. The solutions to these problems are derived in a step-by-step manner to provide the option of self-study. Most of the results are known from the literature, but it would be negligent not to demonstrate that the power of the graphical method enables one to transcend the set of known results, as we do in a few cases.

The final, and rather difficult, chapter is devoted to an application of the graphical theory within a calculation relevant in the area of atomic physics. There we demonstrate the elegance and superiority of the graphical method in deriving a result for which the algebraic method is only a theoretical option.

This book contains, perhaps, less written text than might be expected but many more figures. It is thus obvious that the diagrams represent an essential part of the presentation and merit detailed study. We have tried hard to provide enough explanatory figure captions to soften the impact of complex diagrams.

Only the future will tell whether we were successful in compiling a useful and reader-friendly book, as was our intention. Although great care was exercised in the preparation of text and diagrams, we have to accept the blame for all errors escaping unnoticed in the final copy.

Vienna and Wantage
June 2009

Ewald Balcar
Stephen W. Lovesey

PS: Looking back, it could be said, this project may have started a long time ago, when we had first contact with coupling coefficients in quantum mechanics. For one of us (E.B.) it was his teacher, mentor and friend Prof. Otto Hittmair, who introduced him to these strange entities, that govern even stranger nuclear reactions. In later years both authors had the privilege of working as postdocs with Walter Marshall (later Lord Marshall of Goring), and again, we encountered these vector coupling coefficients as essential ingredients of many-electron wave functions describing magnetic neutron scattering.

On and off through the years, due to our mutual influence and collaboration, familiarity with the vector coupling coefficients of higher rank increased and culminated in the developments of a program[1] for the algebraic values of these coefficients as a tool for theoretical physicists working in this field.

It was only in recent past, when the focus of our work shifted from neutrons to x-rays, that we became aware of the potential and promise offered by mastering the graphical technique for manipulating vector coupling coefficients. In order to profit from this potential one has, however, to suffer the labours of the plain acquiring enough skills before attempting to climb. The urge to facilitate the hard work by providing a 'how-to' manual for mastering the graphical theory of angular momentum supplied the primary momentum for assembling the material collected in this book.

[1] In order to obtain this Windows program please contact one of the authors by e-mail.

Contents

1

Introduction

1.1 Historical Background

The theory of angular momentum in quantum mechanics is central to atomic and nuclear physics. In applications of the theory, one deals with methods of coupling and recoupling states described by quantum numbers which contain, in most cases, angular momenta (spin and orbital) and and their projections. Pioneer work on the subject has been done during the first part of the 20th century and a useful collection of reprints can be found in Biedenharn and Van Dam [1]. Of special note are applications of the theory to the interpretation of atomic spectra underpinned by the definitive contribution of Condon and Shortley [2] and, later, seminal work by Racah [3] that now is a bedrock of atomic and nuclear physics.

In the second half of the 20th century a rising number of books contributed to the continuous evolution of the theory of angular momentum, see Edmonds [4], Rose [5] or Brink and Satchler [6], and its application to atomic spectroscopy, Judd [7], and the nuclear shell-model de Shalit and Talmi [8], to cite just an incomplete selection.

The rising power of electronic computing provided the basis for the creation of extensive tabulations of Rotenberg et al. [9]. At the same time the increased complexity of algebraic calculations within the theory of angular momentum lead to attempts to assist these calculations by the development of a graphical technique.

Work by Yutsis[1] et al. using graphical representations was first published in Russian at the Lithuanian Academy of Science in 1960 and soon translated into English, Yutsis, Levinson and Vanagas [10]. This original source may be quite difficult to obtain. The concepts of graphical representation were soon extended by El Baz and Castel [11] with detailed suggestions of possible physical applications. However, a more recent book, Varshalovich, Moskalev and Khersonskii [12], provides a comprehensive reference work for all aspects of the

[1] Note that Jucys is an alternative spelling used instead of Yutsis.

E. Balcar, S.W. Lovesey, *Introduction to the Graphical Theory of Angular Momentum*, STMP 234, 1–6, DOI 10.1007/978-3-642-03118-2_1,

quantum theory of angular momentum including two chapters of an extensive presentation of the graphical technique. A major part of our introduction to the use of graphical methods is based on the results given in [12]. Although our graphical representations deviate from the diagrams in [12] in some details this should not pose a difficulty. A short introduction to the graphical method can also be found in Brink and Satchler [13].

1.2 Intent and Motivation

In a calculation of a matrix element for a transition amplitude, say, Clebsch–Gordan coefficients, from application of the Wigner–Eckart Theorem, arise together with higher-order coefficients, in associated reduced matrix elements. Typically, the raw result of the calculation can be re-expressed with advantage. One aims to make visible the physical content of the product of coefficients by exposing an effective quantum-mechanical operator that encapsulates the essence of the physical processes. This enlightenment comes from an introduction of additional variables that recouple initial variables. In practice, recoupling is actually achieved by identities for products of coefficients, and an extensive collection of identities has been compiled in the past few decades. Further progress with this type of application is more likely achieved with powerful graphical techniques that are the subject of this book. The techniques permit both an analysis of expressions into physically relevant components and the construction of identities to execute a recoupling agenda.

To express the advantage of the graphical method in a more concrete form, the usual approach would be to split all terms in a sum into an, often considerable, number of basic elements which are all linked by further summations, and then attempt to proceed to a new form by regrouping, manipulations and eliminations. The complexity of this undertaking is partly due to the symmetry properties of the basic elements. These involve, in view of the physical background, angular momenta fulfilling triangle conditions. The graphical representation of the basic elements provides an ideal tool to visualize and control the triangle conditions and to combine the angular momenta in diagrams, also called coupling diagrams.

If an exact solution of a problem is not of prime importance, the graphical method can, very quickly, indicate the general properties of the solution, usually on the back of an envelope or paper napkin if physicists are involved.

This book has been compiled by two physicists aiming to provide familiarity with the concepts and techniques of the graphical method. Our primary intent is to discuss enough worked examples such that the reader gradually becomes proficient in applying the graphical method. It is perhaps worthwhile to note that, the graphical technique enables one to transcend beyond published results and to derive new relations as demonstrated in a few places in this book.

The graphical theory offers a powerful method to treat problems involving $3jm$- and $3nj$-symbols for which, except for the simplest examples, the equivalent algebraic procedures may not be applied easily. Because the two descriptors $3jm$ and $3nj$ seem, at first sight, quite similar it is worth noting that $3jm$ refers to a mathematical quantity defined by three angular momenta j and corresponding projections m. With loss of precision, these symbols are often in the literature just named $3j$-symbols. We use, therefore, $3jm$ to avoid confusion with the second type of symbol described by the term $3nj$. Here, the number n takes the values 1,2,3,. . . , where $3n$ is the number of angular momentum variables in the corresponding symbols and diagrams.

1.3 Layout and Contents

In the following chapters we endeavour to guide the reader through a step by step introduction to the graphical technique, starting from the simplest building blocks and gradually proceeding to more complicated graphs and diagrams.

It is, of course, of advantage to be familiar with the mathematical theory of the coupling coefficients and related quantities which is to be complemented by the graphical method. The following material recapitulates, therefore, some well-known properties of $3jm$-symbols.

In Chap. 2 we provide the reader with a collection of definitions, relations and formulae relevant to $3jm$-symbols and frequently used in the later development of the subject. Only a minimal discussion of $6j$- and $9j$-symbols and their relation to the coupling of three and four angular momenta, respectively, is given. Further details of the properties of these elements of angular momentum theory are reserved for later chapters.

In the subsequent chapters we will, as long as possible, display the algebraic elements and expressions along with the graphical diagrams to define the elements of the graphical theory by comparison with their algebraic counterparts. An attempt at proving by algebra some complicated identities might be motivation to invest in mastering the powerful graphical technique that makes light work of such proofs.

In Chap. 3 we present the basic building blocks and tools of the graphical theory. The basic concepts, nodes and angular momentum lines, are employed to represent a $3jm$-symbol by a node with three attached angular momentum lines, where the direction of a line is linked to the sign of the corresponding projection.

To facilitate treatment of the following examples, we introduce, at an early stage, five special diagrams representing simple factors, leaving their derivation for a later stage. Nodes have a node sign that derives from the order of columns in the corresponding $3jm$-symbol. If the lines attached to a node change their cyclic order then the node sign changes, and this is equivalent to an exchange of columns in the $3jm$-symbol. This has to be contrasted by the explicit change of node sign without rearranging the angular momentum lines. The simplest examples refer to $3jm$-symbols with one angular momentum equal to zero. At this point we advance the concepts of zero-line and two-line nodes which are closely related.

An essential element of the theory is the equivalence of a sum over a projection, present in two $3jm$-symbols, and the corresponding angular momentum line connecting the two nodes. Equally important and frequently used is the option of reversing the direction of an internal angular momentum line. For completeness, the chapter includes also the discussion of line reversal for an external line.

Equipped with these tools we devote Chap. 4 to a discussion of identities obeyed by $3jm$-symbols. One of these identities involves not only summation over a projection but also over the angular momentum itself. This example provides us with the line contraction tool which is most useful for the reduction of angular momentum lines. A new symbol found in this chapter is the simplest closed diagram, the $3j$-symbol. This symbol is the first of the group of closed diagrams called $3nj$-symbols. Depending on the value of n we have $3j$-, $6j$-, $9j$-symbols and symbols of even higher rank.

Chapters 5 and 6 contain definitions and transformations of $6j$- and $9j$-symbols, respectively. We discuss some symmetry properties and special cases with zero angular momentum. Both $3nj$-symbols are of considerable importance in the theory of atomic spectra. The $6j$-symbols represent essential properties of reduced matrix elements of many-electron states. Transformations between different coupling systems, e.g. from Russell–Saunders or SL-coupling to jj-coupling, are defined by $9j$-symbols.

All the manipulations and conventions established up to this point are then reformulated into general rules for treating and transforming diagrams and collected in Chap. 7. This chapter repeats, and then generalizes results and procedures encountered in preceding chapters. The central emphasis is put on methods of separation that allow to split a complicated diagram into smaller symbols. The basic building blocks are the treatment of single lines and the contraction of two angular momentum lines. From these two procedures we are able to derive the conditions for three-, four- and five-line separations. In all cases, the rules for a separation may also be applied in reverse direction and describe the joining of two diagrams.

Equipped with an arsenal of tools and prescriptions for the manipulation of diagrams by deformation, transformation, line contraction, separation etc., we devote our attention in Chaps. 8 and 9 to the treatment of closed and open diagrams, respectively.

The central objective in the presentation of Chap. 8 concerns the summation over one angular momentum in various $3nj$-symbols. Note that, this summation is on top of the summations over all projections implicit to the connecting lines between nodes. The problems are arranged in chapters of rising complexity, starting with sums of single $3nj$-symbols and then progressing to sums of two and three $3nj$-symbols. Each part starts with an algebraic expression to be evaluated. Assuming that the result cannot be derived by algebraic methods, we take recourse to the graphical theory and obtain the result by diagrammatic means. With a few exceptions, these results are, of course, not new and can be found in the literature. Our emphasis lies on the in-between steps for solving the problem by graphical methods. The reader is, thus, encouraged to use the posed problems as exercises for self-study. It may well turn out that the graphical steps described here are not the only way to arrive at a certain solution. In Chap. 8 we are also confronted with the fact that a sum may be performed, but the result is not simple but leads instead to a $3nj$-symbol of higher order, and two kinds of $12j$-symbols are encountered.

The same arrangement of posing an algebraic problem for which a solution is obtained via graphical methods is used in Chap. 9, which deals with open diagrams. The diagrams have, therefore, external lines attached to only one node, in contrast to internal lines which have both ends attached to a node. In many cases the result is found as a transformation of the original expression. Often we will proceed by separating a closed diagram from the sub-diagram containing the external lines.

Open diagrams are, in a way, more general than closed diagrams because we know from atomic physics that any matrix element of a spherical tensor operator can be written as a product between a $3jm$-symbol and a factor that is independent of the projections (Wigner–Eckart Theorem). In graphical terms we say that, a node with three lines, the $3jm$-symbol, accompanies the closed diagram representing the reduced matrix element.

Since open diagrams involving two $3jm$-symbols have been exhaustively discussed in the introductory chapters, we focus here on sums involving three or more $3jm$-symbols. A new formula for a simple product of two $3jm$-symbols is presented in Chap. 9 and this formula is then verified by deriving from it a result given in a preceding chapter.

1.4 Application of the Graphical Theory

Before entering into a discussion of the contents of Chap. 10 we highlight a few, necessarily incomplete, steps describing the passage of the graphical theory into the areas of physics and mathematics.

Very early on Briggs [14] and also El Baz and Castel [11] presented most ambitious attempts of applying graphical methods to the *complete* calculation of matrix elements of various interaction operators. Although many diagrams will be familiar from this book, the notation there covers additional symbols for Clebsch–Gordan coefficients and even fractional parentage coefficients and is thus more comprehensive and more complicated compared to what we use here.

In the books by Judd [7] and Cowan [15], both directed at atomic spectroscopy, there are only few diagrams and hints of the graphical theory because all results could be derived algebraically. However, in the late eighties of the 20th century the definitive work on the *Quantum Theory of Angular Momentum* by Varshalovich et al. [12] contains an extensive presentation of the rules and methods of the graphical approach, together with a wealth of formulae derived with this technique. At the same time work by Thole and van der Laan [16] demonstrates the use of $12j$-symbols and the application of graphical methods.

The use of the graphical method is evident in several publications by van der Laan and Thole [17–22] and in papers by Carra et al. [23–26]. The role of Theo Thole in propagating the diagrammatic technique is emphasized by Brouder and Brinkmann [27]. The latter publication also displays diagrams of coupling coefficients that have caught the attention of mathematicians with interest in graph theory. As examples we cite a few publications [28–31] in which cubic graphs or Yutsis's graphs, as they are also called, are used in the context of calculating or reducing general recoupling coefficients.

Resuming our discussion of the contents of Chap. 10, the graphical method is applied to a realistic example from the theory of resonant x-ray scattering. This, unfortunately, requires a certain amount of familiarity with the theory of atomic spectra and with the properties of spherical tensor operators and their many-electron matrix elements. The basic concepts and definitions for the coupling of tensor operators and the structure of the reduced matrix elements of unit tensor operators are collected in Appendix A.

The specific problem treated in Chap. 10 concerns a product of two matrix elements of the electric dipole operator. This product is part of the scattering amplitude describing resonance enhanced x-ray scattering. From a graphical point of view, the reduced matrix elements are represented by $12j$-symbols containing the angular momentum quantum numbers of the atomic states in question. Resonance-enhanced x-ray scattering is a second-order process in which a photo-absorbing electron is excited to an intermediate state and returns by emission to the original electron shell. Thus, the starting point for the graphical treatment is defined as a product of two dipole matrix elements, and the corresponding diagrams consist of two $3jm$- and two $12j$-symbols.

The main target in Chap. 10 is the demonstration by graphical techniques, that the product of matrix elements can be replaced by a *single* matrix element of an *equivalent* atomic tensor operator. This significant result elucidates several selection rules governing the underlying electronic process. Three cases of decreasing complexity are considered which differ by the amount of physical information retained from the intermediate state visited by the electron. In all three cases the diagrammatic transformations are able to reach the characteristic structure defining the equivalent atomic tensor in question. The large number of figures in this chapter may give the impression of great difficulty, but it is, in fact, due to our documenting every single step in the transformation process. We hope that passing from figure to figure with few steps in the development will facilitate recognition and appreciation. Of course, we cannot deny that a prior knowledge of the properties of unit tensor operators is helpful in directing the transformations to the desired result.

2

Properties of $3jm$- and $3nj$-Symbols

2.1 Properties of $3jm$-Symbols

Basic elements in the theory of angular momentum are the Clebsch–Gordan coefficients for coupling two states characterized by j_1, m_1 and j_2, m_2 into a new state with quantum numbers J, M. The numbers j and their projections, or magnetic quantum numbers[1], m have integer or half-odd integer values. Hence, in an exponent to (-1) the quantum numbers must combine to integers as in $j \pm m$ or $j_1 + j_2 + J$, for example. The notation used for the corresponding Clebsch–Gordan coefficient is $(j_1 m_1 j_2 m_2 | JM)$ and it can be non-zero only if j_1, j_2, J fulfil the triangle condition $\Delta\{j_1, j_2, J\}$ equivalent to the condition $j_1 + j_2 \geq J \geq |j_1 - j_2|$ and, furthermore, provided that $m_1 + m_2 = M$. The Clebsch–Gordan coefficients are chosen to be purely real and constitute a unitary transformation.

The coupling coefficients introduced in the previous paragraph are closely related to the so-called Wigner coefficients, $3j$-symbols or, more precisely, $3jm$-symbols:

$$\begin{pmatrix} j_1 & j_2 & J \\ m_1 & m_2 & -M \end{pmatrix} = \frac{(-1)^{j_1 - j_2 + M}}{\sqrt{(2J+1)}} (j_1 m_1 j_2 m_2 | JM) \; .$$

$3jm$-symbols are also purely real and this convention is adopted for all $3nj$-symbols encountered later. Note that, the exponent $j_1 - j_2 + M$ is integer. Although the $3jm$-symbols, or Wigner symbols, are usually referred to as just $3j$-symbols in the literature, we have chosen the term $3jm$-symbol here in order to distinguish these from the $3j$-, $6j$- and $9j$-symbol etc. in the theory. The Clebsch–Gordan coefficients correspond to the notation and phase used in Condon and Shortley [2], although, unlike them, we do not repeat j_1, j_2 in the right-hand part of the coefficient. The definition above can be inverted to give

[1] The terms 'projection' and 'magnetic quantum number' are used interchangeably in the text, with the first term favoured overall.

E. Balcar, S.W. Lovesey, *Introduction to the Graphical Theory of Angular Momentum*, STMP 234, 7–10, DOI 10.1007/978-3-642-03118-2_2,

$$(j_1m_1j_2m_2|JM) = (-1)^{j_2-j_1-M}\sqrt{2J+1}\begin{pmatrix} j_1 & j_2 & J \\ m_1 & m_2 & -M \end{pmatrix} . \tag{2.1}$$

From this definition follows the transformation from plain product states $|j_1m_1\rangle|j_2m_2\rangle$ to the coupled state $|j_1j_2JM\rangle$ and the inverse transformation, with the Clebsch–Gordan coefficients replaced by $3jm$-symbols as in (2.1).

$$|j_1j_2JM\rangle = \sum_{m_1m_2} (-1)^{j_2-j_1-M}\sqrt{2J+1}\begin{pmatrix} j_1 & j_2 & J \\ m_1 & m_2 & -M \end{pmatrix} |j_1m_1\rangle|j_2m_2\rangle ,$$

$$|j_1m_1\rangle|j_2m_2\rangle = \sum_{JM}(-1)^{j_2-j_1-M}\sqrt{2J+1}\begin{pmatrix} j_1 & j_2 & J \\ m_1 & m_2 & -M \end{pmatrix} |j_1j_2JM\rangle .$$

As mentioned before, the sum of the arguments $j_1 + j_2 + J$ is integer and the triangle condition $\triangle\{j_1, j_2, J\}$ must be fulfilled.

An advantage of the $3jm$-symbols over the coupling coefficients lies in their neat symmetry properties. E.g., even permutation of the columns leaves a $3jm$-symbol invariant,

$$\begin{pmatrix} j_1 & j_2 & j_3 \\ m_1 & m_2 & m_3 \end{pmatrix} = \begin{pmatrix} j_2 & j_3 & j_1 \\ m_2 & m_3 & m_1 \end{pmatrix} = \begin{pmatrix} j_3 & j_1 & j_2 \\ m_3 & m_1 & m_2 \end{pmatrix} , \tag{2.2}$$

while an odd permutation introduces a phase factor $(-1)^{j_1+j_2+j_3}$

$$\begin{pmatrix} j_1 & j_2 & j_3 \\ m_1 & m_2 & m_3 \end{pmatrix} = (-1)^{j_1+j_2+j_3}\begin{pmatrix} j_2 & j_1 & j_3 \\ m_2 & m_1 & m_3 \end{pmatrix} = \text{etc.} \tag{2.3}$$

The same phase factor is required for a sign change of m_1, m_2 and m_3

$$\begin{pmatrix} j_1 & j_2 & j_3 \\ m_1 & m_2 & m_3 \end{pmatrix} = (-1)^{j_1+j_2+j_3}\begin{pmatrix} j_1 & j_2 & j_3 \\ -m_1 & -m_2 & -m_3 \end{pmatrix} . \tag{2.4}$$

The $3jm$-symbol can be non-zero only if $m_1 + m_2 + m_3 = 0$ and the triangle condition $\triangle\{j_1, j_3, j_3\}$ for j_1, j_2 and j_3 is fulfilled. At least one of the three angular momenta in a $3jm$-symbol has to be integer and the sum of all three must also be an integer. From this we see that, two of the angular momenta in a $3jm$-symbol may be half-integer and this has the following consequences for the simplification of a general phase factor: every phase $(-1)^{4j_i} \equiv 1$ or $(-1)^{2j_i \pm 2m_i} \equiv 1$ or, for the j_i of a $3jm$-symbol, $(-1)^{2j_1+2j_2+2j_3} \equiv 1$ can thus be removed or added because the exponents are even integers in all these cases.

The following identities will be very useful in subsequent derivations. From the properties of the Clebsch–Gordan coefficients follow the orthogonality properties of the $3jm$-symbols

$$\sum_{m_1,m_2} \begin{pmatrix} j_1 & j_2 & j_3 \\ m_1 & m_2 & m_3 \end{pmatrix}\begin{pmatrix} j_1 & j_2 & j_3' \\ m_1 & m_2 & m_3' \end{pmatrix} = \frac{\delta_{j_3,j_3'}\delta_{m_3,m_3'}}{(2j_3+1)} , \tag{2.5}$$

and

$$\sum_{j_3,m_3}(2j_3+1)\begin{pmatrix} j_1 & j_2 & j_3 \\ m_1 & m_2 & m_3 \end{pmatrix}\begin{pmatrix} j_1 & j_2 & j_3 \\ m_1' & m_2' & m_3 \end{pmatrix} = \delta_{m_1,m_1'}\delta_{m_2,m_2'} \;. \tag{2.6}$$

These identities are so important, they merit an entire section in our subsequent presentation. Because the sum of the m_i in a $3jm$-symbol is zero by definition, some of the summations over magnetic quantum numbers may be purely formal and this can be seen for a special case of (2.5):

$$\sum_{m_1,m_2,m_3}\begin{pmatrix} j_1 & j_2 & j_3 \\ m_1 & m_2 & m_3 \end{pmatrix}\begin{pmatrix} j_1 & j_2 & j_3 \\ m_1 & m_2 & m_3 \end{pmatrix} = 1 \;. \tag{2.7}$$

The following equality uses the cyclic permutation symmetry, (2.2), and shows the value of this special $3jm$-symbol with one j and its projection m set zero:

$$\begin{pmatrix} j & 0 & j \\ -m & 0 & m \end{pmatrix} = \begin{pmatrix} j & j & 0 \\ m & -m & 0 \end{pmatrix} = (-1)^{j-m}\frac{1}{\sqrt{2j+1}} \;. \tag{2.8}$$

This result will be frequently used in the more general form

$$\begin{aligned}&(-1)^{j_1-m_1}\begin{pmatrix} j_1 & 0 & j_2 \\ -m_1 & 0 & m_2 \end{pmatrix} \\ &= (-1)^{j_1-m_1}\begin{pmatrix} j_2 & j_1 & 0 \\ m_2 & -m_1 & 0 \end{pmatrix} = \frac{\delta_{j_1,j_2}\delta_{m_1,m_2}}{\sqrt{2j_1+1}} \;,\end{aligned} \tag{2.9}$$

or

$$\begin{pmatrix} j_1 & 0 & j_2 \\ -m_1 & 0 & m_2 \end{pmatrix} = \begin{pmatrix} j_2 & j_1 & 0 \\ m_2 & -m_1 & 0 \end{pmatrix} = (-1)^{j_1-m_1}\frac{\delta_{j_1,j_2}\delta_{m_1,m_2}}{\sqrt{2j_1+1}} \;. \tag{2.10}$$

where the Kronecker-δ take care of the triangle condition $\triangle\{j_1, j_2, 0\}$ and the condition $m_2 - m_1 = 0$ for this special case.

The algebraic formula used by Rotenberg et al. [9] for the calculation of $3jm$-symbols is

$$\begin{aligned}\begin{pmatrix} j_1 & j_2 & j_3 \\ m_1 & m_2 & m_3 \end{pmatrix} &= (-1)^{j_1-j_2-m_3} \\ &\times \left[\frac{(j_1-j_2-j_3)!(j_1-j_2+j_3)!(-j_1+j_2+j_3)!}{(j_1+j_2+j_3+1)!}\right]^{1/2} \\ &\times \left[(j_1+m_1)!(j_1-m_1)!(j_2+m_2)!(j_2-m_2)!(j_3+m_3)!(j_3-m_3)!\right]^{1/2} \\ &\times \left[\sum_k \frac{(-1)^k}{k!(j_1+j_2-j_3-k)!(j_1-m_1-k)!(j_2+m_2-k)!} \right. \\ &\qquad \left. \times \frac{1}{(j_3-j_2+m_1+k)!(j_3-j_1-m_2+k)!}\right] \;.\end{aligned}$$

Tables of $3jm$-symbols have been given by Edmonds [4], Rotenberg et al. [9] and Varshalovich et al. [12], to cite just a few. Of course, nowadays computer programs will provide values for $3jm$-symbols, in some cases even in algebraic form.

2.2 Properties of $3nj$-Symbols

If the theory of angular momentum is extended to consider more than two angular momenta one encounters several possible coupling schemes. For three angular momenta j_1, j_2 and j_3 the coupling into a state of total angular momentum J could proceed via two routes. As a first step one could couple j_1 and j_2 to j_{12} and then j_{12} with j_3 to obtain the total J. Alternatively, one could first couple j_2 and j_3 to j_{23} and then j_1 with j_{23} for J. The states belonging to the two coupling schemes can be transformed into each other [7] and the elements of the transformation matrix are proportional to a $6j$-symbol. A symbolic relation for these couplings is,

$$\langle[(j_1, j_2)j_{12}, j_3]J\|[j_1, (j_2, j_3)j_{23}]J\rangle \propto \begin{Bmatrix} j_3 & J & j_{12} \\ j_1 & j_2 & j_{23} \end{Bmatrix} .$$

The symmetry properties of $6j$-symbols will be discussed later in section 5.2.

The next natural extension considers the coupling schemes in which four angular momenta are involved. These two coupling schemes are well known in atomic physics and referred to as SL and jj-coupling. In this case one finds [7] the transformation matrix is proportional to a $9j$-symbol, and symbolically,

$$\langle[(j_1, j_2)j_{12}, (j_3, j_4)j_{34}]J\|[(j_1, j_3)j_{13}, (j_2, j_4)j_{24}]J\rangle \propto \begin{Bmatrix} j_1 & j_2 & j_{12} \\ j_3 & j_4 & j_{34} \\ j_{13} & j_{24} & J \end{Bmatrix} .$$

Again, the discussion of the symmetry properties of the $9j$-symbol will be postponed.

In the following sections we will present a graphical technique equivalent to the algebraic treatment of complex expressions built from $3jm$-symbols, $3nj$-symbols like $6j$-, $9j$-symbols and symbols of higher order.

2.3 Comment on Notation

To simplify the notation, the subscripted variables $j_1, j_2, j_3, \ldots$ will be replaced by $a, b, c, \ldots$ and the magnetic quantum numbers, or projections, $m_1, m_2, m_3, \ldots$ will be represented by corresponding Greek letters $\alpha, \beta, \gamma, \ldots$ but we will continue to refer to these as j- and m-values, respectively.

3

Basic Tools for the Graphical Method

3.1 Graphical Representation of $3jm$-Symbols

The graphical representation of $3jm$-symbols is based on angular momentum lines which have a direction indicated by points and feathers of an arrow. Three angular momentum lines corresponding to the angular momenta in the $3jm$-symbol join at a common point called a node. A negative projection is represented by a line entering the node while lines with a positive projection are pointed away from the node. Each node is characterized by a node sign, which, in turn, depends on the arrangement of columns in the corresponding $3jm$-symbol. If the sequence $a \rightarrow b \rightarrow c$ corresponds, as in Fig. 3.1, to a clockwise arrangement in the diagram then this is signified by a negative node sign within the circle representing the node point. For a counter-clockwise arrangement the positive node sign may be associated with the positive direction of the z-axis in a right-handed Cartesian coordinate system[1].

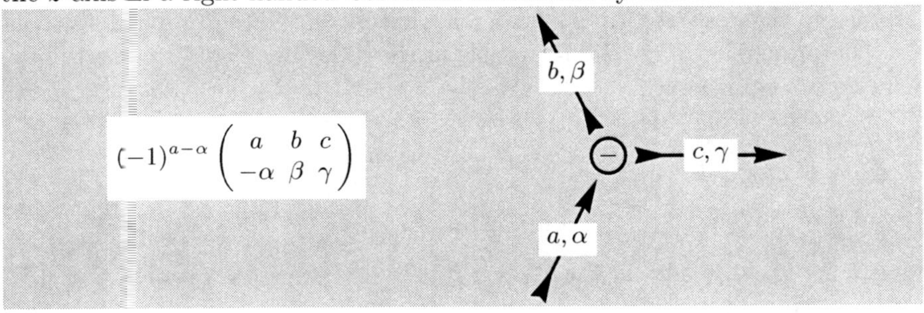

Fig. 3.1. Note that ingoing arrows correspond to negative and outgoing lines to positive m quantum numbers in the $3jm$-symbol, and that ingoing lines require an appropriate phase factor to accompany the $3jm$-symbol

Figure 3.1 shows a layout used often in the text. The algebraic symbol has a graphical equivalent consisting of a signed node and three directed lines.

---

[1] Such a positive node sign is found, f.e., in Fig. 3.10.

E. Balcar, S.W. Lovesey, *Introduction to the Graphical Theory of Angular Momentum*, STMP 234, 11–30, DOI 10.1007/978-3-642-03118-2_3,

Outgoing and ingoing arrows imply positive and negative projections, respectively, where the latter require a phase factor in the algebraic expression.

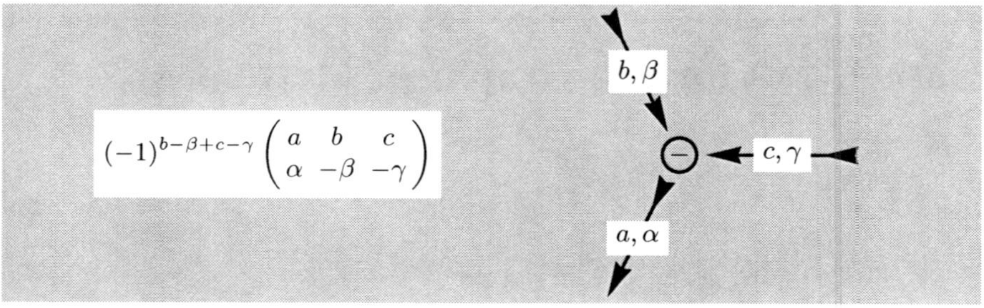

Fig. 3.2. The arrangement of angular momentum lines is the same as in Fig. 3.1 but the directions are different here. Note that the node sign is, therefore, still negative but the phase is different because of the two ingoing lines

We repeat that, a negative projection $-\alpha$ in the $3jm$-symbol requires a phase $(-1)^{a-\alpha}$ represented by an ingoing a-line in the diagram. The two $3jm$-symbols in Figs. 3.1 and 3.2 are not equal and differ by a phase factor $(-1)^{2a}$ and this can be demonstrated by using $(-1)^{\alpha-\beta-\gamma} = 1$ because the m-values in a $3jm$-symbol must sum to zero by definition.

$$(-1)^{a-\alpha}\begin{pmatrix} a & b & c \\ -\alpha & \beta & \gamma \end{pmatrix} = (-1)^{a+b+c}(-1)^{\alpha-\beta-\gamma}(-1)^{a-\alpha}\begin{pmatrix} a & b & c \\ \alpha & -\beta & -\gamma \end{pmatrix}$$
$$= (-1)^{2a}(-1)^{b-\beta+c-\gamma}\begin{pmatrix} a & b & c \\ \alpha & -\beta & -\gamma \end{pmatrix} .$$

The sign change of all m-values in a $3jm$-symbol introduces the first phase $(-1)^{a+b+c}$, see (2.4). Because $-\alpha + \beta + \gamma = 0$ is true by definition here, the phase $(-1)^{\alpha-\beta-\gamma} = 1$ may be added without affecting the value, thus producing the desired form suitable for transcription into graphical symbols.

The properties of the $3jm$-symbols in the following Fig. 3.3 make the two expressions equivalent.

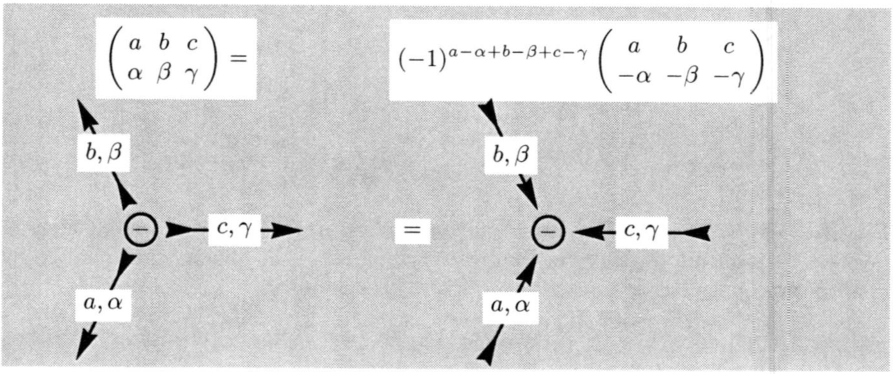

Fig. 3.3. Graphical representation of two equivalent $3jm$-symbols. One has $\alpha+\beta+\gamma=0$ and identity (2.4)

We see from Fig. 3.3 that, for $3jm$-diagrams with three angular momentum lines of the same kind, all directions may be reversed without a change in the value or phase of the diagram.

It should be mentioned that, although our diagrams have been drawn with a symmetric arrangement of angular momentum lines, the orientation is in no way essential, the lines may be deformed, but *only if* the cyclic arrangement remains unchanged. If, however, two lines change places, then this operation requires a change of node sign and this will be indicated by an '!' next to the node[2]. Of course, the whole diagram may be rotated without any effect.

The connection between the sequence $a \to b \to c$ in the $3jm$-symbol and the sign of the node may be remembered by a cork-screw rule: a clockwise turn takes the cork-screw down and the node sign is negative, the upward movement corresponds to a positive node sign. This convention is also adopted by crystallographers to label right- and left-handed screw axes.

3.2 Special Elements of the Graphical Method

The diagrams in Fig. 3.4 represent various factors which occur frequently in the application of the graphical method. This is the reason for introducing these special diagrams at an early stage.

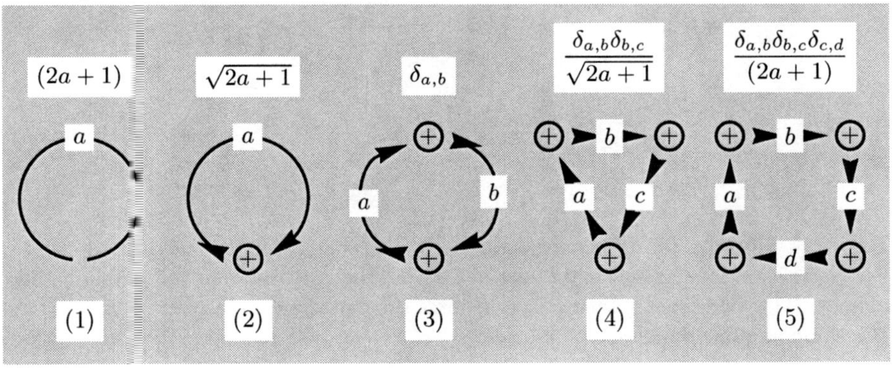

Fig. 3.4. Analytical expressions for the factors are shown above each diagram. The gap in diagram (1) indicates a removed node. These diagrams are referred to as circular, one-node, two-node, triangular and quadrangular diagram, respectively

All diagrams in Fig. 3.4 are closed diagrams with no external lines. It has to be emphasized that, these diagrams are special insofar as they contain nodes which are connected to two lines only and we will refer to these as *two-line* nodes to underline the difference to the *zero-line* nodes of the next section. Except for the first diagram, which has no direction and no node, all others have the lines arranged in clockwise order. The genesis of the factors listed in Fig. 3.4 and their relations will be discussed at a later stage.

[2] Skip forward to Fig. 4.2 for a first example.

3.3 Treatment of $3jm$-Symbols with a Zero-line

The value of a $3jm$-symbol is particularly simple if one of the three angular momenta equals zero as we see from (2.8).

$$\begin{pmatrix} j & j & 0 \\ m & -m & 0 \end{pmatrix} = \begin{pmatrix} j & 0 & j \\ -m & 0 & m \end{pmatrix} = \frac{(-1)^{j-m}}{\sqrt{2j+1}} .$$

From this result and from (2.9) we deduce the more general formula

$$(-1)^{a-\alpha} \begin{pmatrix} a & 0 & c \\ -\alpha & 0 & \gamma \end{pmatrix} = \frac{\delta_{a,c}\delta_{\alpha,\gamma}}{\sqrt{2a+1}} ,$$

which is now transcribed into the graphical representation. The diagram will consist of a node with two angular momentum lines and one zero-line. Provided that the cyclic arrangement of the lines agrees with the node sign, the node with the 0-line, zero-line node for short, may be removed and a triangular diagram (see Fig. 3.4, element (4)) representing the factor is inserted.

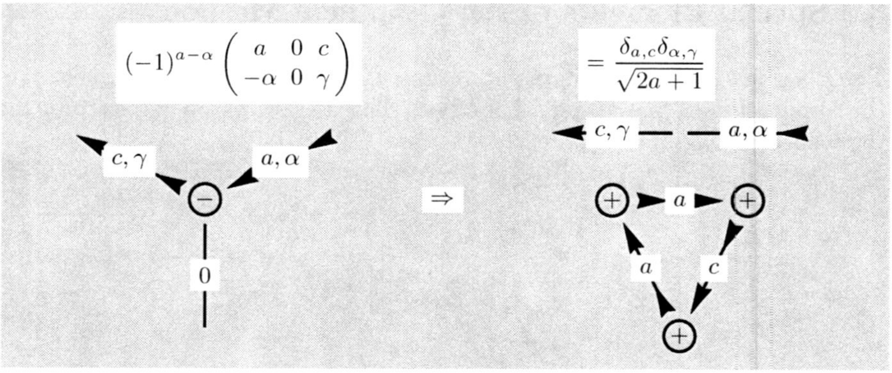

Fig. 3.5. Since $a \to 0 \to c$ corresponds to a clockwise arrangement of the lines, the node sign − is correct and node together with 0-line may be removed without phase change. The two lines entering and leaving the zero-line node will, of course have the same angular momentum values

In this diagram we have introduced a new element, a single free line representing the two Kronecker-δ.

All angular momenta in the triangular diagram have to be the same and it is not significant whether a or c appears twice in the triangular diagram in Fig. 3.5. In the general case of diagram (4) in Fig. 3.4 the triangular diagram contains three angular momenta which must have the same values as expressed two Kronecker-δ.

In the next figure we draw the $3jm$-symbol with a positive node sign and arrive, of course, at the same result. The direction of this free line is not significant. Because the triangular diagram represents a factor with a single Kronecker-δ here, namely $\delta_{a,c}$, the angular momentum lines have been adjusted accordingly.

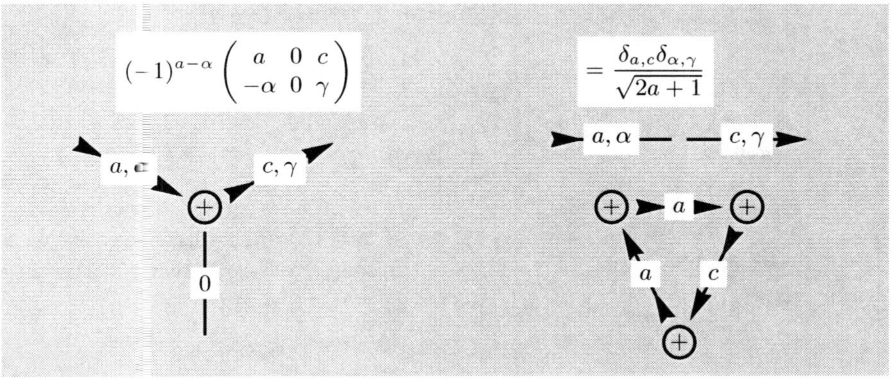

Fig. 3.6. If $a \rightarrow 0 \rightarrow c$ corresponds to an counter-clockwise arrangement the node sign is +. Again the node together with the 0-line may be removed without phase change. The two lines entering and leaving the zero-line node will, of course have the same angular momentum values

Comparing the diagrams for the $3jm$-symbols in left-hand sides of Figs. 3.5 and 3.6 one notices that the symbols are mirror images of each other and the mirror operation implies a change of node sign since the cyclic arrangement of the angular momentum lines is changed. Since the nodes in Figs. 3.5 and 3.6 are both removable, the triangular diagrams are the same in both cases.

■ **Removal of a zero-line node:**
The term zero-line node refers to a node with two angular momentum lines and one zero-line attached to it. As seen in this section, provided the node sign agrees with the cyclic order passing from ingoing through the zero- to the outgoing line, then the node together with the zero-line can be removed. The resulting free line ensures the same direction and angular momentum for the joined lines.

The case when the node sign does not agree with the cyclic order of the three lines requires a node sign change.

3.4 Direct Change of a Node Sign

It is possible to change a node sign *without* altering the arrangement of the lines. Such a change then requires an explicit or direct[3] change of phase because the different node sign implies an exchange of two columns in the $3jm$-symbol. Compared to the $3jm$-symbol of Fig. 3.5 columns are exchanged and the corresponding phase factor has to be taken into account.

[3] Our choice of words indicates the fact that manipulations or transformations may lead to indirect or implicit changes of node sign.

$$(-1)^{a-\alpha}\begin{pmatrix} a & c & 0 \\ -\alpha & \gamma & 0 \end{pmatrix} = (-1)^{a+0+c}(-1)^{a-\alpha}\begin{pmatrix} a & 0 & c \\ -\alpha & 0 & \gamma \end{pmatrix}$$
$$= (-1)^{a+c}\frac{\delta_{a,c}\delta_{\alpha,\gamma}}{\sqrt{2a+1}}$$
$$= (-1)^{2a}\frac{\delta_{a,c}\delta_{\alpha,\gamma}}{\sqrt{2a+1}} \, .$$

From (2.3) we know that the exchange of two columns introduces a phase $(-1)^{a+0+c}$ which remains as $(-1)^{2a}$ in the last equality. Beware, variable a can be half-odd integer, in which case $(-1)^{2a} = -1$ is important.

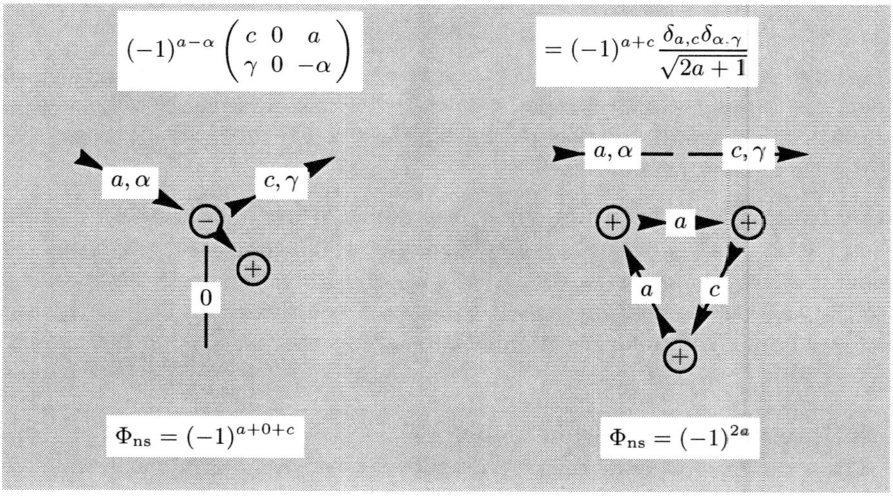

Fig. 3.7. The arrangement $a \to 0 \to c$ is counter-clockwise but the node sign is negative here. Thus, the node sign has to be changed to achieve the correct arrangement of in- and outgoing lines before the node together with the 0-line may be removed. A corresponding phase change is the consequence

The direct node change is indicated by a single arrow head pointing from the original node to the node with the different node sign. Due to the change in the node sign a phase $\Phi_{ns} = (-1)^{2a}$ becomes attached to the diagram. We have seen here that it is possible to change a node sign without rearranging the angular momentum lines at the price of an extra phase factor. Keeping track of the phase factors is an important part of the graphical technique.

■ **Direct change of a node sign:**
The direct change of a node sign without alteration in the cyclic order of the angular momentum lines introduces an explicit phase Φ_{ns} for the diagram. In general such a phase factor for a direct node sign change will have the form $\Phi_{ns} = (-1)^{a+b+c}$, where a, b and c represent the three lines connected to the node in question.

3.5 Single External Line

In the context of discussing the methods applicable to the $j = 0$ case, we will consider two simple examples in the following. The starting point is a general relation, (2.5), valid for $3jm$-symbols,

$$\sum_{\alpha,\beta} \begin{pmatrix} a & b & c \\ \alpha & \beta & \gamma \end{pmatrix} \begin{pmatrix} a & b & c' \\ \alpha & \beta & \gamma' \end{pmatrix} = \frac{\delta_{c,c'}\delta_{\gamma,\gamma'}}{(2c+1)} .$$

We consider the special case $c' = \gamma' = 0$ here

$$\sum_{\alpha,\beta} \begin{pmatrix} a & b & c \\ \alpha & \beta & \gamma \end{pmatrix} \begin{pmatrix} a & b & 0 \\ \alpha & \beta & 0 \end{pmatrix} = \frac{\delta_{c,0}\delta_{\gamma,0}}{\sqrt{2c+1}} ,$$

and use the result given earlier in (2.10) for a $3jm$-symbol with a zero column.

$$\sum_{\alpha,\beta} \begin{pmatrix} a & b & c \\ \alpha & \beta & \gamma \end{pmatrix} (-1)^{a-\alpha} \frac{\delta_{a,b}\delta_{-\alpha,\beta}}{\sqrt{(2a+1)}} = \delta_{c,0}\delta_{\gamma,0} .$$

The Kronecker-δ forces $a = b$ and the sum over β leads to the following relation

$$\sum_{\alpha} (-1)^{a-\alpha} \begin{pmatrix} a & a & c \\ \alpha & -\alpha & \gamma \end{pmatrix} = \delta_{c,0}\delta_{\gamma,0}\sqrt{2a+1} ,$$

and this result is now expressed in graphical form.

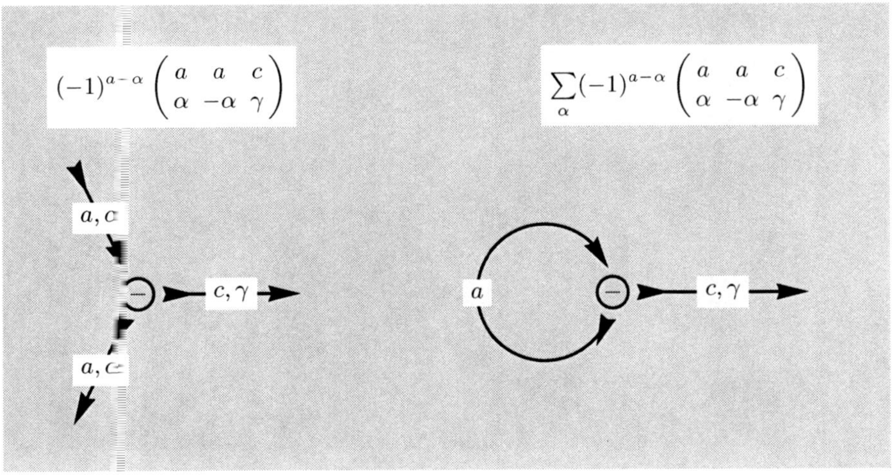

Fig. 3.8. Joining the two lines with angular momentum a is the graphical equivalent for the sum over α

The resulting diagram has only one external line belonging to the angular momentum c. We have, however, seen from the algebraic result given above, that the only possible value for c is zero and we deduce that the single external line must have $c = 0$ and this result will also be true for a single line connecting two diagrams[4] if at least one of the diagrams is closed with no external lines.

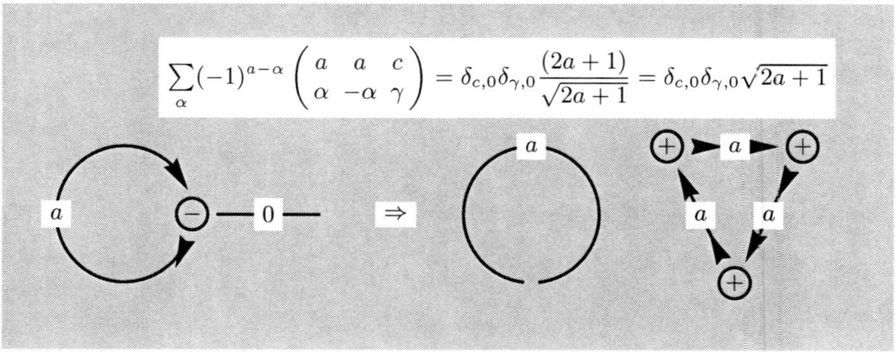

Fig. 3.9. If the $c = 0$ line and the node are removed in accordance with the procedure specified on p. 19 we obtain diagrams given in Fig. 3.4

Here the 0-line is removed together with the remaining node in accordance with the rule specified on page 19 and the diagrams given in Fig. 3.4.

Compared to Fig. 3.5 one and the same line a enters and leaves the zero-line node here and, consequently, the triangular diagram in Fig. 3.9 contains three entries a and requires no Kronecker-δ.

We will now consider a slight variation of this analysis starting with a positive node sign.

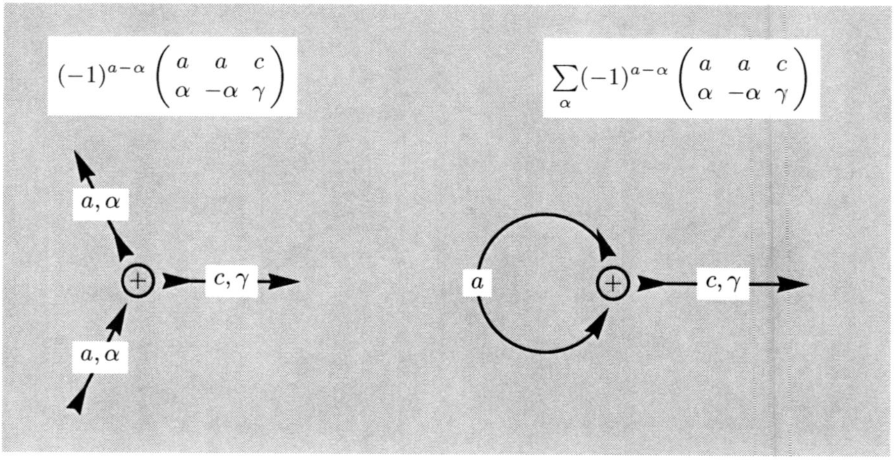

Fig. 3.10. Compared with Fig. 3.8 the in- and outgoing a-lines have changed place and the node sign is now +. The a-line leaving and entering the node symbolizes the sum over α

With arguments following the same lines as given above we find

---

[4] Because Kronecker-δ are inferred from the algebraic result, there is no graphical symbol in this case.

$$\sum_{\alpha}(-1)^{a-\alpha}\begin{pmatrix} a & a & c \\ \alpha & -\alpha & \gamma \end{pmatrix} = \delta_{c,0}\delta_{\gamma,0}\frac{(2a+1)}{\sqrt{2a+1}} = \delta_{c,0}\delta_{\gamma,0}\sqrt{2a+1}$$

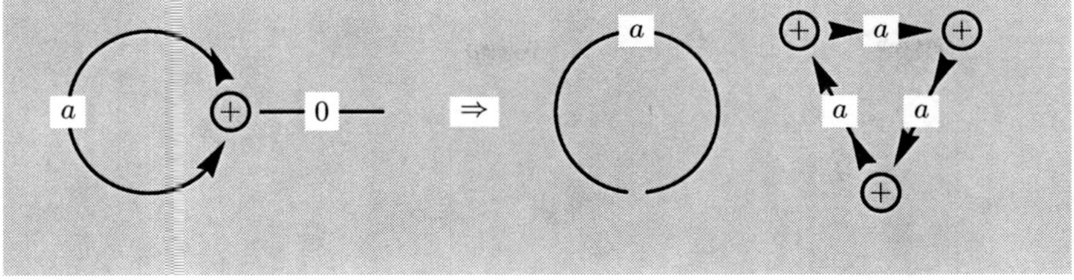

Fig. 3.11. If the 0-line and the node are removed, the remaining circular and triangular diagram can be found from Fig. 3.4

Both results, Fig. 3.9 and Fig. 3.11, are identical because in both cases the cyclic arrangement of ingoing, zero line and outgoing line was as required for the removal of the node.

From these results it is clear that the case of a single external line is closely related to that of a zero-line node discussed on p. 15. In the following box conditions for treating a single external line are formulated with general angular momentum descriptors j and j_1, j_2 in place of specific variables used in the diagrams.

> ■ **Single external line:**
> In a diagram with a single external line this external line can only have the value $j = 0$ and can be removed with the node to which it is attached. The node can be removed if the node sign corresponds to the cyclic arrangement of ingoing j_1-line, zero value line and outgoing j_2-line. The non-zero lines must have the same angular momentum value $j_1 = j_2$ and retain their directions when joined. The node removal introduces a factor $1/\sqrt{2j_1+1}$.

In other cases, when the cyclic arrangement does not agree with the node sign, the node sign has to be changed and this affects the phase of the diagram. Note that the presence of the 0-line in the diagram is essential for the decision whether a phase change is necessary or not.

3.6 Two-line Nodes

Once just the zero-line is removed from a diagram the situation becomes ambiguous with respect to a necessary phase change. To remove this ambiguity for the two-line nodes we have to adopt an additional convention.

There is, as we will see, an alternative option of closing the a-line in Fig. 3.11 if we keep the $j = 0$ line *inside* the closed loop:

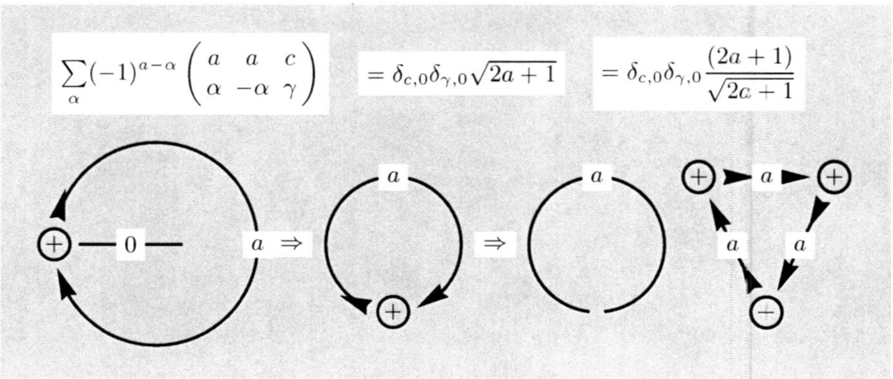

Fig. 3.12. If only the 0-line is removed, the remaining diagram with just one two-line node equals entry (2) in Fig. 3.4

We note that the special case of the two-line node diagram (2) given in Fig. 3.4 must be obtained with the zero-line enclosed by the angular momentum line to guarantee the node removal without phase change. The same convention holds for the other diagrams (3)-(5) of Fig. 3.4. The single node diagram (2) in Fig. 3.12 agrees with the diagram given by Varshalovich et al. [12] for the factor $\sqrt{2a+1}$.

With the zero-line inside the diagram it becomes a matter of taste whether we prefer to express the result by the two-line node diagram (2) or the product of diagram (1) and (4) in Fig. 3.4.

We are now in a position to derive diagram (4) given in Fig. 3.4 which contains only two-line nodes. Following our convention we assume that the nodes are remnants of removed zero-lines *inside* the closed diagram. Consider a product of three special $3jm$-symbols:

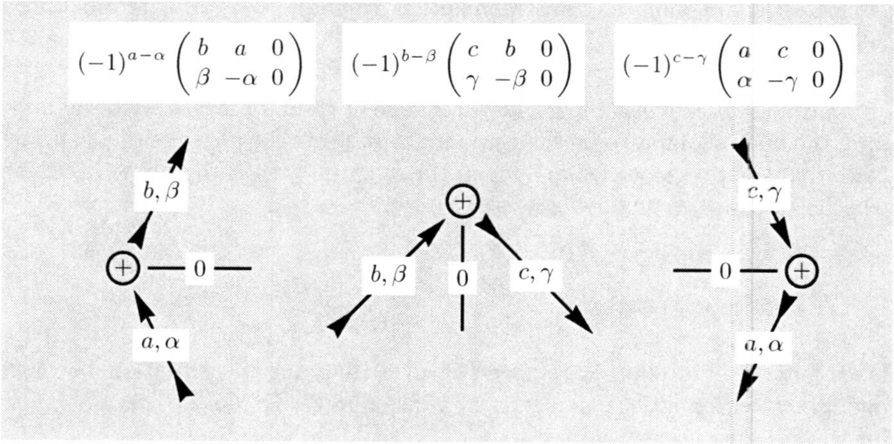

Fig. 3.13. Three $3jm$-symbols, each with one angular momentum zero, are written and drawn here

Each $3jm$-symbol has the value given in (2.9) and we introduce now sums over α, β and γ. The sums over β and γ remove two Kronecker-δ and lead to $\delta_{\alpha,\alpha} = 1$. The remaining sum over α adds just the factor $(2a+1)$.

$$\sum_{\alpha,\beta,\gamma} \frac{\delta_{a,b}\delta_{\alpha,\beta}}{\sqrt{2a+1}} \frac{\delta_{b,c}\delta_{\beta,\gamma}}{\sqrt{2b+1}} \frac{\delta_{c,a}\delta_{\gamma,\alpha}}{\sqrt{2c+1}} = \delta_{a,b}\delta_{b,c}\delta_{c,a} \frac{(2a+1)}{(2a+1)^{3/2}} = \frac{\delta_{a,b}\delta_{b,c}}{\sqrt{2a+1}} .$$

For the graphical representation corresponding angular momentum lines have to be connected and we arrive at a triangular diagram.

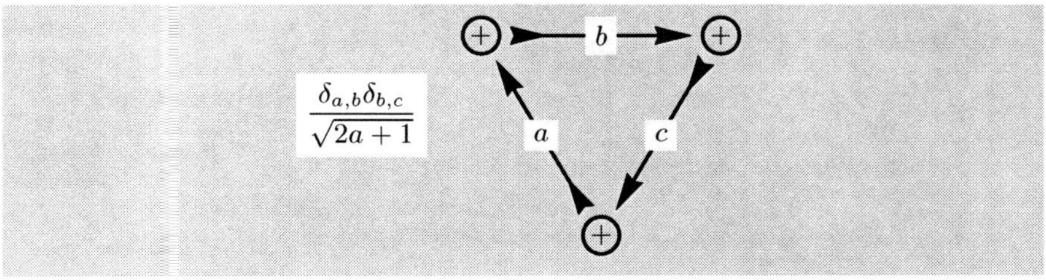

Fig. 3.14. The three $3jm$-symbols combine into one closed diagram and the angular momentum lines between the nodes indicate the summations over α, β and γ. The result for the diagram is, therefore, independent of projections and agrees with diagram (4) in Fig. 3.4

The three removed zero-lines were all *inside* the closed diagram.

It is now fairly obvious that a similar procedure with four $3jm$-symbols will lead us to the quadrangular diagram in Fig. 3.4, and the same is true for the diagram with two nodes.

In principle one could wish to remove all the two-line nodes from the triangular diagram in Fig. 3.14. The result would be a product of a closed circle, $(2a-1)$, and three new triangular diagrams each having the value $1/\sqrt{2a+1}$ and this agrees with the value given to the triangular diagram in Figs. 3.4 and 3.14. Except as a demonstration of consistency this exercise is of limited value.

3.7 A Simple Closed Diagram

For this example we use the relation given above in Fig. 3.9

$$\sum_{\alpha} (-1)^{a-\alpha} \begin{pmatrix} a & a & c \\ \alpha & -\alpha & \gamma \end{pmatrix} = \delta_{c,0}\delta_{\gamma,0}\sqrt{2a+1} ,$$

together with a slight variation of the same equality

$$\sum_{\beta} (-1)^{b-\beta}(-1)^{c-\gamma} \begin{pmatrix} b & b & c \\ \beta & -\beta & -\gamma \end{pmatrix} = \delta_{c,0}\delta_{\gamma,0}\sqrt{2b+1} .$$

Of course, the product of these relations represents another equality which may later additionally also be summed over γ.

$$\sum_{\alpha,\beta} (-1)^{a-\alpha} \begin{pmatrix} a & a & c \\ \alpha & -\alpha & \gamma \end{pmatrix} (-1)^{b-\beta}(-1)^{c-\gamma} \begin{pmatrix} b & b & c \\ \beta & -\beta & -\gamma \end{pmatrix} =$$

$$= \delta_{c,0}\delta_{\gamma,0}\sqrt{2a+1}\sqrt{2b+1} \, .$$

A sum over γ will remove the Kronecker $\delta_{\gamma,0}$. The transcription into the graphical formalism is now used to derive the same result by the graphical method.

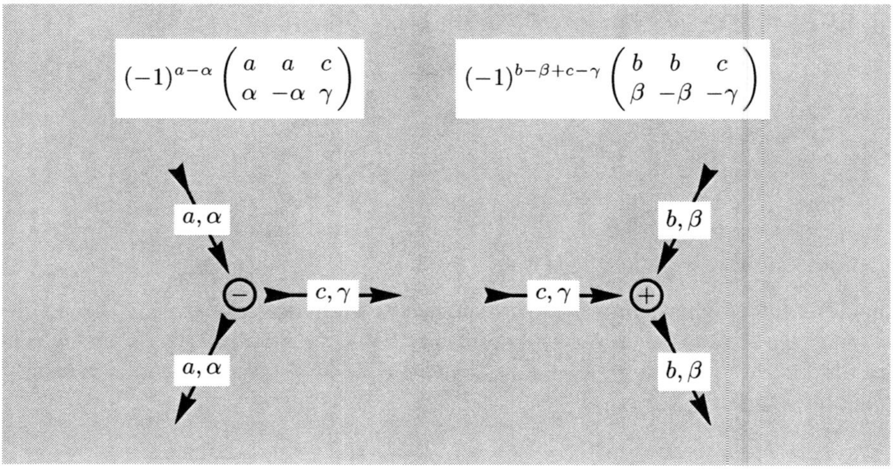

Fig. 3.15. The diagram of the second $3jm$-symbol has been rotated to prepare for the sum over γ

The introduction of the sums over α, β and γ will result in a closed diagram without any external line.

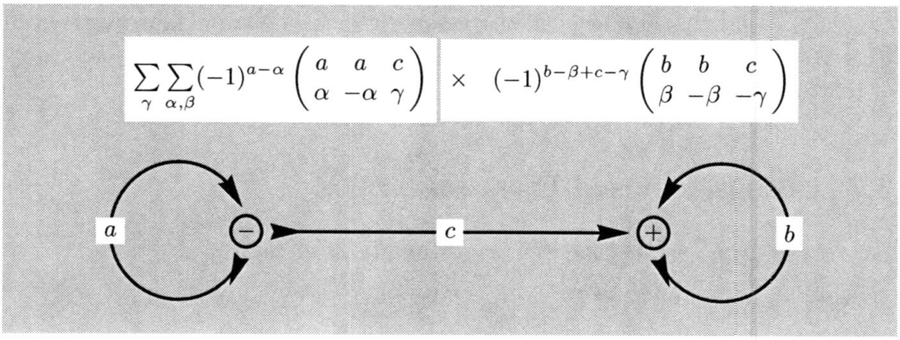

Fig. 3.16. The sum over γ is represented by a line between nodes and the sums over α and β are lines which leave and enter the same node

This is the simplest case of a single line connecting two diagrams. From the algebraic result we know already that the only possible value is given by $c = 0$.

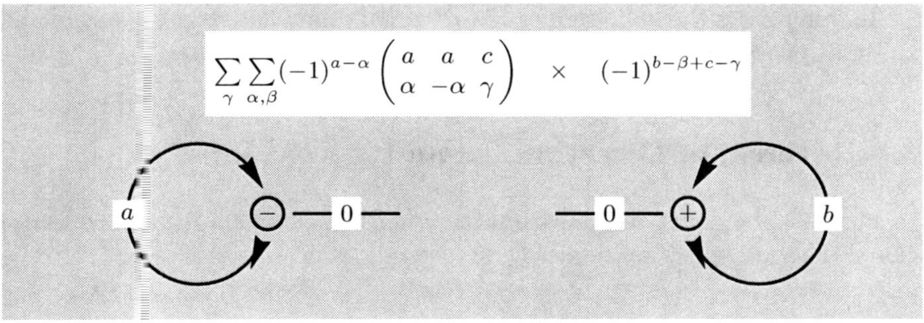

Fig. 3.17. Setting $c = 0$ leaves behind two nodes with attached 0-lines which both have the correct node sign in relation to the arrangement of ingoing, 0-line and outgoing lines

Once the $c = 0$ angular momentum line has been reduced to 0-lines, the remaining diagrams can be further simplified by removing the nodes together with the 0-lines and by introducing the corresponding factors as in Figs. 3.9 and 3.11.

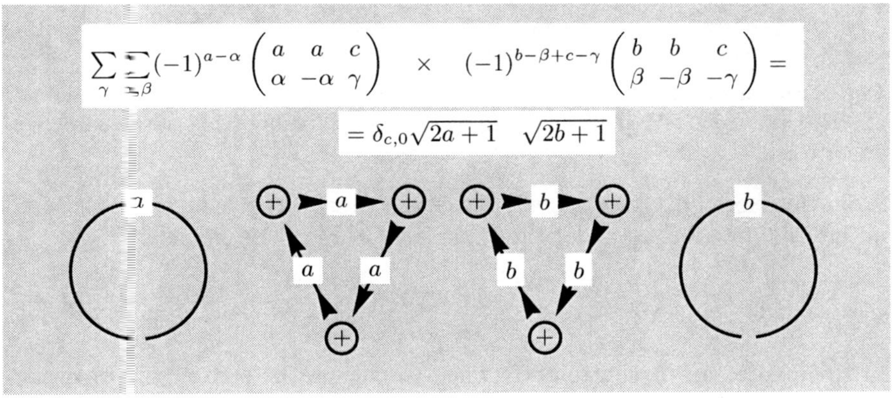

Fig. 3.18. The removal of both nodes with their 0-lines resulted in two circular and two triangular diagrams

Except for the sum over γ which has removed one Kronecker $\delta_{\gamma,0}$ the result of the graphical method agrees with the result given above.

> ■ **Single internal line:**
> A single internal line connecting two diagrams, where at least one of these must be a closed diagram, can only have the value $j = 0$ and results in two nodes with a 0-line and two non-zero lines. These two lines must have the same angular momentum value $j_1 = j_2$ and the node removal introduces a factor $1/\sqrt{2j_1+1}$. The node with the 0-line can be removed if the node sign corresponds to the cyclic arrangement of ingoing j_1-line, zero value line and outgoing j_2-line.

In other cases the node sign or line directions may have to be changed and an extra phase factor can be the result.

3.8 Reversal of Direction for an Internal Line

In Fig. 3.16 the sums over all magnetic quantum numbers have produced a closed diagram which we repeat here:

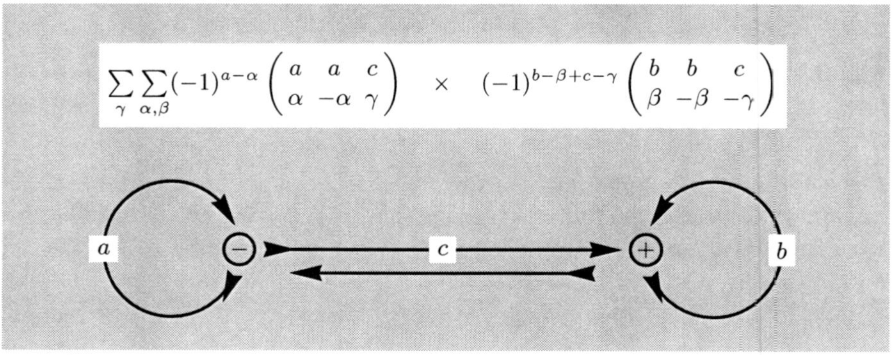

Fig. 3.19. In the sum over γ the summation index can be changed to $-\gamma$ without affecting the value. We indicate intention to change the direction of the c-line with an antiparallel arrow

From this closed diagram we will derive the rule for reversing the direction of an internal line. We consider the expression in Fig. 3.19, namely,

$$\sum_\gamma \sum_{\alpha,\beta} (-1)^{a-\alpha} \begin{pmatrix} a & a & c \\ \alpha & -\alpha & \gamma \end{pmatrix} (-1)^{b-\beta+c-\gamma} \begin{pmatrix} b & b & c \\ \beta & -\beta & -\gamma \end{pmatrix} .$$

In this expression it is straightforward to change the sign of the projection $\gamma \to -\gamma$ and this change can have no influence on the value of the expression.

$$\sum_\gamma \sum_{\alpha,\beta} (-1)^{a-\alpha} \begin{pmatrix} a & a & c \\ \alpha & -\alpha & -\gamma \end{pmatrix} (-1)^{b-\beta+c+\gamma} \begin{pmatrix} b & b & c \\ \beta & -\beta & \gamma \end{pmatrix} .$$

This sum, however, is no longer directly translatable into a graphical expression because of the phase factor $(-1)^{c+\gamma}$. This phase can be transformed

$$(-1)^{c+\gamma} \cdot 1 = (-1)^{c+\gamma} \cdot (-1)^{2c-2\gamma} = (-1)^{c-\gamma} \cdot (-1)^{2c} ,$$

and introduced into the algebraic expression

$$\sum_\gamma \sum_{\alpha,\beta} (-1)^{a-\alpha} (-1)^{c-\gamma} \begin{pmatrix} a & a & c \\ \alpha & -\alpha & -\gamma \end{pmatrix} (-1)^{b-\beta} \begin{pmatrix} b & b & c \\ \beta & -\beta & \gamma \end{pmatrix} \cdot (-1)^{2c} ,$$

which has now the form required for transcription into graphical symbols.

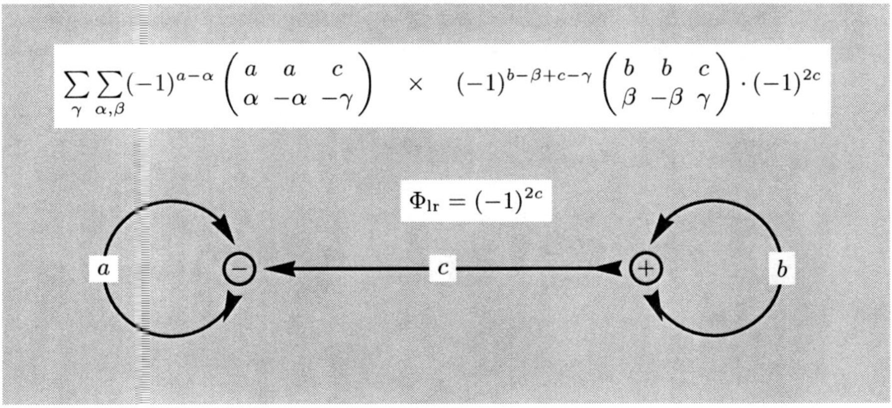

Fig. 3.20. The diagram has acquired an extra phase Φ_{lr} through the reversal of the c-line

We note that the algebraic expression in Fig. 3.20, as we have seen above, is equivalent to the expression in Fig. 3.19. In passing from that diagram to Fig. 3.20 the direction of the c-line has been reversed adding the extra phase factor Φ_{lr} in the process.

■ **Line reversal of an internal line:**
The change of direction of an internal j-line leads to the introduction of an additional phase factor $\Phi_{\mathrm{lr}} = (-1)^{2j}$, where the subscript is short for 'line reversal'. If all three angular momentum lines belonging to one node are reversed the resulting phase factor is $\Phi_{\mathrm{lr}} = (-1)^{2j_1+2j_2+2j_3} = 1$.

Up to now we have encountered two kinds of phases discernible by the subscripts in Φ_{lr}, for *line reversal*, and Φ_{ns}, for *node sign* change. In all cases, the direct change of node sign introduces a phase $(-1)^{j_1+j_2+j_3}$ deriving from the three angular momentum lines joined at the node, see, for example, Fig. 3.7. Line reversals of internal lines, on the other hand, lead to phases with factors of 2 in the exponent as in Fig. 3.20.

As noted before, it is an *essential* element of the graphical technique to keep track of the phases generated by line reversals as well as direct changes of node sign. Equally important it is not to overlook node sign changes encountered in transformations of diagrams.

3.9 Line Reversal of an External Line

In some cases it is desirable to change the direction of an external line. The corresponding formal procedure is included here for completeness although it may seem artificial. It is applied transiently in later sections (see e.g. Figs. 4.11

and 4.12) to facilitate the application of a diagrammatic technique or prescription[5]. Again, for completeness, we include special cases for the removal of nodes with zero-lines.

3.9.1 Reversal of an Outgoing Line

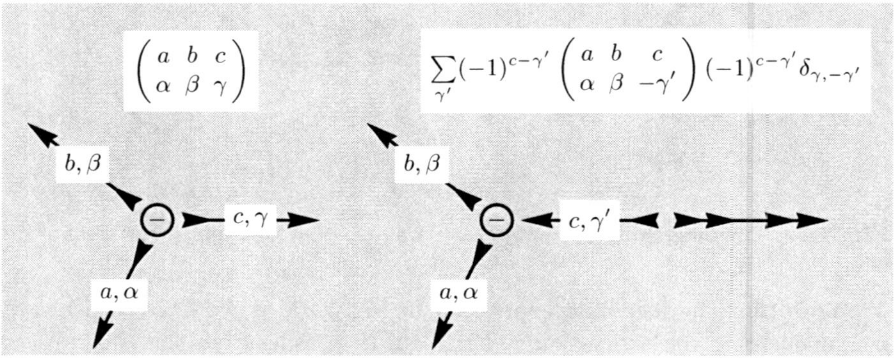

Fig. 3.21. Graphical representation of two equivalent $3jm$-symbols, where one outgoing line is reversed in the right-hand diagram. The double pointed arrow retains the original direction

This demonstrates that the change in direction of an outgoing external line involves a sum over the projection, a Kronecker-δ and a phase factor to cancel the phase factor of the new ingoing line if the sum is performed.

3.9.2 Reversal of an Ingoing Line

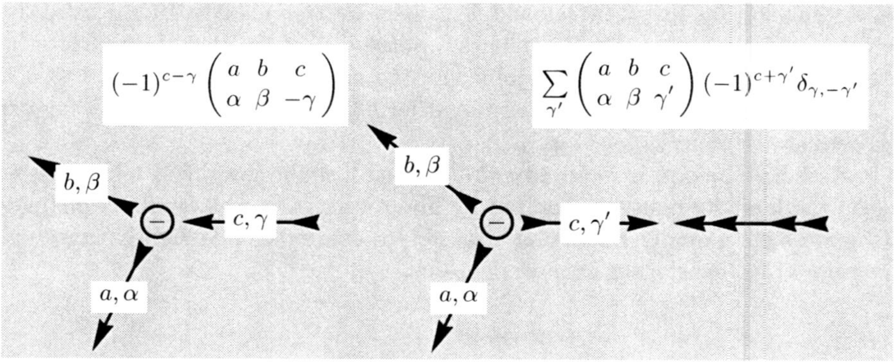

Fig. 3.22. Graphical representation of two equivalent $3jm$-symbols, where one ingoing line is reversed in the right-hand diagram. The double pointed arrow retains the original direction

---

[5] An alternative and less elegant approach would be to introduce (slightly) different sets of rules covering the cases for external lines of various directions.

The situation here is similar to an outgoing line, except that in this case the phase has a different sign for γ' to reproduce the original phase after summation.

3.9.3 Reversal of an Outgoing Line and a $j = 0$-Line

We return to Fig. 3.21 and consider a special case with $a = 0$

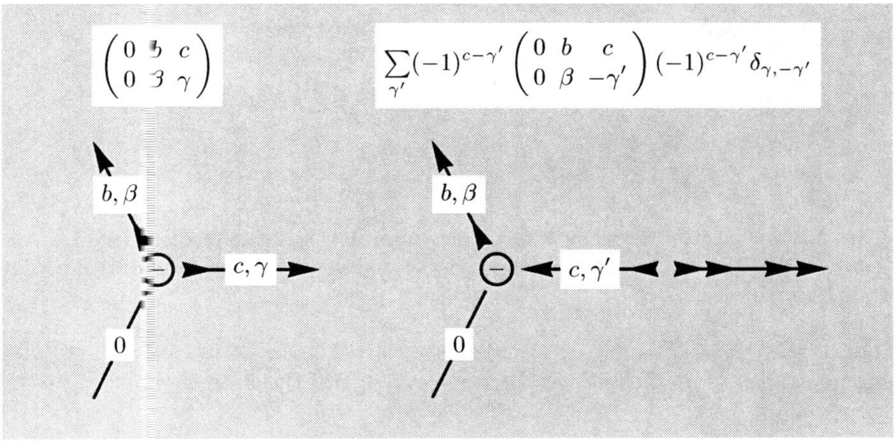

Fig. 3.23. Setting $a = 0$ creates a zero-line node with two lines, but both non-zero lines are directed outwards. If the c-line were reversed, the arrangement of ingoing, zero and outgoing line would be appropriate for node removal

The double arrow that returns a line to its outgoing direction corresponds to the algebraic form $(-1)^{c-\gamma'}\delta_{\gamma,-\gamma'}$ in correspondence with Sect. 3.9.1.

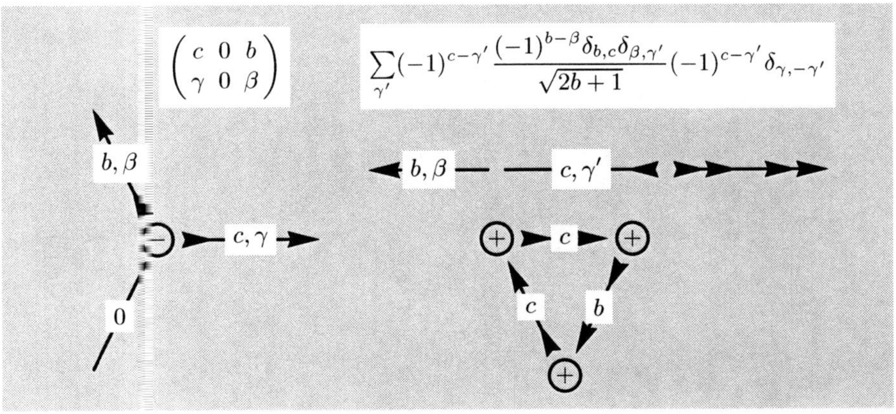

Fig. 3.24. The left part of the picture is the same as in Fig. 3.23 except for a rearrangement of the columns in the $3jm$-symbol. On the right-hand side the node has been removed together with the 0-line and (2.8) is used for the $3jm$-symbol

In the following step the c-line is returned to its original direction and this leads to a single line with arrows in both directions.

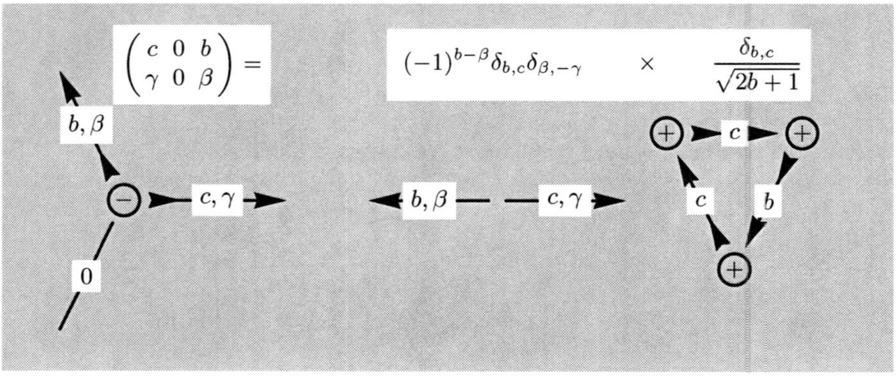

Fig. 3.25. The new diagram with arrows pointing in opposite directions has the value $(-1)^{b-\beta}\delta_{b,c}\delta_{\beta,-\gamma}$ and the action of the double arrow has returned the c-line to its original direction

The phase $(-1)^{b-\beta}\delta_{b,c}\delta_{\beta,-\gamma}$ is expressed through the values of the angular momentum line unaffected by the line reversals of the c-line.

3.9.4 Reversal of an Ingoing Line and a $j = 0$-Line

A similar result is obtained if we consider the reversal of an ingoing line. We reverse all angular momentum lines in the $3jm$-symbol of Fig. 3.23, an operation that leaves the value invariant, see Fig. 3.3.

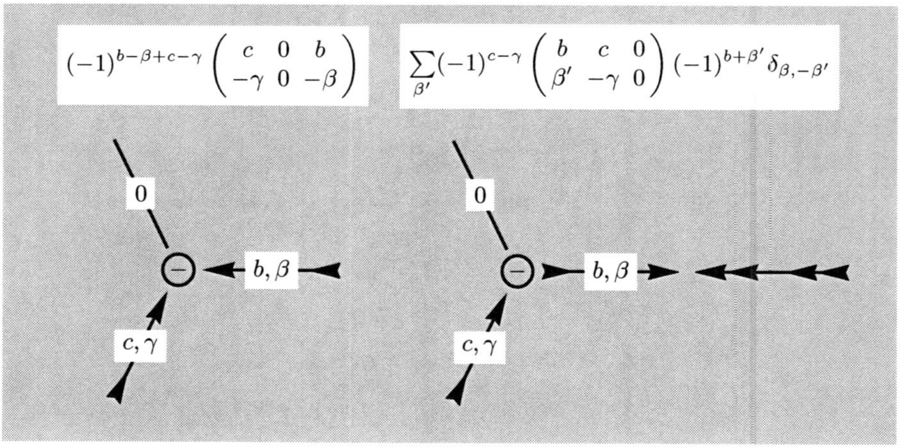

Fig. 3.26. Setting $a = 0$ creates a 0-line node with two non-zero lines, but both lines are directed inwards. If the b-line were reversed, the arrangement of ingoing, zero- and outgoing line would be appropriate for node removal. The value of the $3jm$-symbol is taken from (2.8)

The double arrow that returns a line to its ingoing direction corresponds to the algebraic form $(-1)^{b+\beta'}\delta_{\beta,-\beta'}$ in correspondence with Sect. 3.9.2.

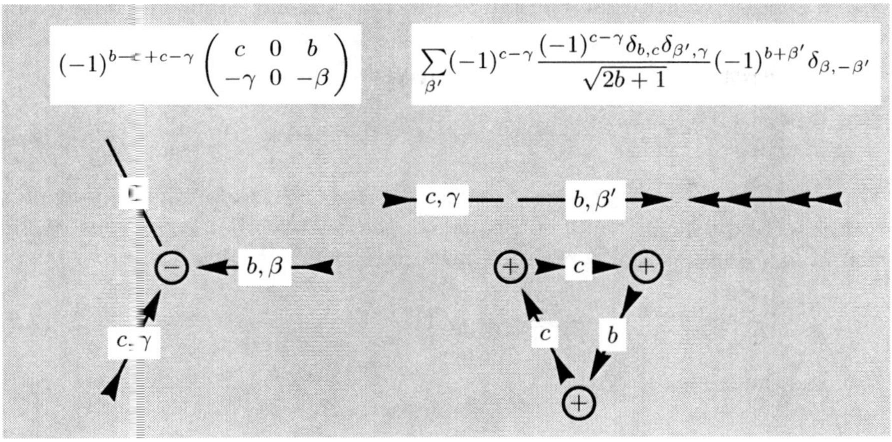

Fig. 3.27. The left part of the picture is the same as in Fig. 3.26 except for a rearrangement of the columns in the $3jm$-symbol. On the right-hand side the node and the 0-line have been removed

In the following step the b-line is returned to its original direction and this leads to a single line with arrows pointing against each other.

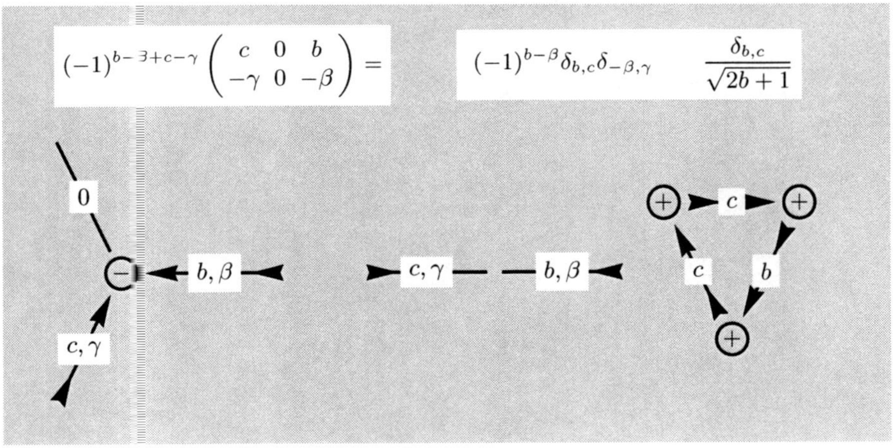

Fig. 3.28. The new diagram with the inward pointing arrows has the value $(-1)^{b-\beta}\delta_{b,c}\delta_{-\beta,\gamma}$ but here it was the b-line which was reversed by the double arrow

The phase $(-1)^{b-\beta}\delta_{b,c}\delta_{\beta,-\gamma}$ is expressed through the values of the angular momentum line affected by the line reversals of the b-line. Because $(-1)^{b-\beta}\delta_{b,c}\delta_{\beta,-\gamma} = (-1)^{c+\gamma}\delta_{b,c}\delta_{\beta,-\gamma}$ the phase would look different if the values of the unreversed c-line were used instead.

In sections 3.9.3 and 3.9.4 we have analysed $3jm$-symbols with one angular momentum and its projection having zero value. The general properties of $3jm$-symbols require that the remaining two angular momenta have equal values, to fulfil the triangle condition, and also that their projections sum to zero. We use (2.10) for the $3jm$-symbol treated in Fig. 3.24,

$$\begin{pmatrix} b & c & 0 \\ \beta & \gamma & 0 \end{pmatrix} = (-1)^{c+\gamma} \frac{\delta_{b,c}\delta_{\beta,-\gamma}}{\sqrt{2c+1}} = (-1)^{b-\beta} \frac{\delta_{b,c}\delta_{\beta,-\gamma}}{\sqrt{2b+1}} , \tag{3.1}$$

where action of the Kronecker-δ makes the two results for the $3jm$-symbol completely equivalent. As a consequence, the $3jm$-symbol invites the definition of a special symbol containing just three values, b, β and γ,

$$\begin{pmatrix} b & c & 0 \\ \beta & \gamma & 0 \end{pmatrix} = \begin{pmatrix} & b & \\ \beta & & \gamma \end{pmatrix} \frac{\delta_{b,c}}{\sqrt{2b+1}} = (-1)^{b-\beta} \delta_{\beta,-\gamma} \frac{\delta_{b,c}}{\sqrt{2b+1}} , \tag{3.2}$$

that has the form

$$\begin{pmatrix} & b & \\ \beta & & \gamma \end{pmatrix} = (-1)^{b-\beta} \delta_{\beta,-\gamma} . \tag{3.3}$$

This new quantity can be interpreted as metric tensor and can, as in Fig. 3.24, be seen to change the sign of a projection. This sign change is related to the change of direction of an external line, and also, in a more general context to the operation of time-reversal. Further analysis of these relationships is reserved for Sect. A.4 of Appendix A.

4

Orthogonality Relations for $3jm$-Symbols

In the preceding chapter the main emphasis rests on the graphical representation and manipulation of single $3jm$-symbols that constitute the basic building blocks for diagrams. The tools and conventions for zero-line nodes, line reversals etc. are discussed. To make further progress in proving that the graphical method is equivalent to the algebra of $3jm$-symbols we make use of valid relations between $3jm$-symbols. In consequence, we gain further insight into graphical methods by considering and re-writing equations (2.5) and (2.6) that represent orthogonality and closure.

4.1 Identity for $3jm$-Symbols I

The first example introduces a new feature, namely the sum of an angular momentum value in addition to the sum of the projection. The orthogonality relation for $3jm$-symbols given in (2.6) has the form,

$$\sum_{c,\gamma}(2c+1)\begin{pmatrix} a & b & c \\ \alpha & \beta & \gamma \end{pmatrix}\begin{pmatrix} a & b & c \\ \alpha' & \beta' & \gamma \end{pmatrix} = \delta_{\alpha,\alpha'}\delta_{\beta,\beta'} \; .$$

The graphical representation of of this product of two $3jm$-symbols consists of two nodes, each with three angular momentum lines directed away from the node. At least for the sum over projection γ one ingoing c-line is required. The expression can be rewritten, with the use of (2.4) and $\alpha' + \beta' + \gamma' = 0$, as

$$\sum_{c,\gamma}(2c+1)\begin{pmatrix} a & b & c \\ \alpha & \beta & \gamma \end{pmatrix}(-1)^{a-\alpha'+b-\beta'+c-\gamma}\begin{pmatrix} a & b & c \\ -\alpha' & -\beta' & -\gamma \end{pmatrix} = \delta_{\alpha,\alpha'}\delta_{\beta,\beta'} \; ,$$

to allow the transcription into graphical symbols[1]. We repeat Fig. 4.4 here for easier reference

[1] Note the factor $(2c+1)$ and similar factors that accompany sums over angular momentum values.

E. Balcar, S.W. Lovesey, *Introduction to the Graphical Theory of Angular Momentum*, STMP 234, 31–40, DOI 10.1007/978-3-642-03118-2_4,

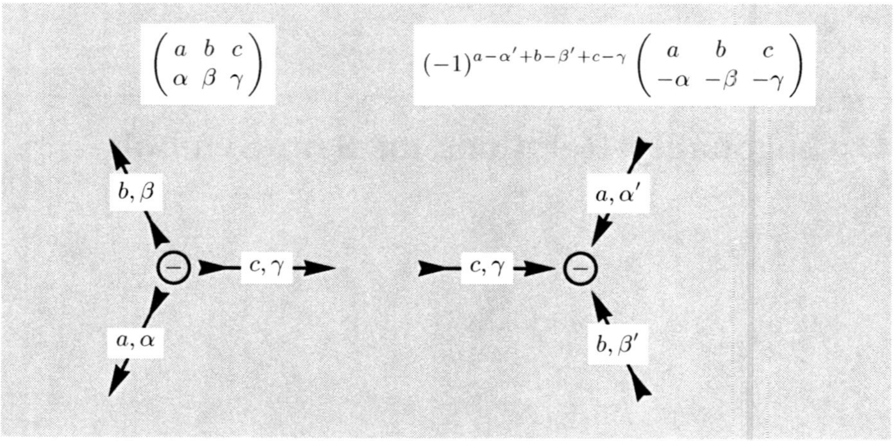

Fig. 4.1. The two diagrams have been arranged such that sums over c, γ can be introduced easily

Summations of projections are represented graphically by a single line connecting two nodes, as in Fig. 3.16. For the first time two nodes in the following Fig. 4.2 are connected by a double line, and this defines in our notation the sum over both c and γ. The use of an exclamation mark ! to emphasize an indirect change of node sign is mentioned in Sect. 3.1 and implemented here for the first time.

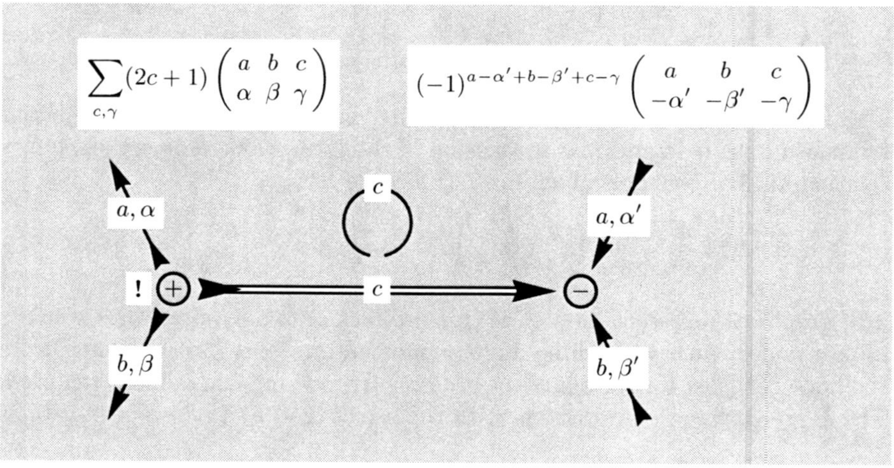

Fig. 4.2. The '!' in the diagram indicates that the external lines at the left node have changed place and the node sign has been altered accordingly. Note that three lines go out from the positive node and three lines go into the negative node

Because there are external lines at both ends of the diagram the single connection rule from p. 23 cannot be applied here. We know, however, from the result for the algebraic expression that the result is very simple, and choose the diagrams correspondingly. Thus, in the graphical representation the sum over c and γ leads to two free lines.

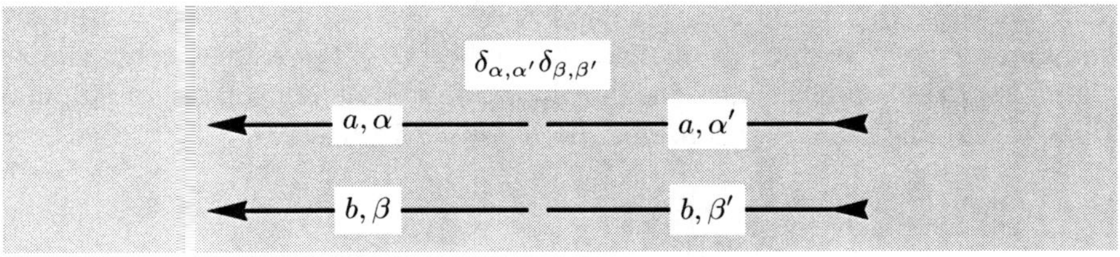

Fig. 4.3. After the removal of the summation line together with the two nodes, two single lines remain, which contribute $\delta_{\alpha,\alpha'}$ and $\delta_{\beta,\beta'}$

The graphical method has thus recovered the algebraic identity set out at the beginning of the section with just a few strokes:

$$\sum_{c,\gamma}(2c+1)\begin{pmatrix} a & b & c \\ \alpha & \beta & \gamma \end{pmatrix}(-1)^{a-\alpha'+b-\beta'+c-\gamma}\begin{pmatrix} a & b & c \\ -\alpha' & -\beta' & -\gamma \end{pmatrix}$$

$$= \sum_{c,\gamma}(2c+1)\begin{pmatrix} a & b & c \\ \alpha & \beta & \gamma \end{pmatrix}\begin{pmatrix} a & b & c \\ \alpha' & \beta' & \gamma \end{pmatrix} = \delta_{\alpha,\alpha'}\delta_{\beta,\beta'}$$

Note by going backwards from Fig. 4.3 to Fig. 4.2 that, the two figures indicate a reversed procedure by which two parallel angular momentum lines may be reduced to just one line at the price of an extra summation over an additional angular momentum and its projection.

■ **Summation convention for an internal line:**
A sum over an internal angular momentum line j, together with the factor $(2j+1)$, can be performed if corresponding lines enter and leave the nodes connected by the j-line. The sum removes the nodes at both ends of j and reconnects the remaining lines as described by the passage from Fig. 4.2 to 4.3. The j-line has to be anti-parallel to the resulting lines.

This procedure of line contraction constitutes an important tool that allows a reduction in the number of angular momentum lines. This technique will frequently be applied in the following sections.

■ **Line contraction method:**
Two lines of the same direction can be contracted to a single line by introducing two new nodes of differing node sign. The two lines enter one node and leave the other in the original directions. The two nodes are connected by a new anti-parallel angular momentum line, say x, and a sum over an internal line x, together with the factor $(2x+1)$ is introduced. The two contracted lines may be free, as in the passage from Fig. 4.3 to 4.2, or internal lines.

4.2 Identity for $3jm$-Symbols II

The following expression is a special case of (2.5) with an additional dummy summation over the third projection. This relation will provide another simple example of the application of the graphical method demonstrating manipulations of the diagrams. We consider the following identity,

$$\sum_{\alpha,\beta,\gamma} \begin{pmatrix} a & b & c \\ \alpha & \beta & \gamma \end{pmatrix} \begin{pmatrix} a & b & c \\ \alpha & \beta & \gamma \end{pmatrix} = 1 ,$$

and rewrite the second $3jm$-symbol in such a way as to allow transcription into graphical symbols,

$$\sum_{\alpha,\beta,\gamma} \begin{pmatrix} a & b & c \\ \alpha & \beta & \gamma \end{pmatrix} (-1)^{a-\alpha+b-\beta+c-\gamma} \begin{pmatrix} a & b & c \\ -\alpha & -\beta & -\gamma \end{pmatrix} = 1 ,$$

where the first $3jm$-symbol has three outgoing lines and the second three ingoing lines in accordance with the corresponding phase factors.

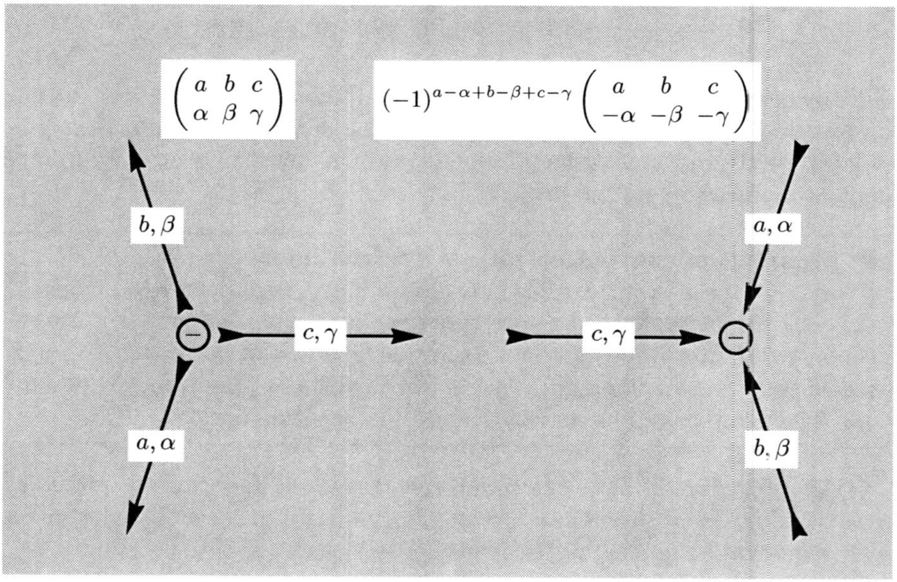

Fig. 4.4. Compared to Fig. 3.3 the diagram of the second $3jm$-symbol has been rotated around the node without changing the cyclic arrangement

For the sums over α and β the corresponding angular momentum lines are to be connected, but the arrangement of a- and b-lines is not yet ideal for that purpose. In the diagram for the $3jm$-symbol on the left-hand side of the following Fig. 4.5 the angular momentum lines for a and b have changed place together with a concomitant change in the node sign.

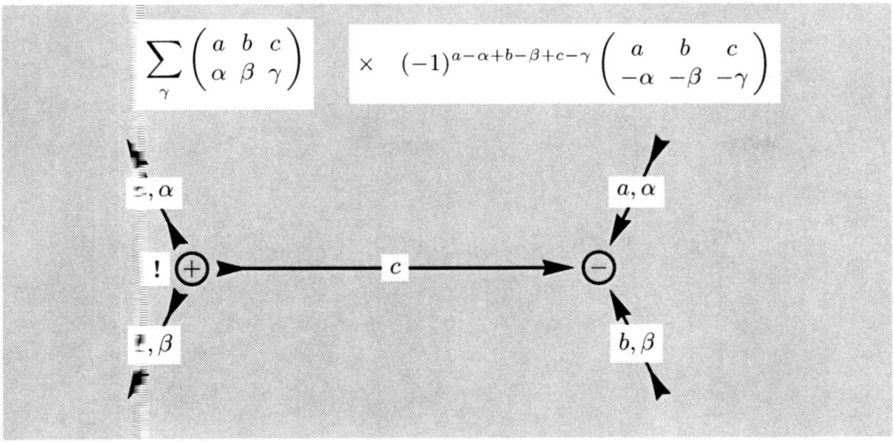

Fig. 4.5. The line joining the two nodes represents the summation over γ

To insert the summations over α and β the angular momentum lines are joined such that they connect the two nodes, as displayed in the following figure:

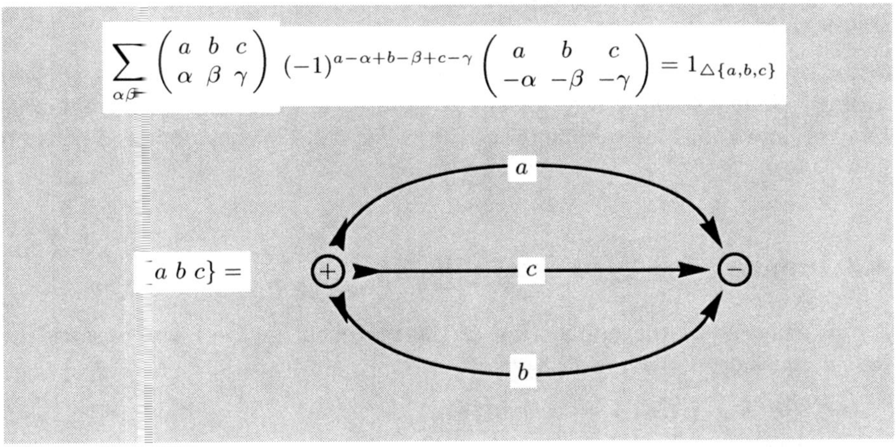

Fig. 4.6. The three lines joining the two nodes represent the summations over α, β and γ in accordance with the algebraic expression

The resulting diagram has the value 1, provided the three angular momenta a, b and c fulfil the triangle condition $\Delta\{a, b, c\}$, and the value is represented by $1_{\Delta\{a,b,c\}}$. This is an example of a *closed diagram* which does no longer depend on α, β and γ. This diagram is the first member of a group of symbols which are called $3nj$-symbols and it belongs to the value $n = 1$ and is referred to as the $3j$-symbol $\{a\ b\ c\}$. It is not difficult to see that $3nj$-symbols have $2n$ nodes, per definition.

It is, perhaps, at this point worthwhile to consider the effect of setting $c = 0$ in the $3j$-symbol displayed in Fig. 4.6.

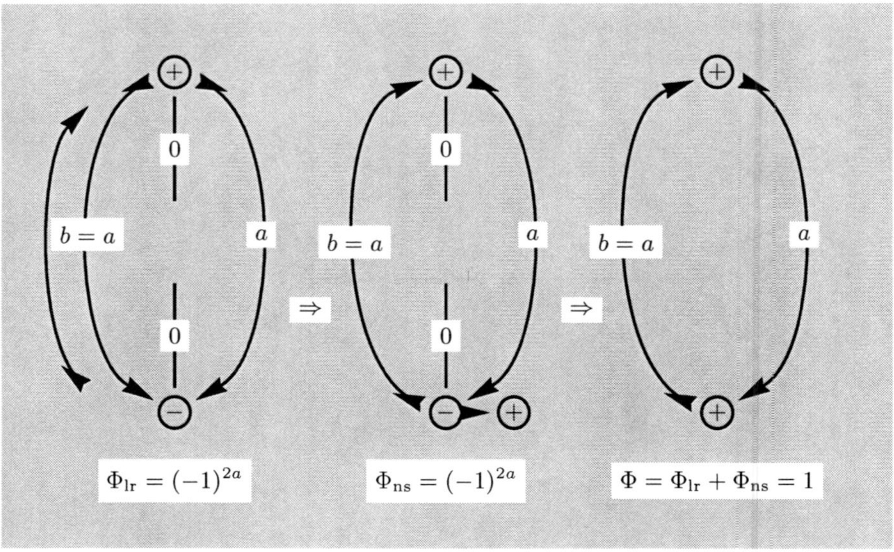

Fig. 4.7. If the c-line is set to zero, the 0-lines are inside the diagram and allow the transition to two-line nodes

After setting $c = 0$ in the leftmost diagram and a line reversal for b, with a direct change of node sign for the bottom node we obtain a diagram with two two-line nodes resembling entry (3) in Fig. 3.4. The accumulated phase is $(-1)^{4a} = 1$.

4.3 Identity for $3jm$-Symbols III

As an example of the application of the graphical method we consider the following relation given in (2.5):

$$\sum_{\alpha,\beta} \begin{pmatrix} a & b & c \\ \alpha & \beta & \gamma \end{pmatrix} \begin{pmatrix} a & b & c' \\ \alpha & \beta & \gamma' \end{pmatrix} = \frac{\delta_{c,c'}\delta_{\gamma,\gamma'}}{(2c+1)} \, .$$

As in previous manipulations of this identity, we rewrite the second $3jm$-symbol in such a way as to create the required phase factors which allow transcription into graphical symbols,

$$\sum_{\alpha,\beta} \begin{pmatrix} a & b & c \\ \alpha & \beta & \gamma \end{pmatrix} (-1)^{a-\alpha+b-\beta+c'-\gamma'} \begin{pmatrix} a & b & c' \\ -\alpha & -\beta & -\gamma' \end{pmatrix} = \frac{\delta_{c,c'}\delta_{\gamma,\gamma'}}{(2c+1)} \, ,$$

where the first $3jm$-symbol has three outgoing lines and the second three ingoing lines. The change to negative m-values introduces a phase $(-1)^{a+b+c}$ which may be augmented with $(-1)^{-\alpha-\beta-\gamma'} = 1$ because $\alpha + \beta + \gamma' = 0$ by definition.

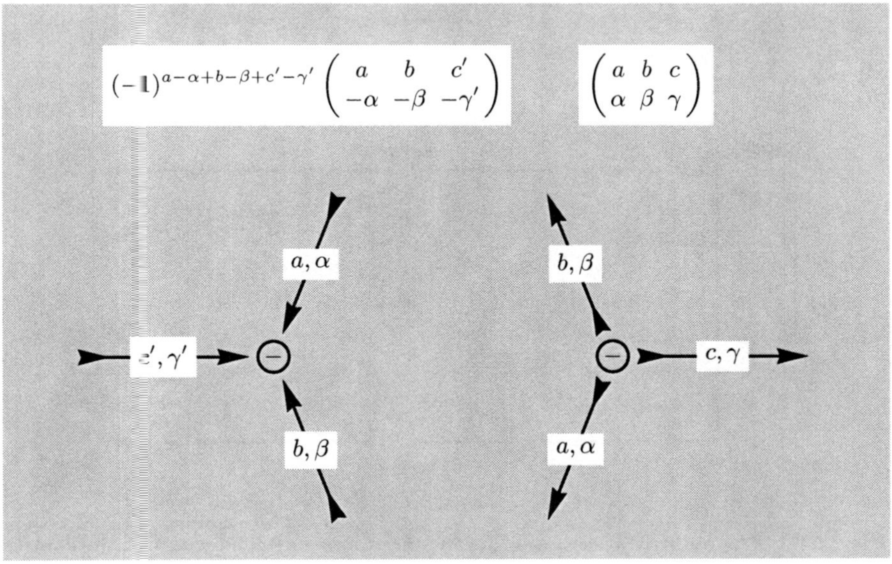

Fig. 4.8. Compared to Fig. 3.3 the diagrams of both $3jm$-symbols have been rotated around the node without changing the cyclic arrangement

To facilitate joining corresponding angular momentum lines the cyclic order of the right-hand side $3jm$-symbol is changed together with the node sign, as required. The next step consists in the introduction of the sums over projections α and β.

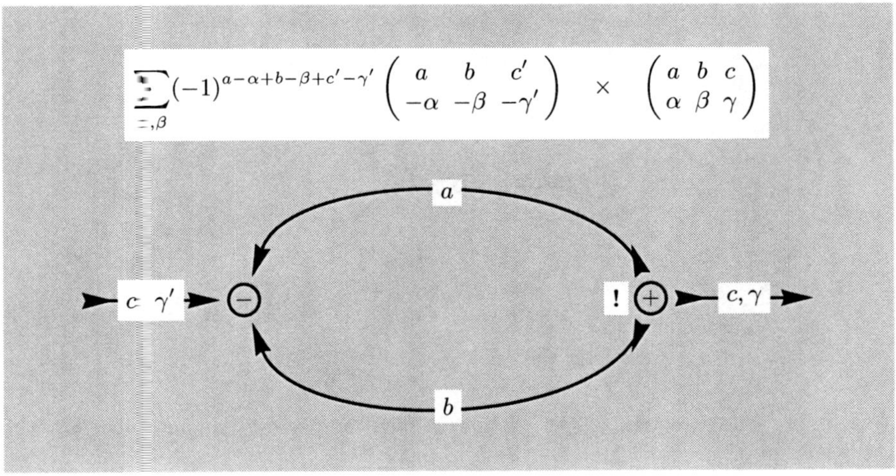

Fig. 4.9. Note that the cyclic order in the right $3jm$-symbol has been changed together with the node sign

The resulting diagram has only two remaining external lines and will now be rearranged without changing its value.

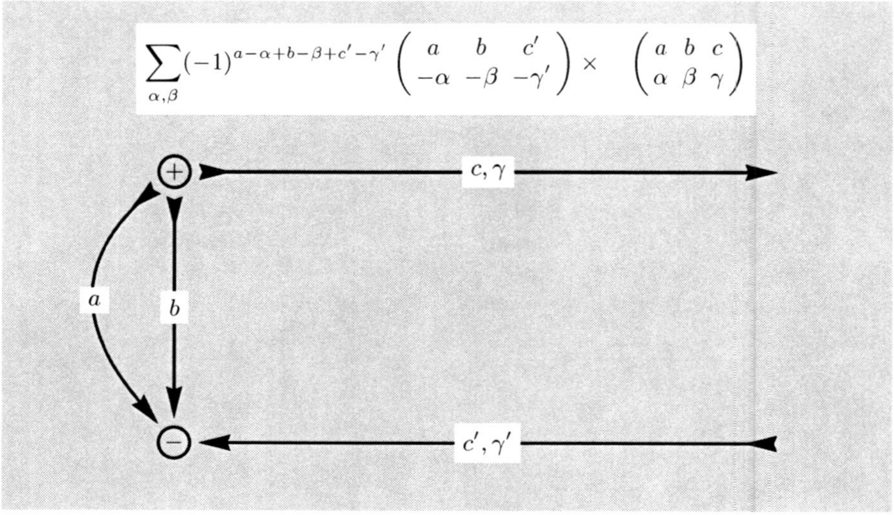

Fig. 4.10. The node positions and the line directions have changed but the content and the characteristics of the diagram remain unchanged

To make further progress we use the method of line contraction, thereby reducing the number of lines with a summation over an extra angular momentum x and its projection ξ, which has been demonstrated in Sect. 4.1. These steps include also a temporary change of direction for an external line to prepare the diagram for sums over x, ξ according to the contraction method, p. 33.

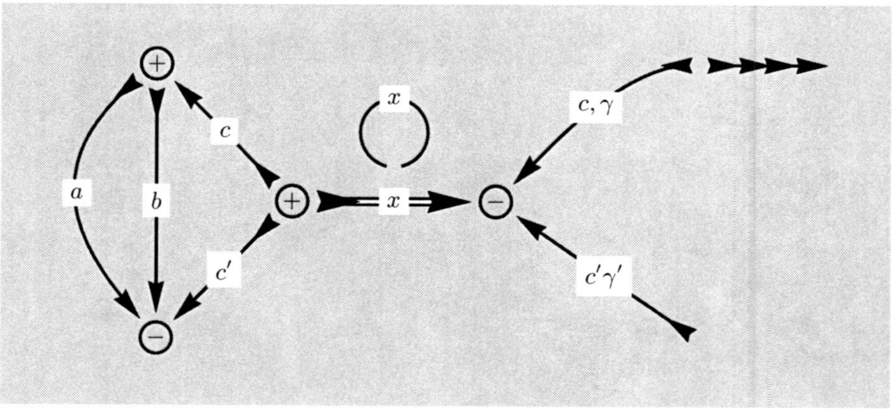

Fig. 4.11. The direction of the c, γ-line is now temporarily changed to allow the introduction of an extra sum over both x and ξ

The single line rule, p. 23, replaces x by zero[2]. In order to create the correct directions for node removal, the direction of the internal c' line is changed.

[2] Note that the double line x connects a closed diagram on the left with the external lines on the right as required by the single line rule.

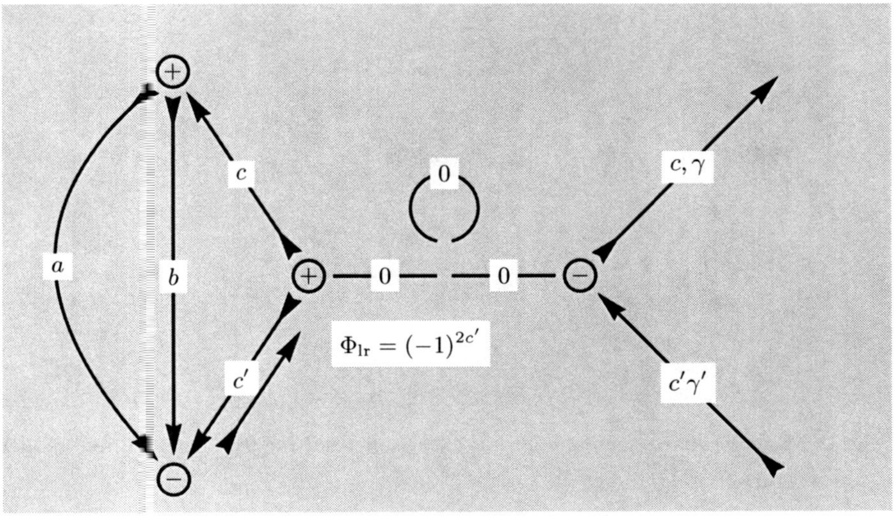

Fig. 4.12. The direction of the c, γ-line has been returned to the previous direction

Both nodes and the 0-lines are now removed, leading to the two triangular diagrams which taken together represent the factor $\delta_{c,c'}/(2c+1)$.

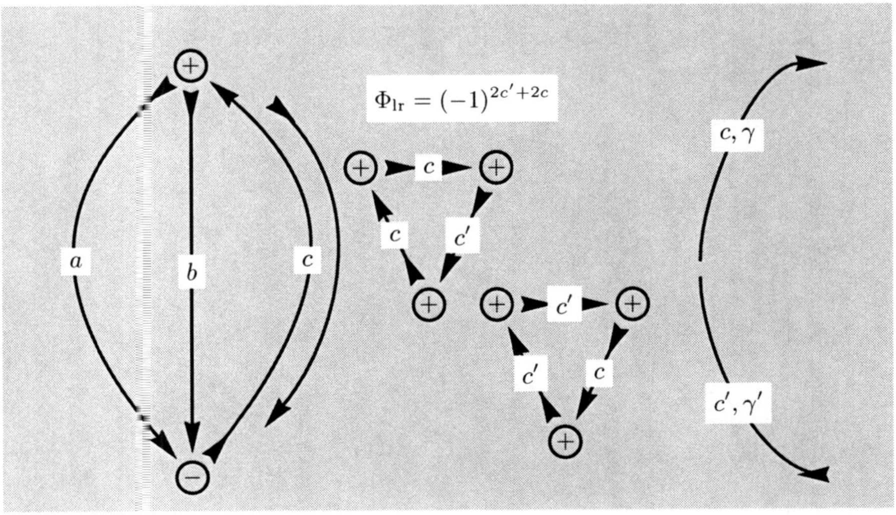

Fig. 4.13. The direction of the c-line in the left diagram is changed to achieve the desired orientation of all three lines The single line on the right-hand side introduces $\delta_{\gamma,\gamma'}$

The two triangular elements can be merged into a single quadrangular diagram shown as entry (5) in Fig. 3.4.

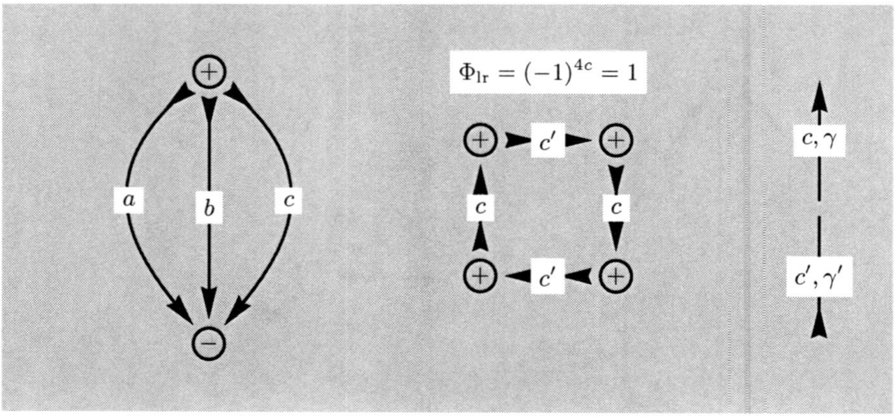

Fig. 4.14. The square element contains the factor resulting from the removed modes

The final result contains the symbol derived in Fig. 4.6, a factor $1/(2c+1)$ and a free line giving two Kronecker-δs required for the relation in question.

$$\sum_{\alpha,\beta}(-1)^{a-\alpha+b-\beta+c'-\gamma'}\begin{pmatrix} a & b & c' \\ -\alpha & -\beta & -\gamma' \end{pmatrix}\begin{pmatrix} a & b & c \\ \alpha & \beta & \gamma \end{pmatrix}$$
$$= \sum_{\alpha,\beta}\begin{pmatrix} a & b & c' \\ \alpha & \beta & \gamma' \end{pmatrix}\begin{pmatrix} a & b & c \\ \alpha & \beta & \gamma \end{pmatrix} = \frac{\delta_{c,c'}\delta_{\gamma,\gamma'}}{(2c+1)} \,.$$

5

Properties of the 6j-Symbol

Introduced in Sect. 2.2, the 6j is a symbol of very high symmetry and many applications in mathematical physics. E.g., sum rules for transition matrix elements. Symmetry properties will be reviewed in Sect. 5.2 after considering the 6j definition first.

5.1 Definition of the 6j-Symbol

The 3nj-symbols in the theory of angular momentum are all constructed from 3jm-symbols and we are going to look now at the 6j-symbol. This symbol has four nodes connected by six angular momentum lines in the graphical representation. An algebraic definition of the 6j-symbol has the following form (see for example [9]):

$$\begin{Bmatrix} a & b & c \\ d & e & f \end{Bmatrix} = \sum_{\substack{\alpha,\beta,\gamma \\ \delta,\varepsilon,\varphi}} (-1)^{d+e+f+\delta+\epsilon+\varphi} \times \begin{pmatrix} a & b & c \\ \alpha & \beta & \gamma \end{pmatrix} \begin{pmatrix} a & e & f \\ \alpha & \varepsilon & -\varphi \end{pmatrix} \times \begin{pmatrix} d & b & f \\ -\delta & \beta & \varphi \end{pmatrix} \begin{pmatrix} d & e & c \\ \delta & -\varepsilon & \gamma \end{pmatrix} .$$

In preparation for the translation into graphical symbols we change the signs of the projections of three symbols, taking into account (2.3) and that the sum of the three projections in a 3jm-symbol is always zero by definition:

$$\begin{Bmatrix} a & b & c \\ d & e & f \end{Bmatrix} = \sum_{\substack{\alpha,\beta,\gamma \\ \delta,\varepsilon,\varphi}} (-1)^{d+e+f+\delta+\epsilon+\varphi} \times \begin{pmatrix} a & b & c \\ \alpha & \beta & \gamma \end{pmatrix} (-1)^{a-\alpha+e-\varepsilon+f+\varphi} \begin{pmatrix} a & e & f \\ -\alpha & -\varepsilon & \varphi \end{pmatrix}$$

E. Balcar, S.W. Lovesey, *Introduction to the Graphical Theory of Angular Momentum*, STMP 234, 41–47, DOI 10.1007/978-3-642-03118-2_5,

$$\times(-1)^{d+\delta+b-\beta+f-\varphi}\begin{pmatrix} d & b & f \\ \delta & -\beta & -\varphi \end{pmatrix}$$
$$\times(-1)^{d-\delta+e+\varepsilon+c-\gamma}\begin{pmatrix} d & e & c \\ -\delta & \varepsilon & -\gamma \end{pmatrix} .$$

Rearranging the quantum numbers in the phase factors we arrive at

$$\begin{Bmatrix} a & b & c \\ d & e & f \end{Bmatrix} = \sum_{\substack{\alpha,\beta,\gamma \\ \delta,\varepsilon,\varphi}} (-1)^{2d+2\delta+2e+2\epsilon+2f+2\varphi}$$
$$\times\begin{pmatrix} a & b & c \\ \alpha & \beta & \gamma \end{pmatrix}(-1)^{a-\alpha+e-\varepsilon}\begin{pmatrix} a & e & f \\ -\alpha & -\varepsilon & \varphi \end{pmatrix}$$
$$\times(-1)^{b-\beta+f-\varphi}\begin{pmatrix} d & b & f \\ \delta & -\beta & -\varphi \end{pmatrix}(-1)^{d-\delta+c-\gamma}\begin{pmatrix} d & e & c \\ -\delta & \varepsilon & -\gamma \end{pmatrix} .$$

With $(-1)^{2d+2\delta} = 1$, $(-1)^{2e+2\epsilon} = 1$ and $(-1)^{2f+2\varphi} = 1$ the first phase factor can now be omitted and all four $3jm$-symbols have the necessary phases and are, in Fig. 5.1, transcribed into diagrammatic form.

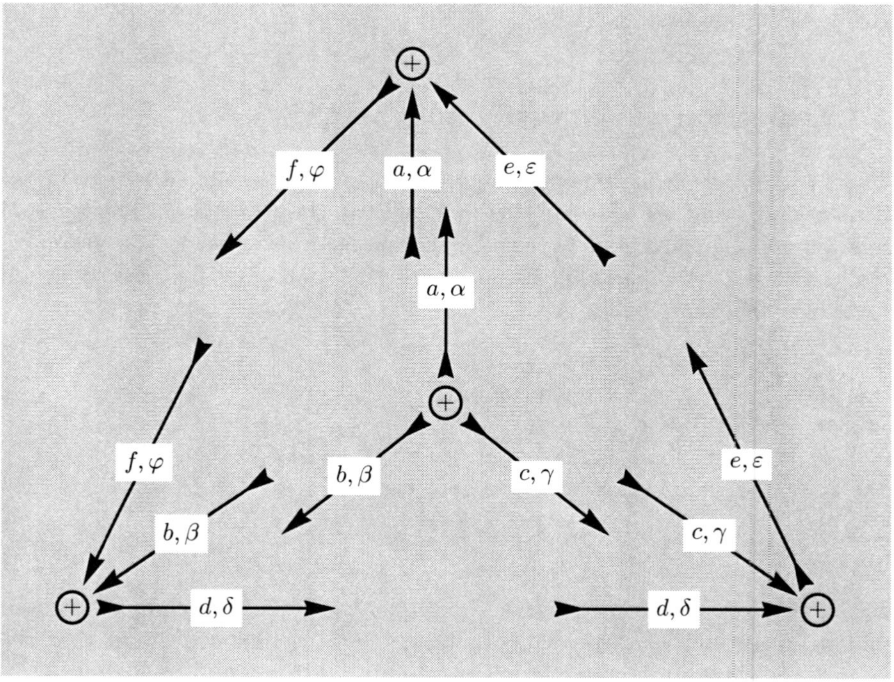

Fig. 5.1. The four $3jm$-symbols in our definition of a $6j$ are arranged such that corresponding in- and outgoing lines are aligned in preparation for sums over all projections

Due to the properties of the $3jm$-symbols not all six projections are independent of each other and, subsequently, some sums are purely formal. The resulting graphical symbol is a closed diagram with no external lines as it must be.

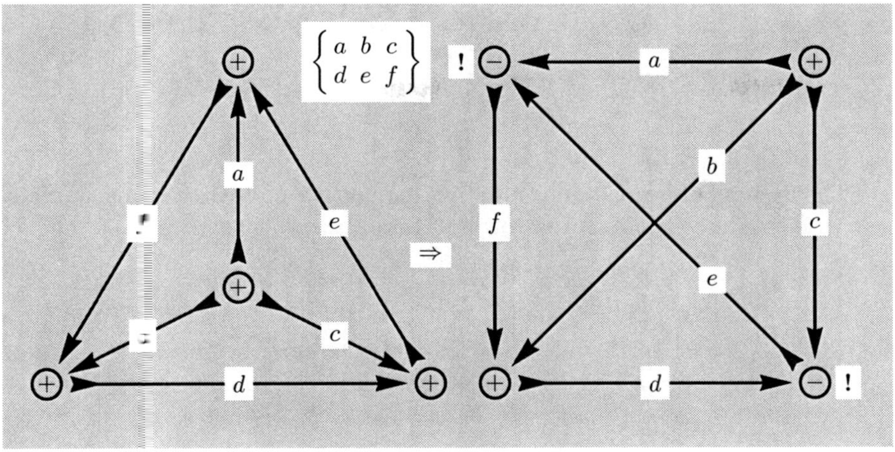

Fig. 5.2. The angular momentum lines connecting the four nodes indicate sums over all projections in the definition of the $6j$

The graphical representation of the $6j$-symbol consists of four positive nodes in a triangular pattern with one central node. The three angular momenta in the top row of the $6j$-symbol are attached to the central node and are directed to the three outer nodes. The three angular momenta from the bottom row of the $6j$-symbol are arranged in counter-clockwise direction as sides of the triangle in such a way that the elements of the columns in the $6j$-symbol are opposites.

The quadratic diagram in the right-hand part of Fig. 5.2 is obtained from the triangular diagram by moving the central node to the right and up across the e-line. This move affects the arrangement of the angular momentum lines at two nodes which change, therefore, to a negative node sign. This is just one example for an allowed deformation of a diagram and we see that there are many more completely equivalent diagrams which can be generated in a similar way. The diagram on the left-hand side of Fig. 5.2 represents the definitive form of a $6j$-symbol and, in some cases, transformed diagrams can acquire a phase when brought into the form of Fig. 5.2 with corresponding line directions and node signs.

5.2 Symmetry Properties of the 6j-Symbol

The $6j$-symbol vanishes unless the four triangle conditions in the four $3jm$-symbols of the definition are satisfied. The triangle conditions for the analytic form of the $6j$-symbol are indicated here:

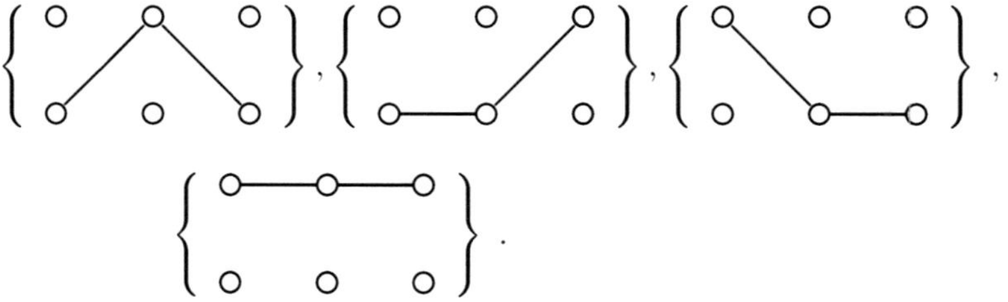

The $6j$ symbol is invariant under any permutation of its columns, and any pairwise exchange between top-row and bottom-row elements

$$\begin{Bmatrix} a & b & c \\ d & e & f \end{Bmatrix} = \begin{Bmatrix} a & e & f \\ d & b & c \end{Bmatrix} = \begin{Bmatrix} d & b & f \\ a & e & c \end{Bmatrix} = \begin{Bmatrix} d & e & c \\ a & b & f \end{Bmatrix} ,$$

for example. It is worthwhile to perform the graphical equivalent of these operations and we consider the last $6j$-symbol in the equality above.

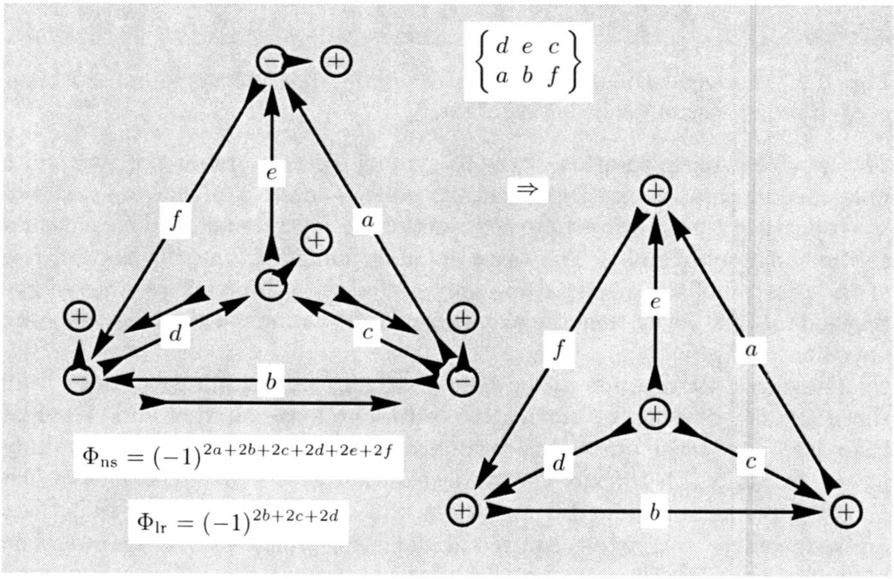

Fig. 5.3. Compared to Fig. 5.2 the central node has changed place with the right-hand bottom node. This operation has altered the cyclic arrangement of lines at all four nodes which have obtained a negative node sign The usual '!' indicator of node sign change is absent here. The left-hand diagram needs several changes in order to assume assume the definitive form of a $6j$-symbol

In order to conform with the diagram given in Fig. 5.2 all node signs and three line directions have to be changed. The corresponding phases result from node sign change where each angular momentum contributes twice to $\Phi_{\rm ns}$, and line reversal contributes $\Phi_{\rm lr}$.

$$\Phi_{\rm ns} = (-1)^{2a+2b+2c+2d+2e+2f} \qquad \Phi_{\rm lr} = (-1)^{2b+2c+2d} .$$

In the first equality we use $(-1)^{2e+2f} = (-1)^{2a}$ and find that the net phase is 1 as expected.

$$\Phi = (-1)^{4a+4b+4c+4d} = 1 \, .$$

Except for an exchange of columns, which leaves a 6j-symbol invariant, the right-hand side diagram in Fig. 5.3 has the definitive form of a 6j-symbol.

5.3 6j-Symbol with a zero element

A special 6j-symbol is obtained by setting $d = 0$, say, and, as a consequence, $e = c$ and $f = b$ to obey the triangle rule.

$$\begin{Bmatrix} a & b & c \\ 0 & c & b \end{Bmatrix} = (-1)^{a+b+c} \frac{1}{\sqrt{(2b+1)(2c+1)}} \, .$$

This result is now derived by the graphical technique:

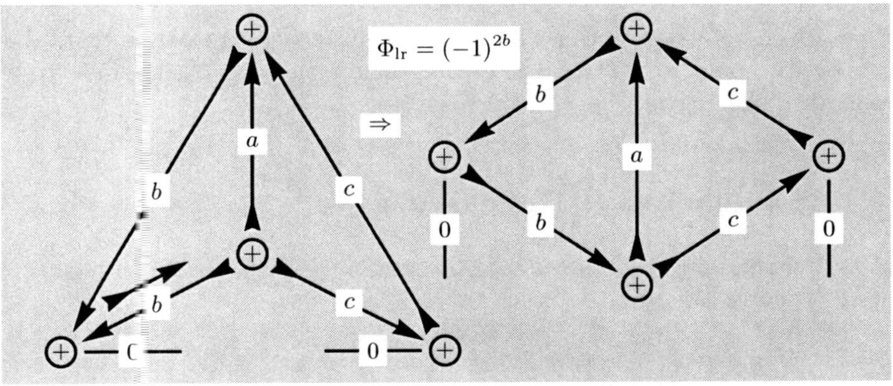

Fig. 5.4. The value for the d-line in Fig. 5.2 is set to zero. The direction of one b-line is reversed to create the appropriate arrangement of lines for node removal of a node with positive node sign

Both nodes with the zero-lines can now be removed and the corresponding factors are taken into account.

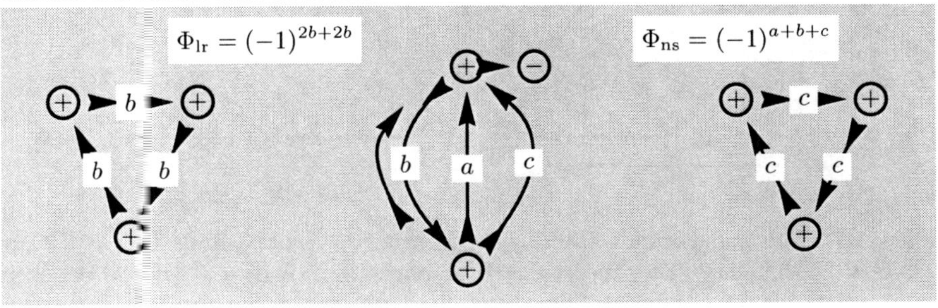

Fig. 5.5. The central diagram needs another line reversal and a sign change of the top node to conform with the 3$j$symbol first given in Fig. 4.6

The phase from line reversals is $\Phi_{\mathrm{lr}} = (-1)^{4b} = 1$ and only the phase from the change of one node sign $\Phi_{\mathrm{ns}} = (-1)^{a+b+c}$ remains.

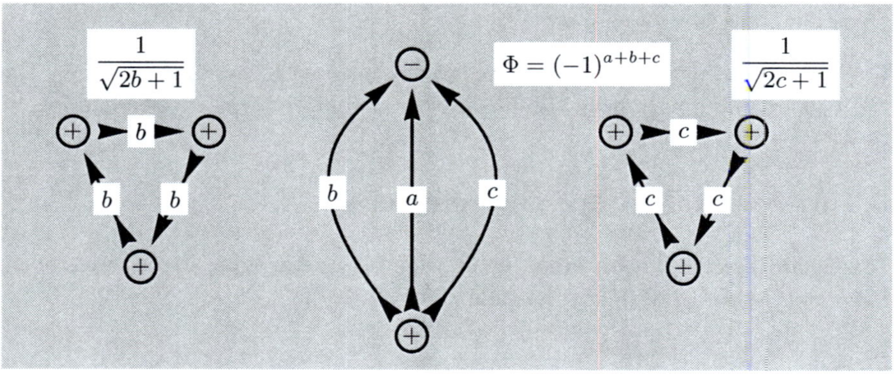

Fig. 5.6. The central diagram resembles now the diagram in Fig. 4.6 except for the order of the lines

To reach complete agreement with Fig. 4.6 would require a rotation that takes the positive node to the top. The angular momentum line b is then moved to the other side, leaving line c in the middle.

5.4 6j-Symbol with Three 0-Lines

A very simple value is obtained for a 6j-symbol in which all angular momenta in the top row are zero.

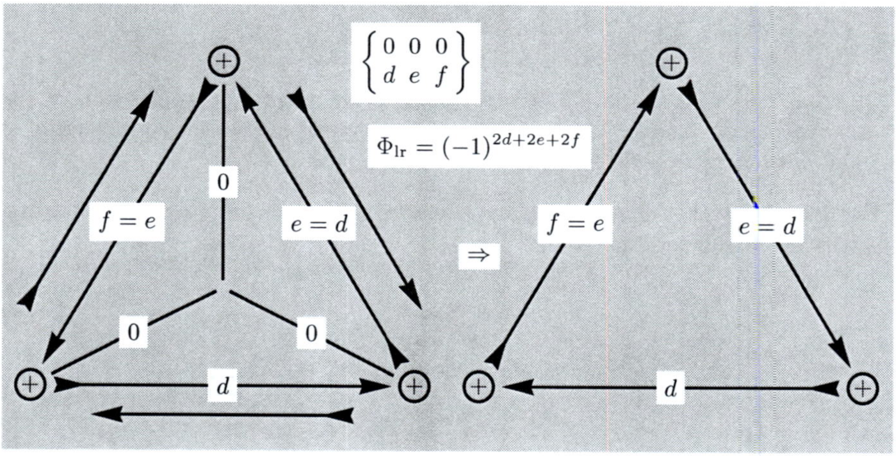

Fig. 5.7. The angular momentum lines leaving the central node in Fig. 5.2 are all zero in this case. The line reversals arrange the 0-lines and the others as is appropriate for node removal. Since the 0-lines are all pointing inside it is possible to change to two-line nodes

The transition to two-line nodes in the right-hand part of Fig. 5.7 reproduces the well known entry (4) from Fig. 3.4 for $1/\sqrt{2d+1}$. Because $d = e = f$ line reversals contribute a phase $\Phi_{\text{lr}} = (-1)^{2d}$. In analytical form we have

$$\begin{Bmatrix} 0 & 0 & 0 \\ d & e & f \end{Bmatrix} = (-1)^{2d} \frac{\delta_{d,e}\delta_{e,f}}{\sqrt{2d+1}} \; .$$

The degenerate central node with 3 zero-lines, equivalent to an $3jm$-symbol with six zeroes and a value of 1, has been left out in Fig. 5.7.

6

Properties of the $9j$-Symbol

Introduced in Sect. 2.2, the $9j$-symbol is, algebraically, less symmetric than the $6j$ but equally useful in mathematical physics. In contrast, the diagrammatic form displays a very high degree of symmetry.

The definition of the $9j$-symbol and its graphical representation is considered in Sect. 6.1 before some of the properties and the symmetries are treated in Sect. 6.2.

6.1 Definition of the $9j$-Symbol

The $3nj$-symbol with $n = 3$ is constructed from six $3jm$-symbols summed over all projections. We note again, that not all projections are independent and so some of the sums are purely formal. The definition of the $9j$-symbol as given in [4] or [9] is

$$\begin{Bmatrix} a & b & c \\ d & e & f \\ k & l & m \end{Bmatrix} = \sum_{\substack{\alpha,\beta,\gamma \\ \delta,\epsilon,\varphi \\ \kappa,\lambda,\mu}} \begin{pmatrix} a & b & c \\ \alpha & \beta & \gamma \end{pmatrix} \begin{pmatrix} d & e & f \\ \delta & \epsilon & \varphi \end{pmatrix} \begin{pmatrix} k & l & m \\ \kappa & \lambda & \mu \end{pmatrix} \times \begin{pmatrix} a & d & k \\ \alpha & \delta & \kappa \end{pmatrix} \begin{pmatrix} b & e & l \\ \beta & \varepsilon & \lambda \end{pmatrix} \begin{pmatrix} c & f & m \\ \gamma & \varphi & \mu \end{pmatrix} .$$

Note the use of an angular momentum m with projection μ in this formula, not to be confused with the projection m of j. In order to prepare this analytic expression for transcription into graphical symbols we change the projections of the last three $3jm$-symbols, from positive to negative according to (2.4), and recall that projections in a $3jm$-symbol add to zero.

$$\begin{Bmatrix} a & b & c \\ d & e & f \\ k & l & m \end{Bmatrix} = \sum_{\substack{\alpha,\beta,\gamma \\ \delta,\epsilon,\varphi \\ \kappa,\lambda,\mu}} \begin{pmatrix} a & b & c \\ \alpha & \beta & \gamma \end{pmatrix} \begin{pmatrix} d & e & f \\ \delta & \epsilon & \varphi \end{pmatrix} \begin{pmatrix} k & l & m \\ \kappa & \lambda & \mu \end{pmatrix}$$

E. Balcar, S.W. Lovesey, *Introduction to the Graphical Theory of Angular Momentum*, STMP 234, 49–53, DOI 10.1007/978-3-642-03118-2_6,

$$\times(-1)^{a-\alpha+d-\delta+k-\kappa}\begin{pmatrix} a & d & k \\ -\alpha & -\delta & -\kappa \end{pmatrix}(-1)^{b-\beta+e-\varepsilon+l-\lambda}\begin{pmatrix} b & e & l \\ -\beta & -\varepsilon & -\lambda \end{pmatrix}$$
$$\times(-1)^{c-\gamma+f-\varphi+m-\mu}\begin{pmatrix} c & f & m \\ -\gamma & -\varphi & -\mu \end{pmatrix} .$$

From this formula one can see that the 9j-symbol is built from three 3jm-symbols with three outgoing lines and three 3jm-symbols with three ingoing lines.

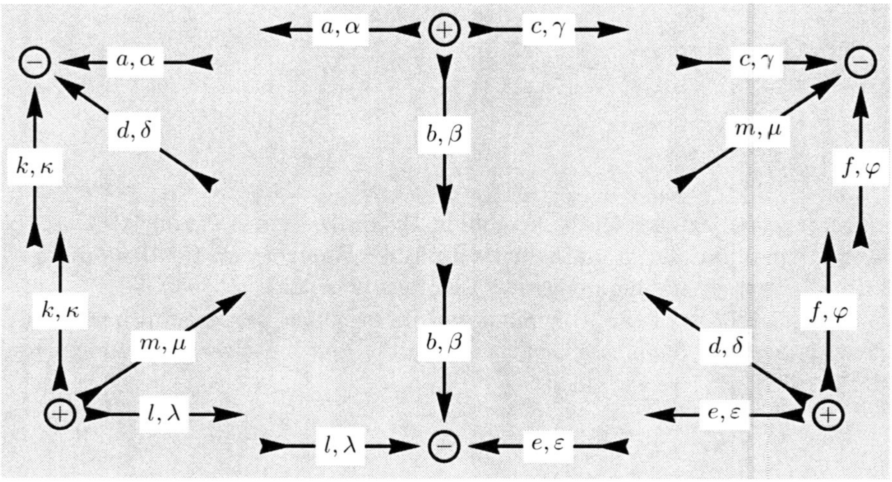

Fig. 6.1. Six 3jm-symbols in our definition of a 9j-symbol are arranged such that the intended line connections are anticipated

All nodes are now connected to indicate the summation over all magnetic quantum numbers.

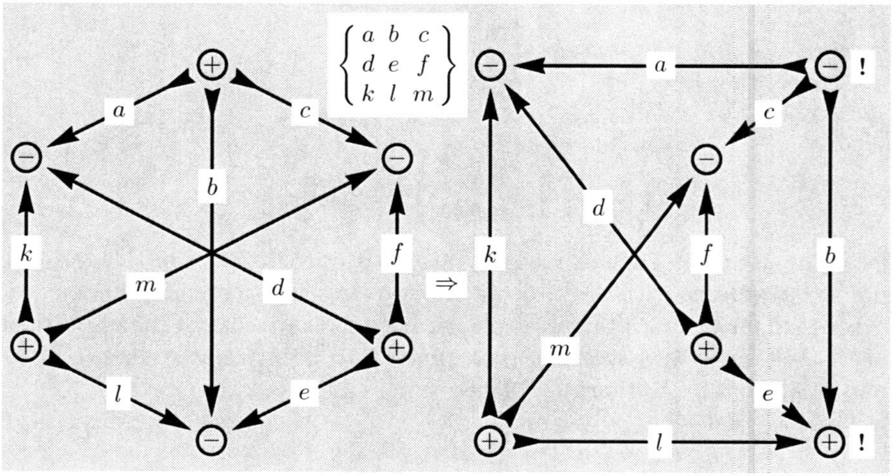

Fig. 6.2. Some of the summations in these diagrams are purely formal. The diagram on the left-hand side is the graphical representation of the 9j-symbol

The graphical representation of the 9j-symbol, on the left-hand side of Fig. 6.2, has nodes arranged in shape of a hexagon with alternating positive and negative node signs. The angular momenta of each row of a hexagon-shaped 9j-symbol go out from one of the positive nodes and end in three negative nodes. All three elements of a column in a 9$j$symbol go into a negative node.

The diagram on the right-hand side of Fig. 6.2 is the result of moving the top and bottom node to the right thereby changing the line arrangement and node sign at both nodes. Both diagrams are completely equivalent with respect to their value.

6.2 Symmetry Properties of the 9j-Symbol

All rows and all columns in a 9j-symbol are subject to a triangle condition as we can see from its definition through six 3jm-symbol. The symmetry properties of the 9j-symbol allow 64 symmetry operations in total. The 9j-symbol is invariant against cyclic (even) permutation of rows and columns and a reflection through the main diagonal. An odd permutation introduces a phase factor. This, of course, is also obtainable from the graphical method.

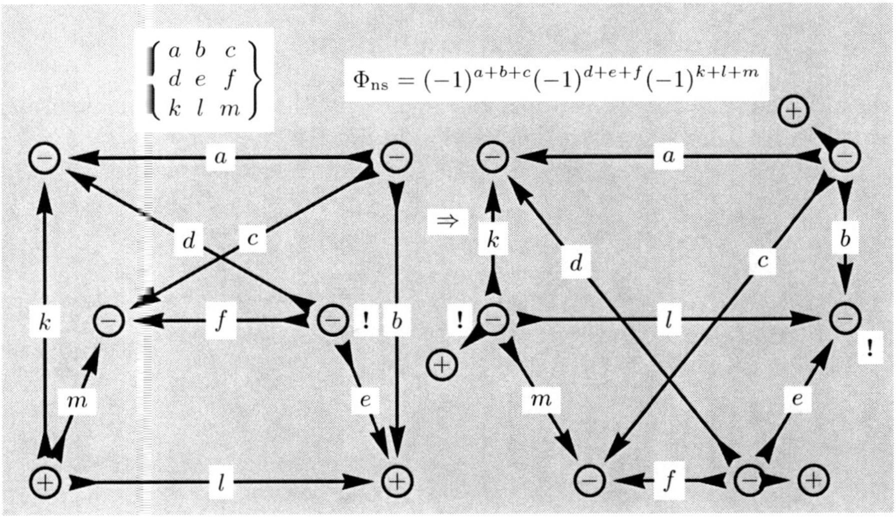

Fig. 6.3. The right-hand side diagram in Fig. 6.2 is rearranged in such a way that the d- and the f-line at the positive inner node change place Here, the horizontal f-line is moved below the l-line in the right-hand side diagram

Note here that in one case an indirect node-sign change, indicated by the '!', is followed by a direct node-sign change. In order to correspond to the graphical form of the 9j-symbol three nodes in Fig. 6.3 need a different node sign and this leads to the phase $\Phi_{\rm ns}$. In each case the outgoing lines belong to positive nodes and enter negative nodes.

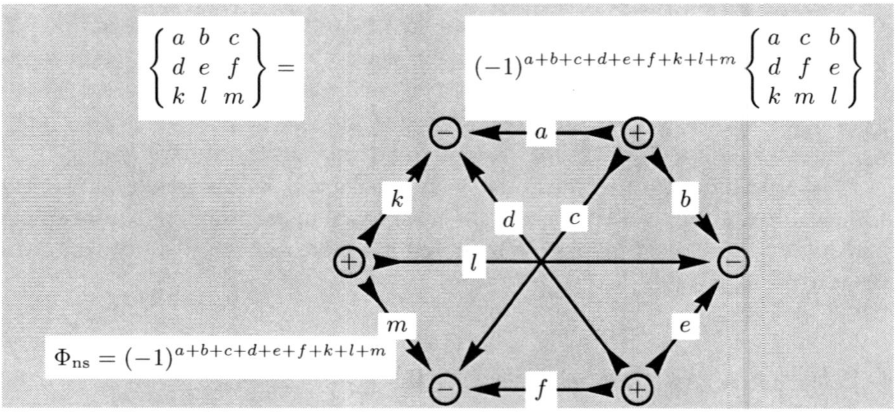

Fig. 6.4. The resulting diagram is, once again, in accordance with the definition of the 9j-symbol but with an additional phase factor

The exchange of the second and third column, which corresponds to an odd permutation, has introduced a phase factor where the exponent is given by the sum over all angular momenta in the 9j-symbol.

6.3 A 9j-Symbol with a zero element

As a further exercise we consider a 9j-symbol with one angular momentum equal to zero and start with the diagram in Fig. 6.2.

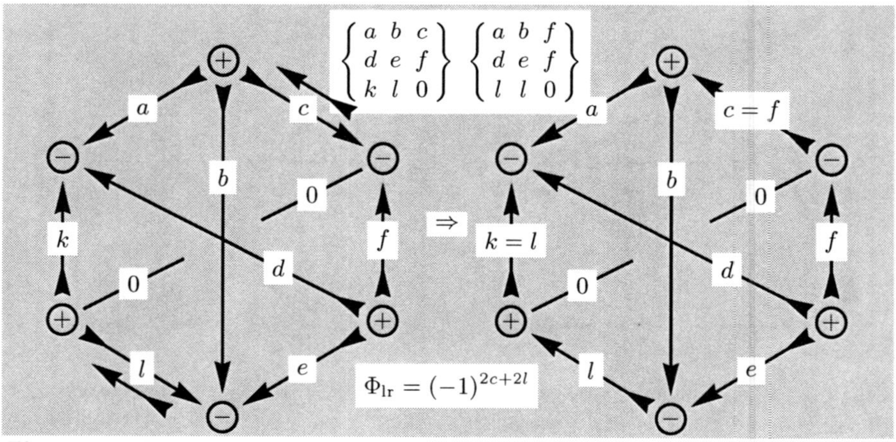

Fig. 6.5. Before the zero-lines and the nodes can be removed, reversal of the c- and the l-line creates the appropriate line arrangement of ingoing, zero- and outgoing line

All rows and columns in a 9j-symbol have to fulfil the triangle rule and this implies $c = f$ and $k = l$ when angular momentum m is set zero. After the removal of the two nodes with their zero-lines we are left with a diagram containing four nodes.

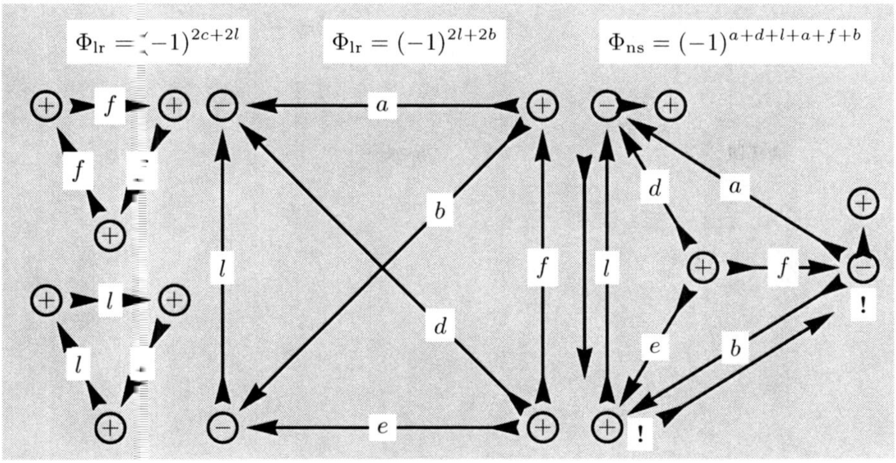

Fig. 6.6. The two factors representing $1/\sqrt{2f+1}$ and $1/\sqrt{2l+1}$ are shown on the left and the Kronecker-δ have been taken into account in the diagrams. To approach the definition of a 6j-symbol the bottom right node is moved into the triangle defined by the a-, b- and l-line

The accumulated phase simplifies because $(-1)^{4l} = 1$ and, furthermore, $(-1)^{2a+2b+2c} = 1$. The remaining phase factor is $(-1)^{b+d+f+l}$.

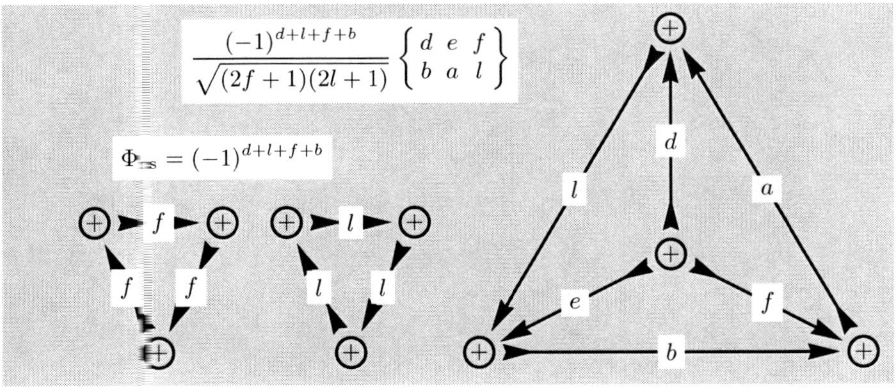

Fig. 6.7. The graphical diagrams are now converted into their analytic expressions

With this result we have concluded the graphical derivation of a formula given, for instance, in [9] for a 9j-symbol with one angular momentum set zero:

$$\left\{ \begin{matrix} a & b & f \\ d & e & f \\ l & l & 0 \end{matrix} \right\} = \frac{(-1)^{d+l+f+b}}{\sqrt{(2f+1)(2l+1)}} \left\{ \begin{matrix} a & b & f \\ e & d & l \end{matrix} \right\} .$$

The first two columns in the 6j-symbol have been rearranged here to facilitate comparison with the relation quoted in the cited reference.

7

General Principles for Diagrams

The methods introduced in Chaps. 3-6 concentrate on the application of the graphical technique to specific examples and simple diagrams. Building on the basic material from these chapters, we are now in a position to establish a set of principles and procedures formulated for general diagrams.

It is worthwhile to remind the reader that several conditions and conventions have been specified in Chap. 3. In addition to the basic facts of using nodes and angular momentum lines, the special methods reserved for zero-lines and the removal of nodes attached to them are covered. Further basic building blocks deal with direct change of a node sign, properties of single external and internal lines, direction reversal and summation over the projection of an internal angular momentum line. Section 3.9 on reversals of external lines is included for completeness although, at that stage, the reason for the development may seem opaque.

Chapter 4 introduces the notation for a sum over both angular momentum and its projection and implements this into a tool for reducing two external lines to a double-line summation. Familiarity with algebraic identities of $3jm$-symbols is used to explain the value of the simplest closed diagram, the $3j$-symbol.

7.1 Multiplication Convention

Some figures in the examples we have seen so far contained diagrams consisting of several unlinked graphs. It was implicitly assumed that such unlinked diagrams translate into a product of the corresponding algebraic expressions. In the same way, an algebraic expression consisting of several factors is converted term by term into the the equivalent graphical symbols. This conversion may, as we have seen in Sect. 5.1, require the introduction of additional phase factors.

E. Balcar, S.W. Lovesey, *Introduction to the Graphical Theory of Angular Momentum*, STMP 234, 55–71, DOI 10.1007/978-3-642-03118-2_7,

7.2 Notation for Graphical Elements

All but a few graphical symbols contain signed nodes represented by a circle with + or − inside and these signs are related, as we have seen, to the cyclic order of the angular momenta belonging to this node. Normally, each node has three angular momenta represented by directed lines going into or coming out from the node. A diagram may be *closed* with only internal lines or *open*, if external lines are present. An external line is connected to only one node with the other end remaining free.

A directed line connecting two nodes represents a sum over projections of the angular momentum. If, in addition, the sum is over both the angular momentum and its projection, then this sum is represented by a directed double line in the diagrams (see for example Fig. 4.2 and note the circular diagram accompanying the double line).

Phase factors due to direct changes of node sign and/or reversals of line direction are represented by Φ_{ns} or Φ_{lr}, respectively, and, in some cases, combined to the total phase factor $\Phi = \Phi_{\mathrm{ns}} + \Phi_{\mathrm{lr}}$.

7.3 Transformation and Deformation

The angular momentum lines connecting to nodes can be straight or curved. They may be extended and deformed without affecting the value of the diagram. If, however, a deformation of a diagram requires a line to pass over another line, thus altering the cyclic arrangement of lines at a node, then the node sign changes with this transformation (see Fig. 5.2, for example). To draw attention to the fact that the node sign has changed due to a diagram transformation a '!' is put next to the node and first seen in Fig. 4.2 and mentioned in Sect. 3.1.

Because the orientation of an angular momentum line has no specific significance, it is clear that, a rotation of a diagram has no effect on the value of the diagram and leads, upon transcription, to an identical algebraic expression. Transforming a diagram into its mirror image, or enantiomorph, has the effect of changing all node signs.

In following sections we will present methods for separating and joining diagrams, changing the direction of angular momentum lines and other possible transformations. The application of the graphical method is usually directed at evaluating a complicated algebraic expression, containing $3jm$- and $3nj$-symbols, which is converted into the equivalent graphical representation. The resulting diagrammatic form may then be treated according to the rules presented here, and, at the end, the result is again transformed into the corresponding algebraic form. When phase factors are neglected, the graphical method can be applied extremely fast by sketching diagrams with undirected lines and unsigned nodes. It is, however, essential, for the exact result, to painstakingly keep track of the phase factors resulting from explicit changes of line directions or node signs.

7.4 Diagrams Connected by a Single Internal Line

Rectangular and similar symbols in the following diagrams represent general symbols with an unspecified number of nodes and, in most cases, at least one of these rectangles with rounded corners has no external lines[1]. The rules are, of course, also applicable to completely closed diagrams, in which case there are no external lines.

In the case of a single line connection the procedure is particularly simple, because from symmetry arguments [10; 12] follows zero as the only possible value for the angular momentum a. Already, a simple example has been encountered in Fig. 3.16.

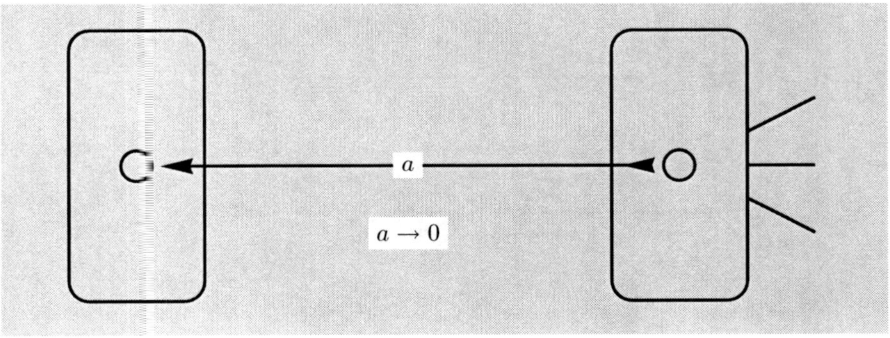

Fig. 7.1. Separation of two diagrams connected by a single line

The single line $a = 0$ is replaced by zero-lines attached to a node in Fig. 7.2. The conditions for removal of the 0-line and each node depend on the non-zero lines (not shown in the diagram). These non-zero lines must have the same angular momentum value.

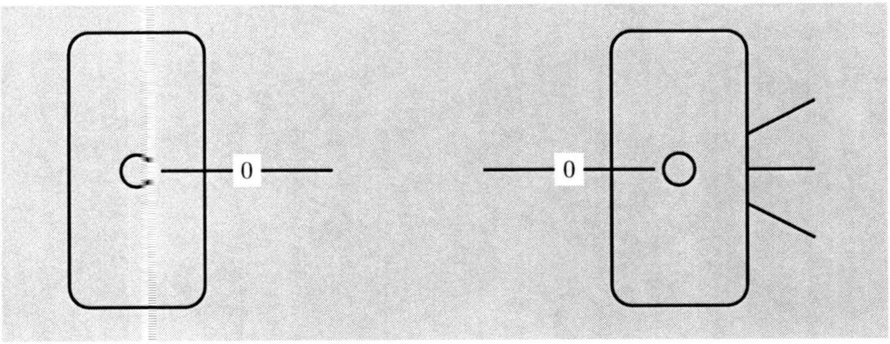

Fig. 7.2. Of the two diagrams at least one has no external lines

The removal of such modes has been demonstrated and discussed in Sects. 3.3 and 3.7.

[1] The three external lines in the general diagrams stand symbolically for any number of external lines.

We have also seen that, a single external line can only have angular momentum zero, see, e.g., Figs. 3.8 and 3.9.

7.5 Separation for Diagrams Connected by Two Lines

Figure 7.3 shows a connection by two internal angular momentum lines of different direction. In order to have the same direction for the lines, we change the direction of a, for example. This action introduces, as we have seen in Fig. 3.20, a phase factor $\Phi_{\mathrm{lr}} = (-1)^{2a}$ for the line reversal of a.

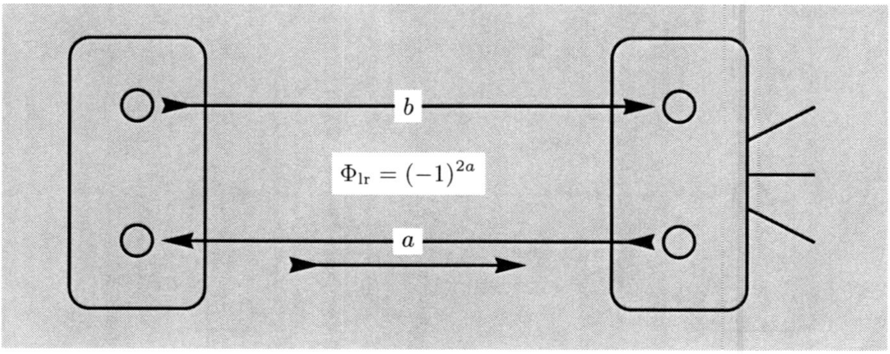

Fig. 7.3. Separation of two diagrams connected by two lines

Two parallel angular momentum lines may be reduced to one line through the introduction of two new nodes of differing node signs and an angular momentum line x, ξ connecting both nodes in the opposite direction. The expression is multiplied by $(2x + 1)$ and summed over x and ξ (see e.g. Fig. 4.11). The technique of reducing the number of lines is also referred to as line contraction, see Sect. 7.6.

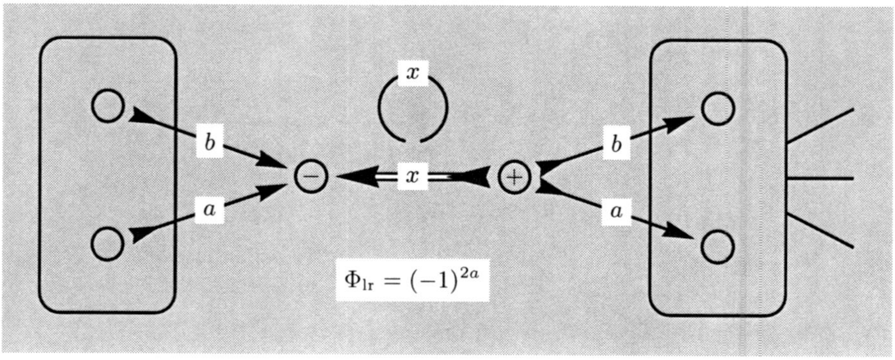

Fig. 7.4. The summation over x and ξ, with a factor $(2x+1)$, reduces the connection to one single line. The double line indicates that both variables x and ξ are to be summed

These two figures can also be read in reverse direction, i.e. such a sum over x, ξ can be performed if corresponding angular momentum lines enter both nodes such that we end up with two parallel lines.

However, since the goal here is to separate the two diagrams we can now appeal to the single line separation procedure as discussed in Sect. 7.4. In order to create the appropriate directions of ingoing and outgoing lines and the required node sign, we change the direction of the a-line in both diagrams and change the right node from + to −. This action leads to additional phase factors for line reversal and $\Phi_{\text{ns}} = (-1)^{a+b}$ for the node sign change.

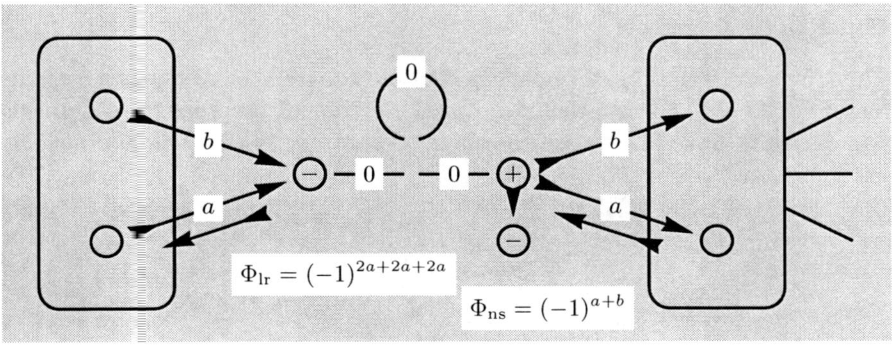

Fig. 7.5. The single line requires $x = 0$ and the line can be replaced by 0-lines. In order to have the correct arrangement of ingoing and outgoing lines at the free nodes two line directions and one node sign are reversed

At both nodes the arrangement of the ingoing and outgoing lines relative to the 0-line is in accordance with the node sign. Thus, we are left with two nodes where two non-zero angular momentum lines go in and out and, of course, the two angular momenta must be the same.

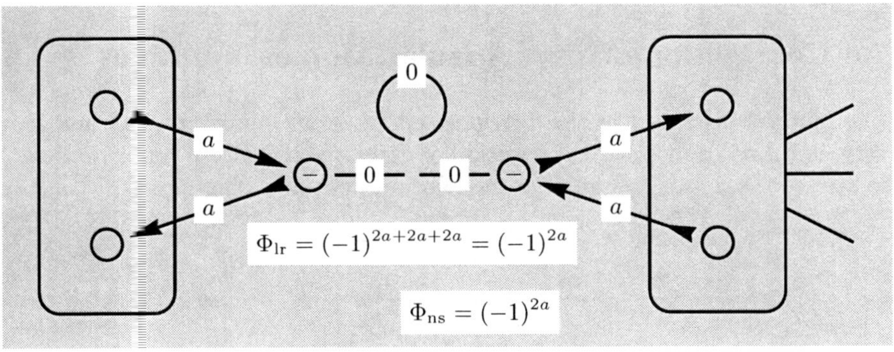

Fig. 7.6. We have $a = b$ for both lines. The circular 0-line corresponds to the factor 1

The removal of each node contributes a factor $1/\sqrt{2a+1}$ represented by a triangular symbol in Fig. 7.7.

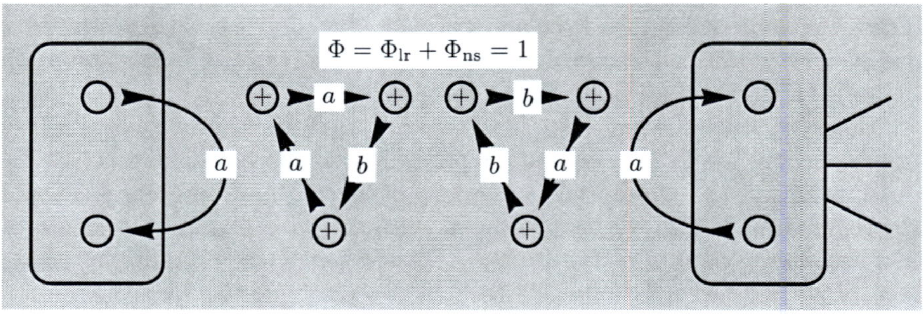

Fig. 7.7. The overall phase is 1

The triangular symbols can be merged into a square symbol representing the factor $1/(2a + 1)$. This concludes our derivation of the rule for separating two diagrams linked by two lines, where at least one diagram should have no external lines.

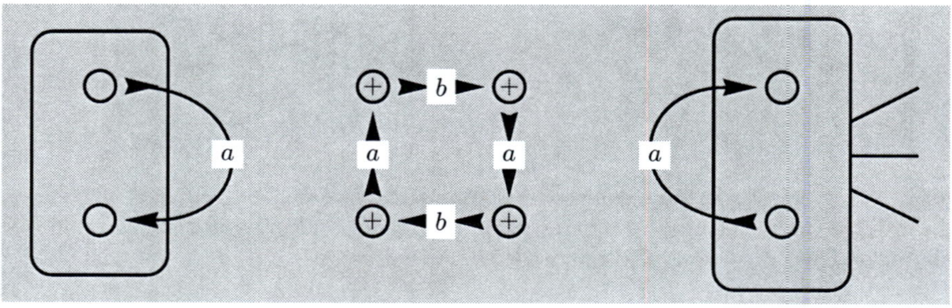

Fig. 7.8. The final result for the separation of diagrams joined by two lines

Note that the quadrangular diagram carries a Kronecker $\delta_{a,b}$.

7.6 Contraction of Two Angular Momentum Lines

The method discussed in the previous section is an important tool and it is very useful not only for the separation of diagrams but also when is necessary to reduce the number of angular momentum lines in a diagram.

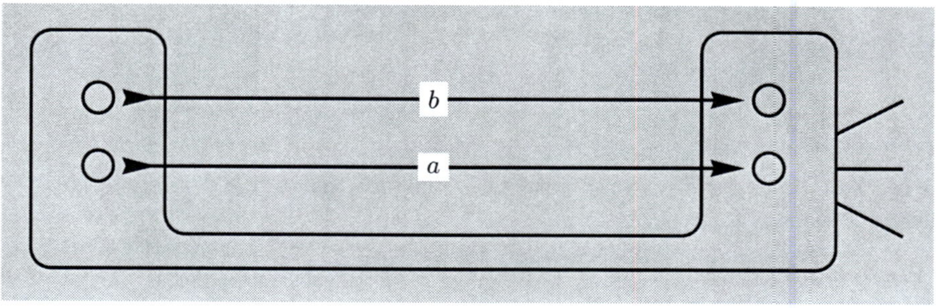

Fig. 7.9. A general diagram with two parallel angular momentum lines

In this case the two parallel lines are part of a bigger diagram and the reduction from two to one line may not immediately lead to a separable diagram.

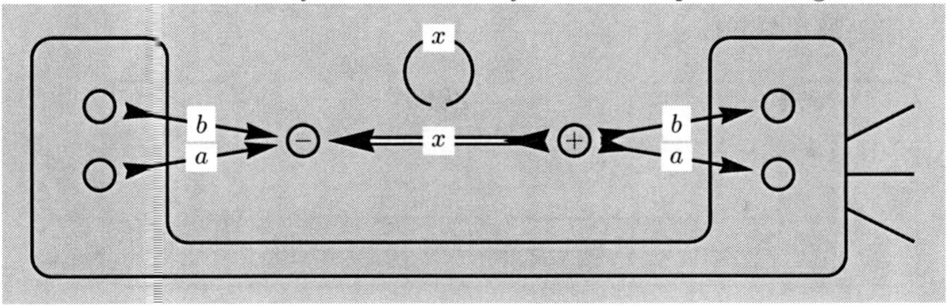

Fig. 7.10. The summation over x and ξ, with a factor $(2x+1)$, reduces two parallel lines to one line. The double line indicates that both variables x and ξ are to be summed

The main point in this section is that, the two lines are not just the connection between two diagrams, as before, but they are angular momentum lines within one single diagram.

It should be emphasized that line contraction illustrated in Fig. 7.10 is the basic procedure for simplification and separation of diagrams which will be applied frequently in following sections.

This method of contracting two lines is also applicable to two external lines, as has been demonstrated in Figs. 4.10 and 4.11.

7.7 Summation Procedure

The graphical method for summing over angular momentum labels, x and ξ, has been used in several examples and also in the preceding section. The sum over x, where the sum contains also the factor $(2x + 1)$, can be performed graphically if both nodes of the double line x are connected to the same angular momenta.

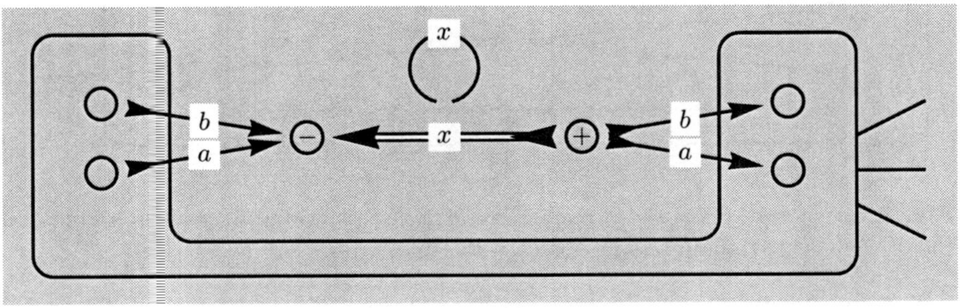

Fig. 7.11. Note that the same lines are found at both nodes, and a and b enter at the left node and go out from the right node The double line indicates that both variables x and ξ are to be summed. As in previous cases, the direction of x opposes a and b

In this case the lines for a, b and x are part of a single diagram and we have assumed that the x-line appears only once in the diagram and not in unlinked associated diagrams.

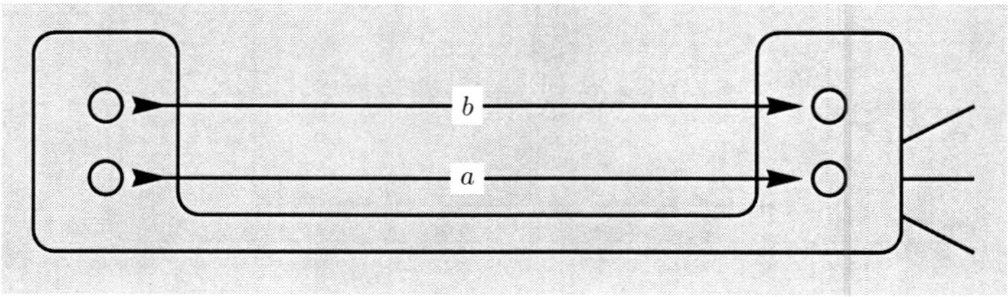

Fig. 7.12. As a result of the summation on x and ξ the nodes connected by x are removed and the a- and b-line remain

We will now consider a variation of the summation procedure.

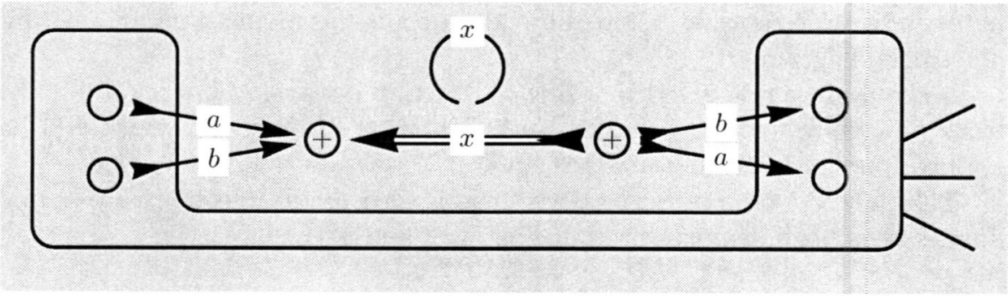

Fig. 7.13. Note that the cyclic arrangement of lines at the left node has changed and both nodes have the same sign. The line directions are the same as in Fig. 7.11

As before the lines for a, b and x are part of a single diagram and we have assumed that the x-line appears only once in the diagram.

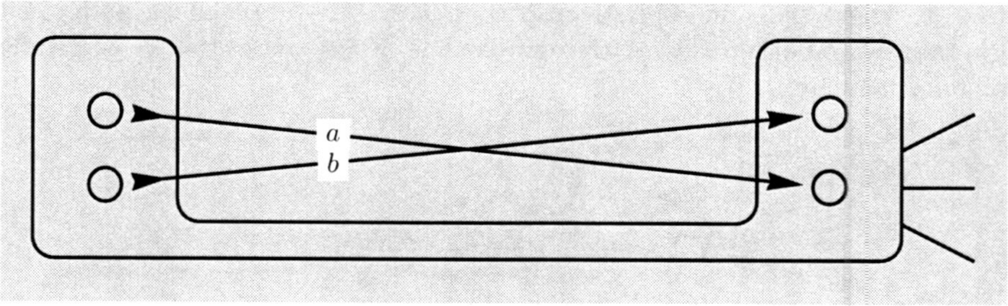

Fig. 7.14. As a result of the summation the nodes connected by x are removed and the a- and b-line remain but in this case the lines cross

There are situations where the angular momentum x is present in several unlinked diagrams. In this case these diagrams have to be linked, by methods introduced in the following sections, until the x-line appears only once and the diagram resembles one of the cases discussed here.

7.8 Separation for Three Lines

The following Fig. 7.15 shows two diagrams connected by three angular momentum lines with directions in common. The prescription for a three-line separation will be derived using line contraction and properties of single internal lines, as encountered in preceding sections.

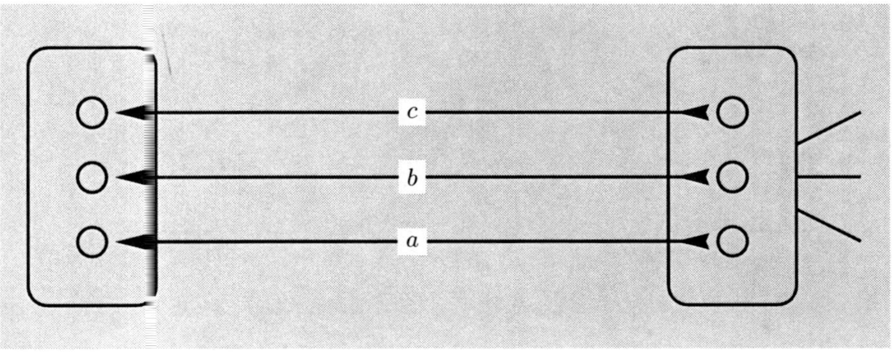

Fig. 7.15. Separation of two diagrams connected by three lines of common direction

The method for reducing two parallel lines to a single line with a corresponding summation has been discussed above. Because the x-line has to be antiparallel to a and b we implement, for convenience, also a direction change for the c-line, thereby introducing the phase $(-1)^{2c}$.

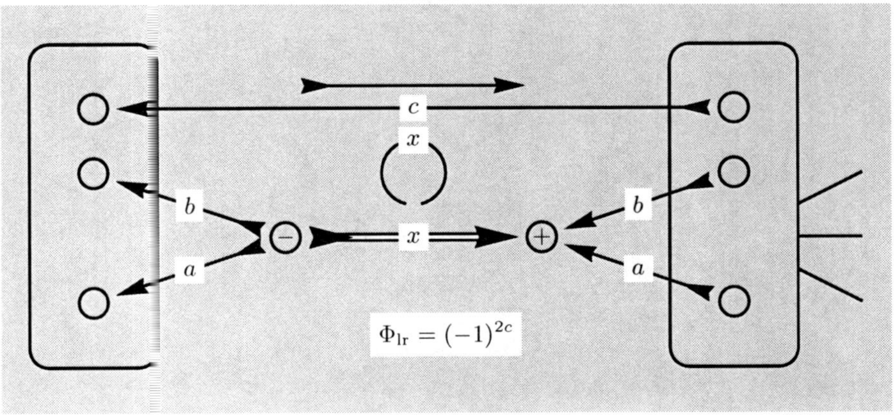

Fig. 7.16. The summation over x and ξ, with a factor $(2x+1)$, reduces the connection to two lines with the same direction, because the direction of c is changed

A further summation over y and η is introduced to reduce the connection to a single line and the directions of both c-lines are changed in preparation for removal of the nodes left after y is set to zero.

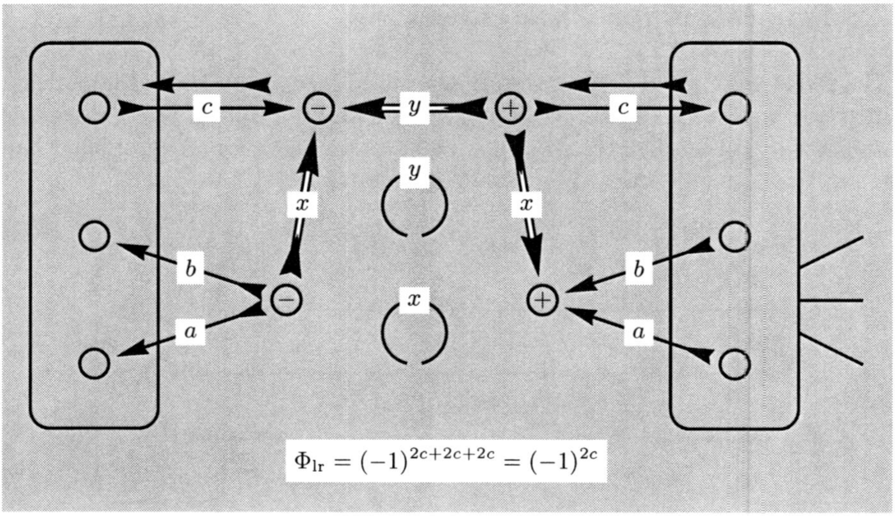

Fig. 7.17. The two lines are reduced to one and we can proceed in the same way as before

In order to achieve the correct arrangement of ingoing and outgoing lines, the node sign for the left node is changed and the corresponding phase taken into account.

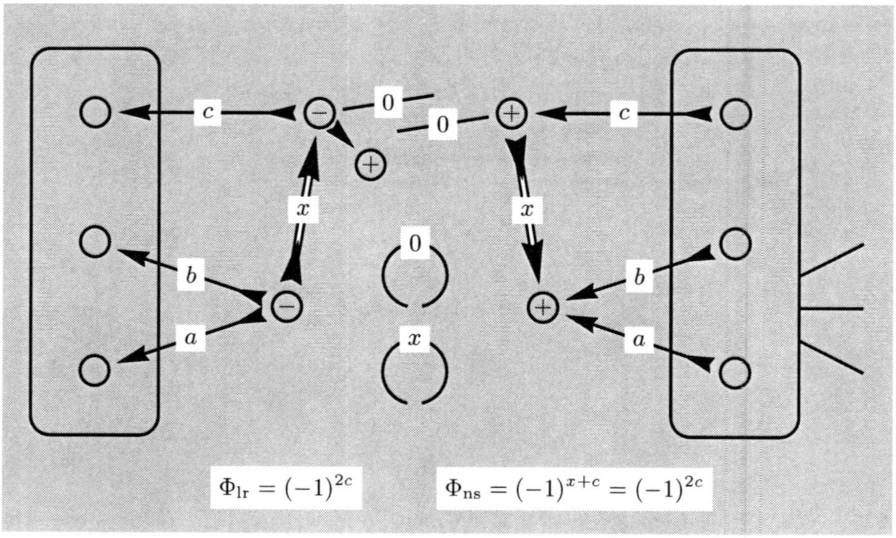

Fig. 7.18. The single line has been removed because $y = 0$. The diagrams are now separated

Because only two non-zero angular momentum lines enter and leave the remaining nodes the angular momentum values have to be the same in both cases.

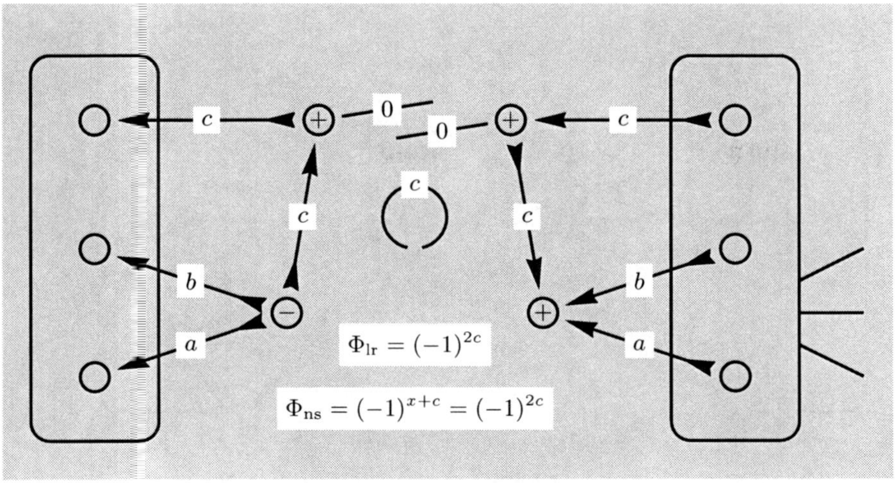

Fig. 7.19. Two nodes with only two non-zero lines remain and introduce $\delta_{x,c}$ such that the sum over x has only one term namely $x = c$

Note that the summation over x collapses to a single term $x = c$.

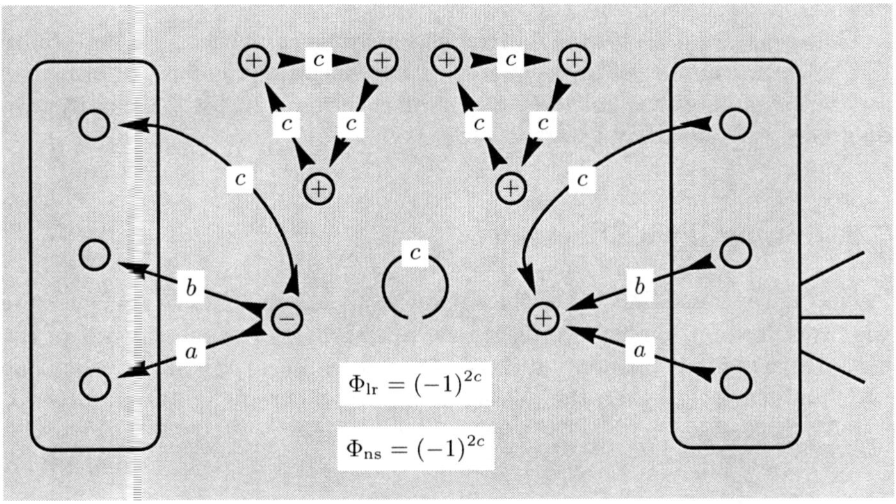

Fig. 7.20. The removal of the nodes from both lines produces two factors, each of value $1/\sqrt{2c+1}$

The product of two factors $1/\sqrt{2c+1}$ cancels $(2c+1)$ which is a remnant of the sum over x. Looking back at our starting point, Fig. 7.15, we find that two diagrams connected by three equidirectional lines can be separated by the introduction of two new nodes of different sign. One node has the three lines as ingoing lines and the other as outgoing lines. At least one diagram must not have external lines.

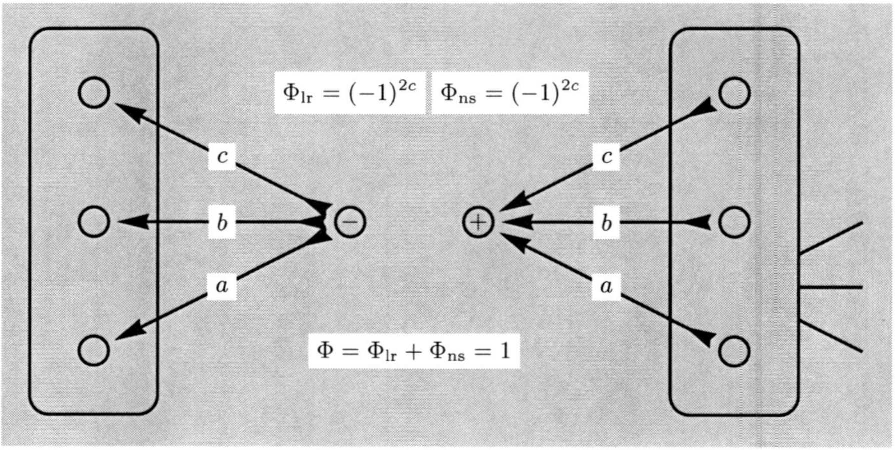

Fig. 7.21. The node signs of the newly created nodes have to be different, but which node is chosen positive and which is chosen negative is a matter of choice

With $\Phi = 1$ the separation of diagrams linked by three angular momentum lines can be performed without additional phase, if the starting point resembles Fig. 7.15.

The separation procedure derived here may be combined with our ability to reduce the number of lines connecting two diagrams to allow, in principle, separation of diagrams linked by any number of lines. At least one of the two diagrams must not have external lines.

7.9 Joining Two Diagrams

Retracing the steps that led to the simple result in Fig. 7.21, we find en route the procedure for joining two diagrams, if both have a node attached to the same three angular momentum lines. If the three lines, a, b and c, enter one node and leave the other, the joining of the two diagrams is straightforward.

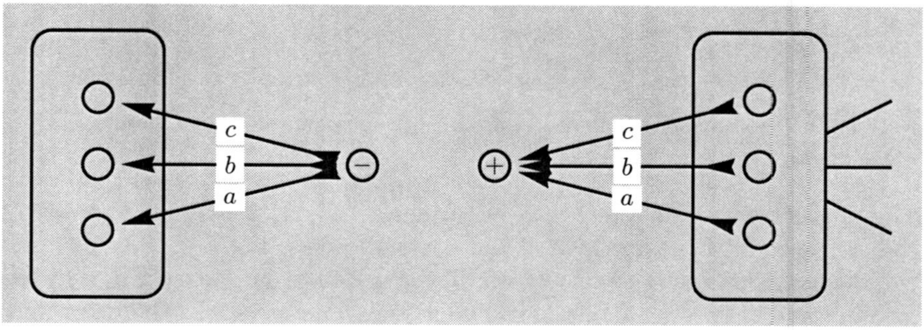

Fig. 7.22. The node signs of the two nodes have to be different, but which node is positive and which negative is, once more, a matter of choice

Changing the node sign for both nodes introduces no phase factor because $\Phi = (-1)^{a+b+c}(-1)^{a+b+c} = 1$. Thus, two separate diagrams can be joined at the two nodes if these have three angular momentum lines in common.

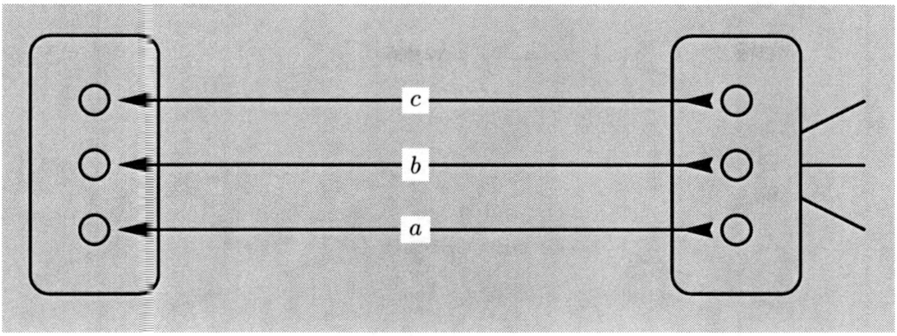

Fig. 7.23. The two previously unlinked diagrams connected by three lines

The procedure described here represents an important and useful tool. It is applied in such cases where an angular momentum is a summation variable present in more than one diagram, see e.g. Sect. 8.2.1.

7.10 Separation for Four Lines

In this case two pairs of lines have the same and opposite direction, respectively. We draw again from the results of previous sections. All being well, it should be routine to perform the following steps.

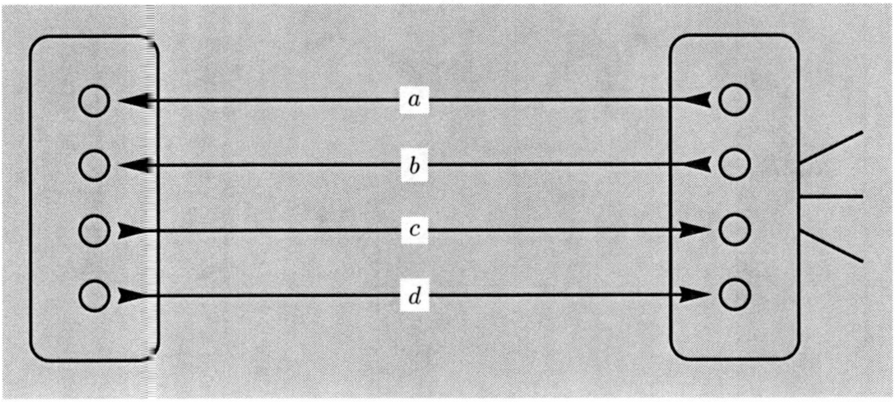

Fig. 7.24. Separation of two diagrams connected by four lines, two of which are pairwise parallel

For each pair of lines two new nodes and a summation over the new angular momentum line is introduced, together with the concomitant factors $(2x+1)$.

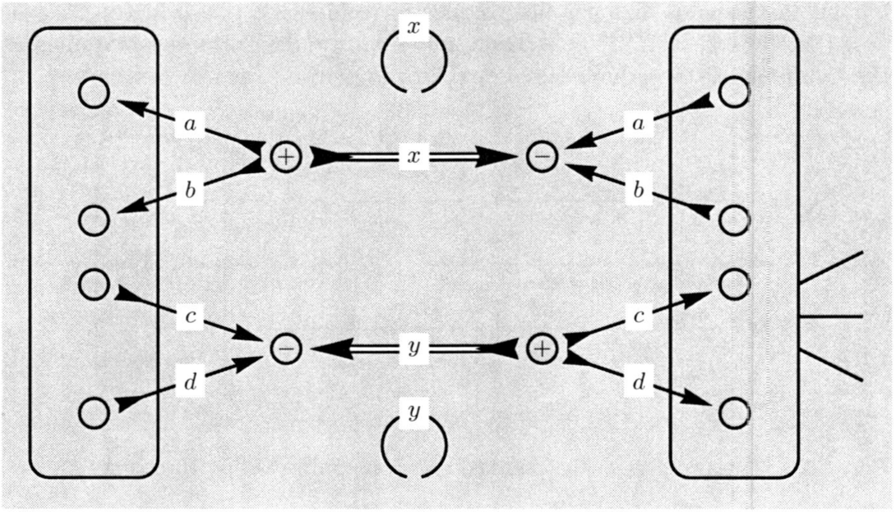

Fig. 7.25. Two additional summations are inserted such that the connections between diagrams are reduced to two

We follow the procedure for diagrams separated by two lines as given in Fig. 7.8.

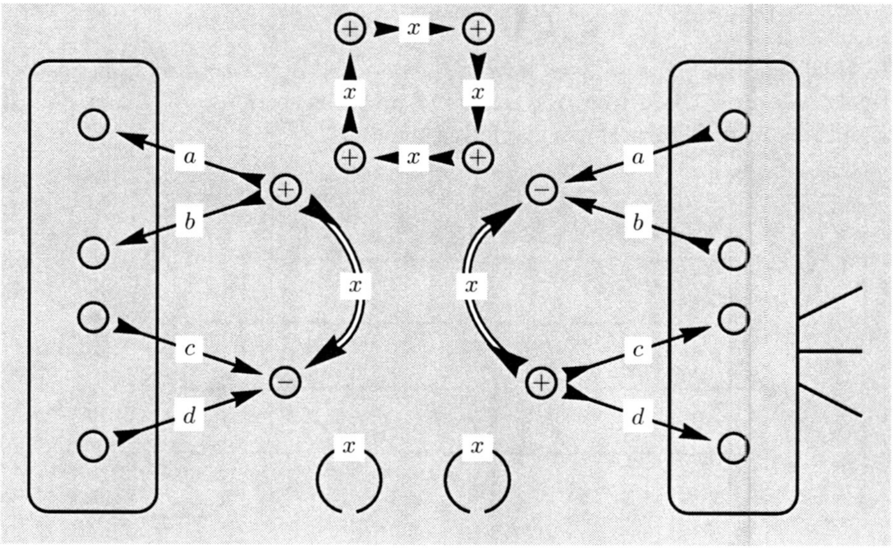

Fig. 7.26. Here we apply the rule for separating diagrams connected by two lines, see Fig. 7.8

This figure simplifies due to the cancellation of one circular and the quadrangular diagram.

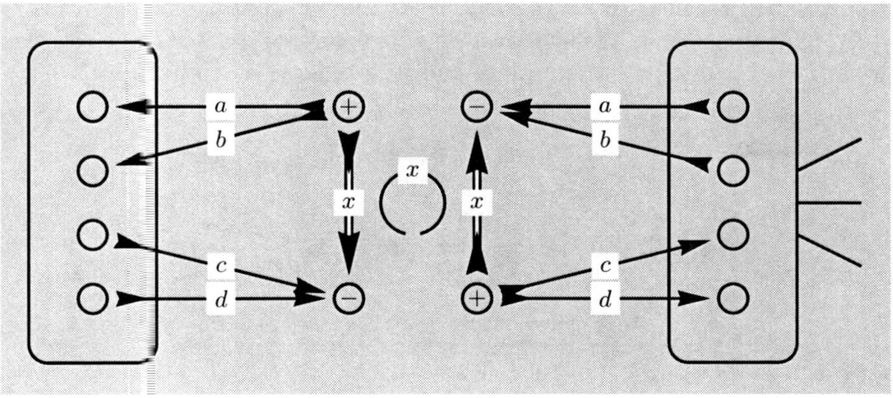

Fig. 7.27. Two of the additional factors cancel and a single sum over x remains

Note that in Fig. 7.27 the summation over the extra angular momentum x remains a part of the solution. Figure 7.27 shows also the way to join two diagrams, which both contain a sum over an angular momentum with the nodes attached to corresponding angular momenta. After joining the two diagrams at two corresponding nodes we return to a situation described in Fig. 7.11, where the double line x appears only once.

7.11 Separation for Five Lines

All five angular momentum lines, in Fig. 7.28, connecting the two diagrams point in the same direction. At least one of the two diagrams has no external lines.

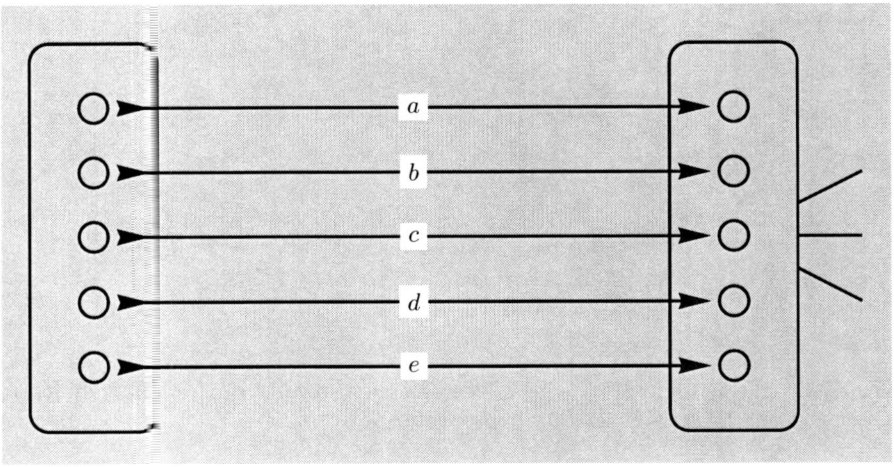

Fig. 7.28. Two diagrams connected by five angular momentum lines

Applying the result from Fig. 7.10 to the a- and b-line, and, separately, to the d- and e-line, demands the introduction of two new variables x and y together with the corresponding summations and factors $(2x+1)$ and $(2y+1)$.

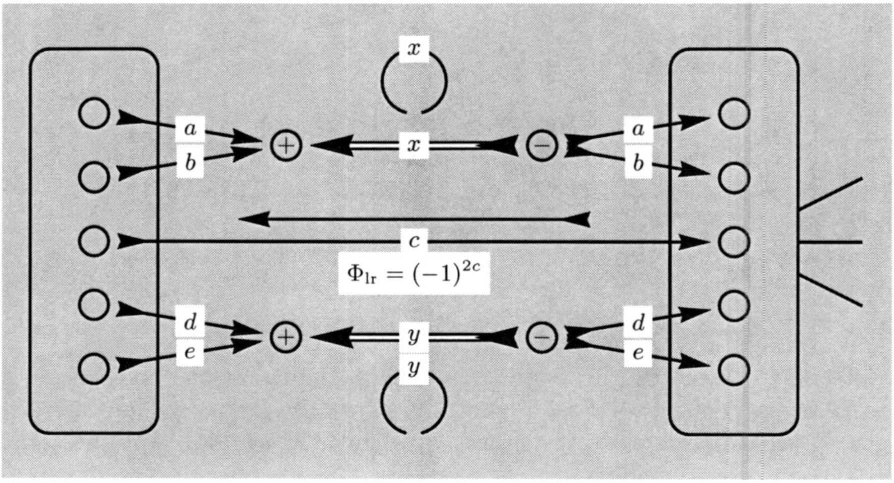

Fig. 7.29. In preparation for the next step the direction of the c-line is reversed and the required phase change is indicated below the c-line

The x-, c- and y-line have all the same direction and we can use the result given in Fig. 7.21 for a three-line separation.

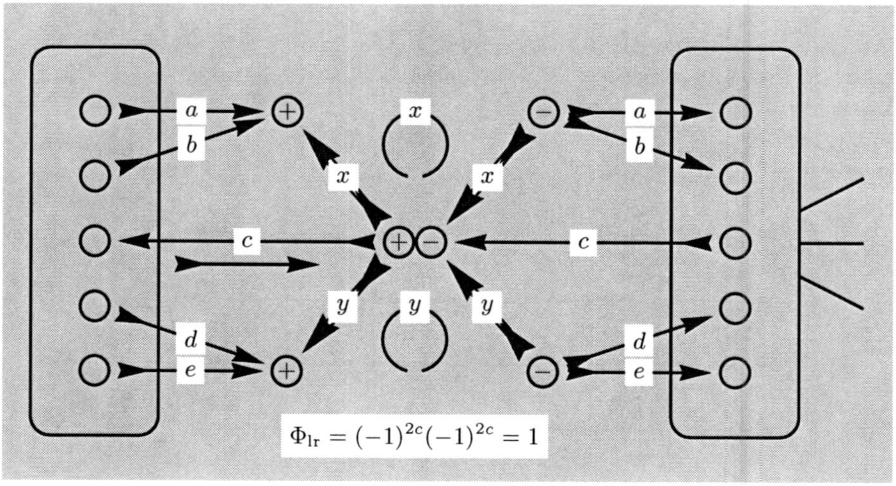

Fig. 7.30. Two new nodes are introduced for the three-line separation. A direction change of one c-line adds another phase factor

Because of $(-1)^{4c} = 1$ there is no extra phase factor.

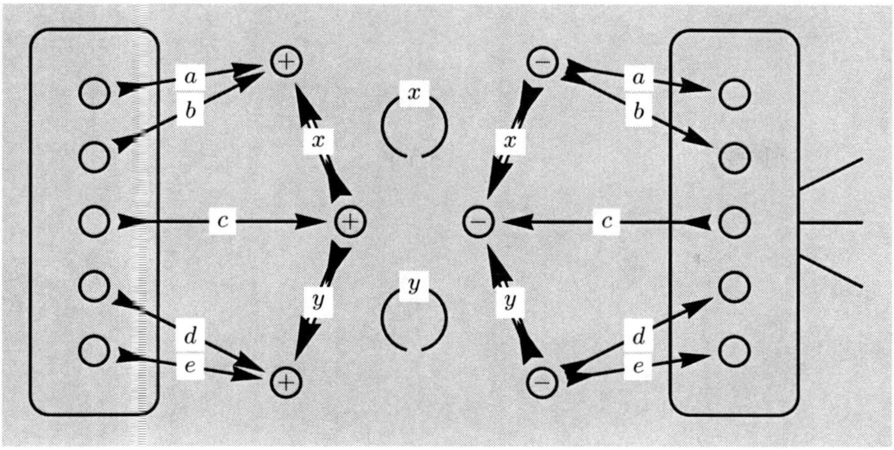

Fig. 7.31. The separation of two diagrams connected by five lines has lead to two extra summations over x and y. Note that one of the two c-lines has no longer the original direction

From this result it is fairly easy to extrapolate to the separation of six lines. The method from Fig. 7.10 will be applied three times and, therefore, three new variables and their sums will be introduced. Once more, the three line separation from Fig. 7.21 produces the desired result.

8

Closed Diagrams

In Chaps. 4–5 the result of an algebraic expression served as the touchstone for demonstration of the equivalent power of the graphical method. In the following sections the emphasis is shifted to posing a problem and deriving the solution by application of the general rules derived so far for the manipulation and transformation of diagrams. In all following sections the summation procedure of Sect. 7.7 will be frequently on call.

8.1 Sums with Single $3nj$-Symbols

All examples in this section analyse sums of single $3nj$-symbols and lead to surprisingly simple results. Even more fascinating, one may find, is the little effort needed in the graphical method to produce outcomes, in comparison with, in most cases, considerable algebraic effort. The results in the following examples are all listed by Varshalovich et al. [12] in Sect. 12.2. To facilitate comparison we have used the same angular momentum variables.

Although only single sums are considered here, the graphical method can, of course, equally well be applied to double sums etc., and many examples are found in [12].

8.1.1 Sum with a $3j$-Symbol

The simplest $3nj$-symbol is the $3j$-symbol of Fig. 4.6 which has the value 1. We use $1_{\triangle\{a,b,x\}}$ to indicate that the three angular momenta must fulfil the triangle condition. In this diagram we replace $c \to x$, to conform with Varshalovich et al. [12], and consider the sum over x. Close inspection reveals that the $3j$-symbol thus limits the range of x-values.

$$\sum_x (2x - 1)\{a\ b\ x\} = \sum_x (2x + 1) 1_{\triangle\{a,b,x\}} = ?$$

E. Balcar, S.W. Lovesey, *Introduction to the Graphical Theory of Angular Momentum*, STMP 234, 73–117, DOI 10.1007/978-3-642-03118-2_8,

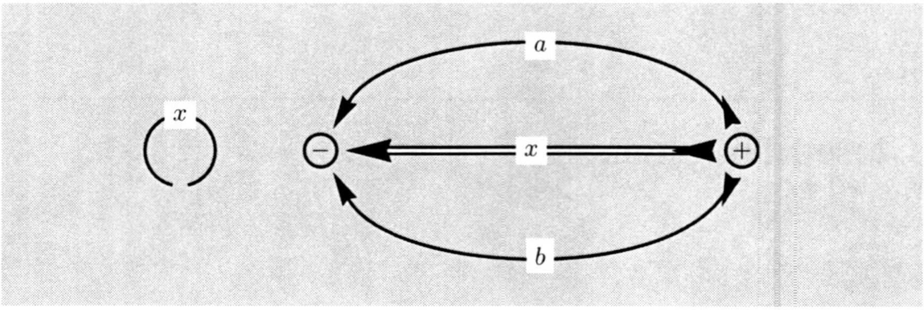

Fig. 8.1. The double line for x indicates that a sum is to be taken over x and this implies a factor $(2x + 1)$ shown in the figure Note that the summation over ξ is part of the definition of the $3j$-symbol but this sum is purely formal here, because ξ is fixed by the values of α and β

It is perhaps not easy to see at once that this sum corresponds to the general diagram shown in Fig. 7.11. To make this connection clearer, we redraw the diagram of Fig. 8.1.

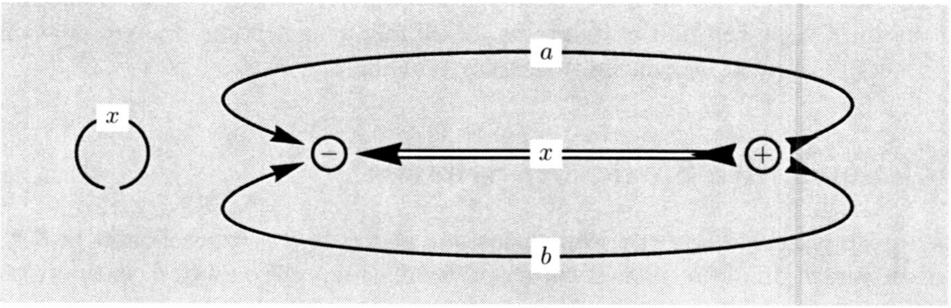

Fig. 8.2. The three angular momenta go out from the +-node and go into the −-node

It is now straightforward to apply the summation as described in Sect. 7.7 and we find

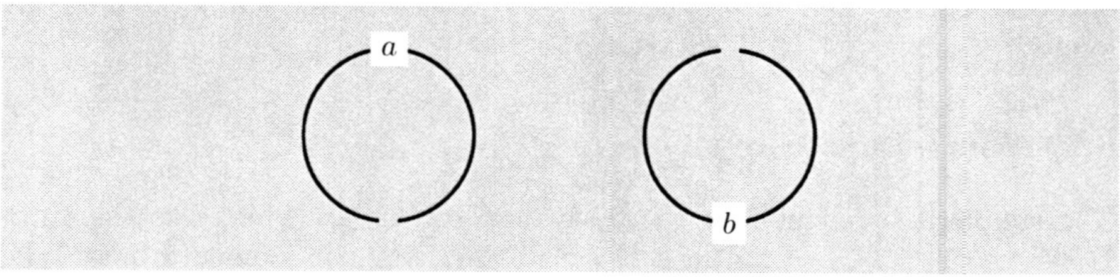

Fig. 8.3. The sum removes the double line and the corresponding factor

The graphical analysis translates into the simple formula

$$\sum_x (2x + 1) 1_{\triangle\{a,b,x\}} = (2a + 1)(2b + 1) ,$$

and this corresponds to the sum of $(2x+1)$ for the values $|a-b| \leq x \leq a+b$ as we can see[1].

8.1.2 Alternating Sum with a $3j$-Symbol

It is instructive to consider a variation of the diagram given in Fig. 8.2 for $a = b$. The sum over x is an example of the summation procedure considered in Fig. 7.13.

$$\sum_x (-1)^{2a+x}(2x+1)1_{\triangle\{a,a,x\}} = ?$$

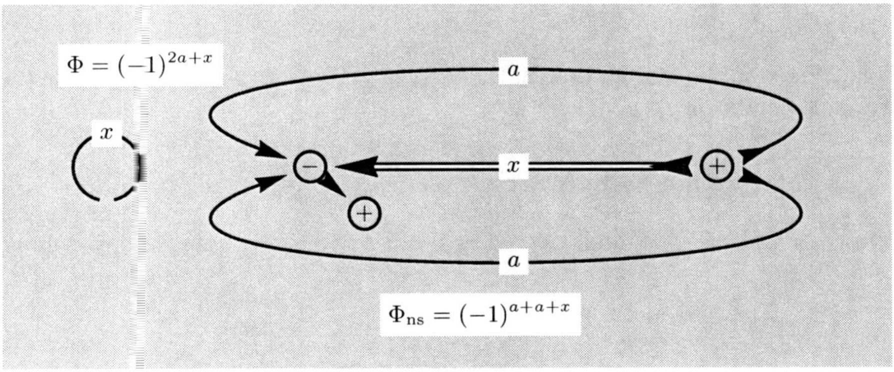

Fig. 8.4. Note the additional phase factor $(-1)^{2a+x}$ which multiplies the terms of the sum. A direct change of node sign prepares the diagram for the summation

Due to the phase factor the terms will have alternating signs and we expect a different result compared to the formula in Sect. 8.1.1. However, this phase is cancelled by the extra phase induced by the direct node sign change in the diagram.

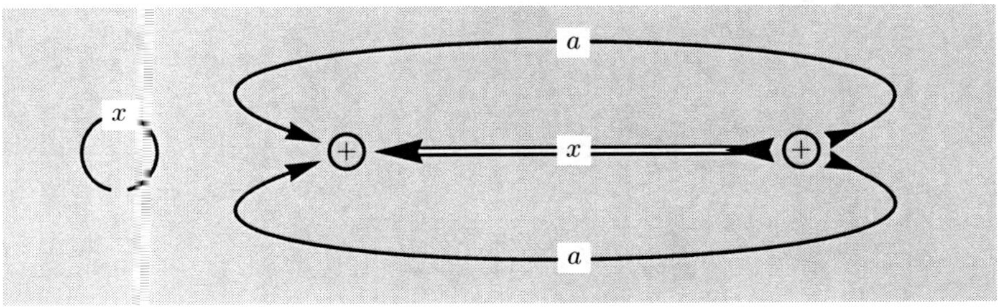

Fig. 8.5. The additional phase factor $(-1)^{a+a+x}$ from change of the node sign in Fig. 8.4 cancels the original phase factor

The sum can now be performed but the lines cross in this case (see Sect. 7.7).

[1] Legend has it that young C. F. Gauß could do similar sums in primary school.

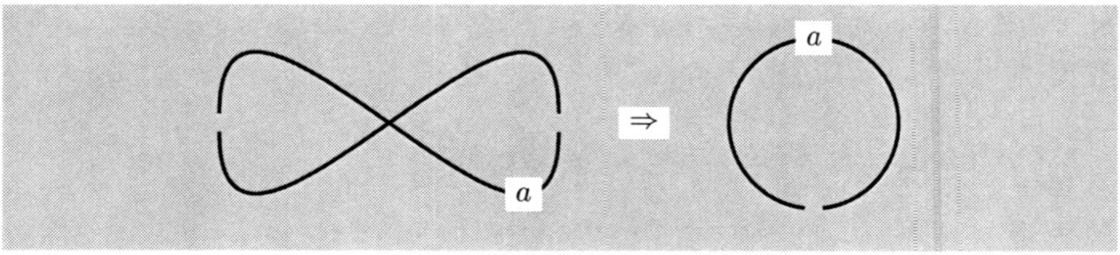

Fig. 8.6. The summation procedure removes the double line together with the corresponding factor

The algebraic result from graphical analysis is,

$$\sum_x (-1)^{2a+x}(2x+1)1_{\triangle\{a,a,x\}} = (2a+1)\,.$$

The range of the summation index $0 \leq x \leq 2a$ here is in accord with the triangle condition $\triangle\{a, a, x\}$.

8.1.3 Sum with a 6*j*-Symbol

On the next level of complexity we turn our attention to sums of 6*j*-symbols.

$$\sum_x (2x+1) \begin{Bmatrix} c & a & b \\ x & a & b \end{Bmatrix} = ?$$

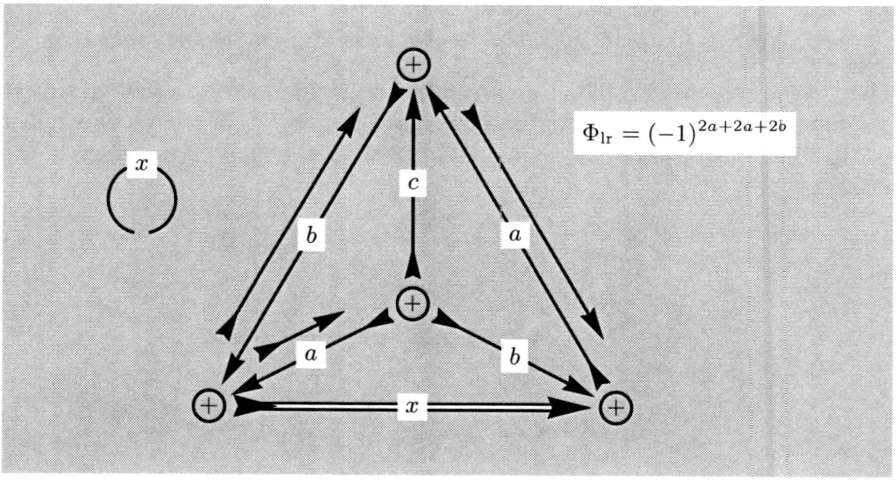

Fig. 8.7. The double line for the angular momentum x is an indication of the sum over x and, as usual, the factor $(2x+1)$ is part of the sum

Changes of line directions are required to prepare the diagram for the application of the summation procedure of Sect. 7.7 and the appropriate phase factor is taken into account.

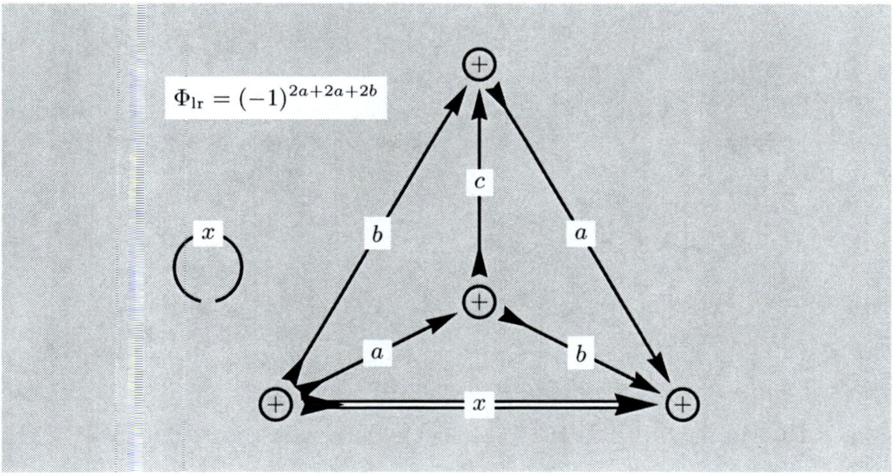

Fig. 8.8. The same angular momentum lines for a and b enter and leave the nodes connected by the double line x

Note that the sum can be performed following the procedure displayed in Figs. 7.13 and 7.14, where the reconnected lines cross.

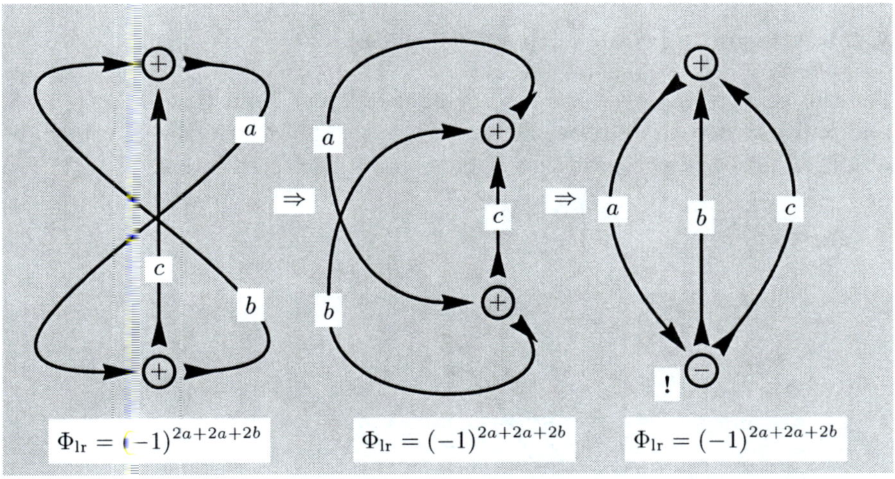

Fig. 8.9. The sum is carried out and the corresponding parts of the a- and b-line have been reconnected

Performing the sum over x has created a rather confusing diagram but a careful check will confirm that the remaining a-, b- and c-lines are very close to a familiar diagram, if the direction of the b-line and the c line are reversed. The additional phase factor gives $\Phi_{\rm lr} = (-1)^{4a+4b+2c} = (-1)^{2c}$. We note also that, the transformation of the diagram has changed the cyclic order at the lower node and the node sign has been changed accordingly.

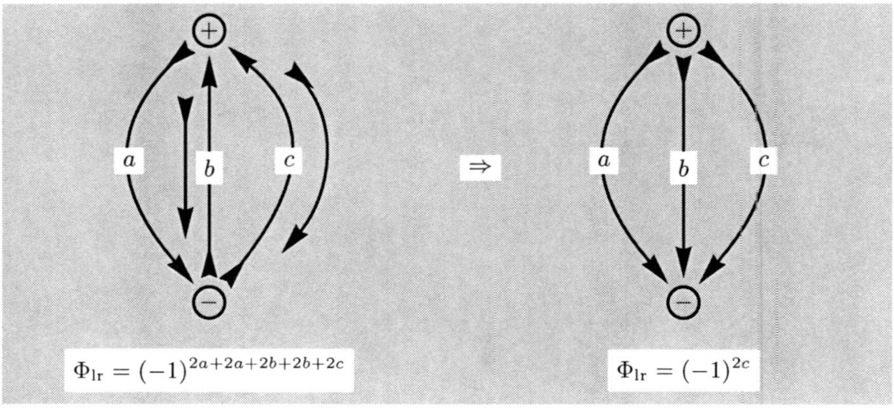

Fig. 8.10. The final result is the $3j$-symbol with the phase $(-1)^{2c}$

The initial diagram from Fig. 8.7 is now translated into algebraic form and from Fig. 8.10 we obtain the following formula:

$$\sum_x (2x+1) \begin{Bmatrix} c & a & b \\ x & a & b \end{Bmatrix} = \sum_x (2x+1) \begin{Bmatrix} a & b & x \\ a & b & c \end{Bmatrix} = (-1)^{2c}\, 1_{\triangle\{a,b,c\}} \,.$$

8.1.4 Alternating Sum with a $6j$-Symbol

For the next example we use a $6j$-symbol different from that in Sect. 8.1.3, and multiply it with an phase factor $\Phi = (-1)^{a+b+x}$ which can, of course, be absorbed into a direct node sign change, as we have seen in Sect. 8.1.2.

$$\sum_x (2x+1)(-1)^{a+b+x} \begin{Bmatrix} c & a & a \\ x & b & b \end{Bmatrix} = ?$$

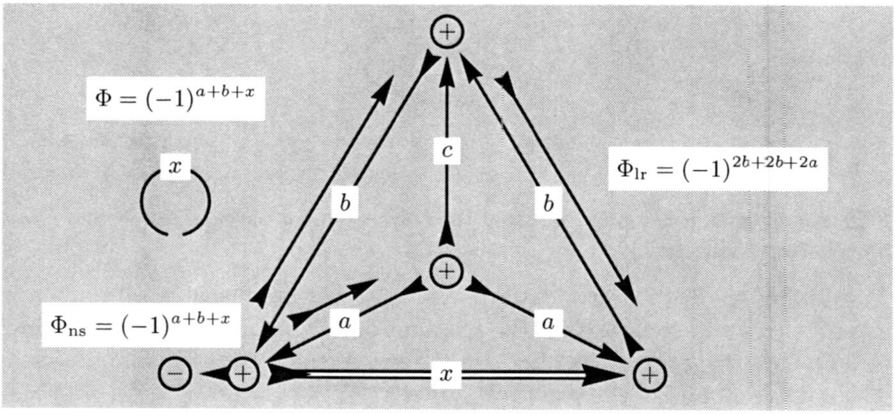

Fig. 8.11. The double line for the angular momentum x is an indication of the sum over x together with the factor $(2x+1)$. Note the extra phase $\Phi = (-1)^{a+b+x}$

The changes of line direction are required to prepare the diagram for the summation procedure and the appropriate phase factor has to be taken into account. The extra phase $\Phi = (-1)^{a+b+x}$ has been cancelled by the node sign change.

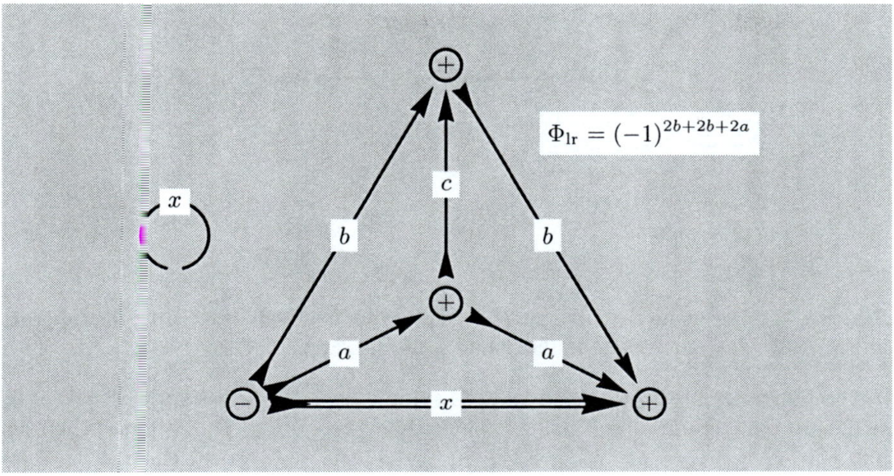

Fig. 8.12. The same angular momentum lines a and b enter and leave the nodes connected by the double line x of opposite direction

Note that the sum can be performed following the procedure displayed in Figs. 7.11 and 7.12.

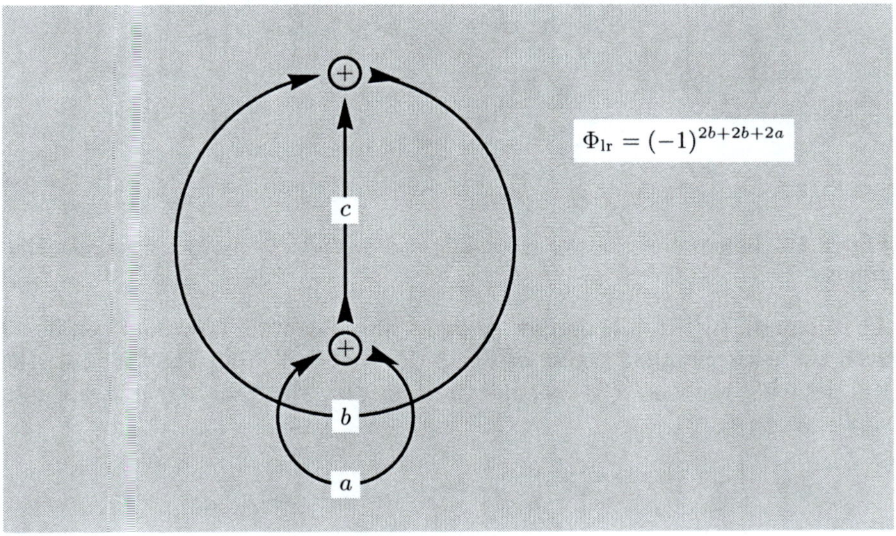

Fig. 8.13. The sum can be performed and the corresponding parts of the a and the b-line are reconnected

It is possible to deform the diagram such that the b-line no longer overlaps with the a-line and the cyclic arrangement at both nodes is not affected.

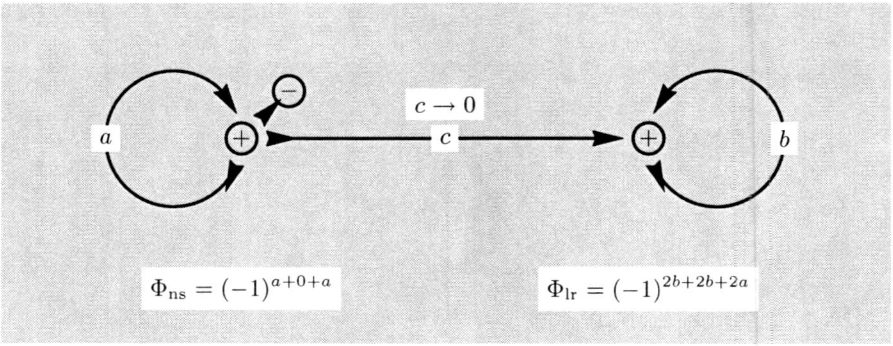

Fig. 8.14. After some rearrangement, the diagram is closely related to the diagram in Fig. 3.16. Note that the circular b-line is at the right

In the figure we have indicated that the single line rule will require $c = 0$. The situation corresponds to Fig. 3.17, with the exception of the left node, which has the wrong sign for node removal. With the phase Φ_{ns} the total phase factor is $\Phi = \Phi_{\text{ns}} + \Phi_{\text{lr}} = (-1)^{4a+4b} = 1$ and the result may be taken from Fig. 3.18.

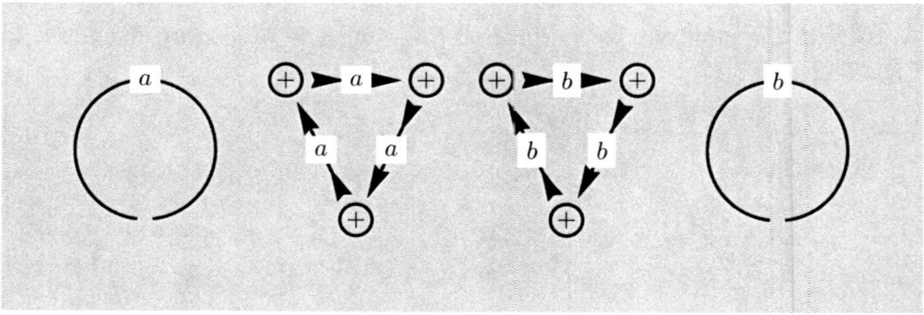

Fig. 8.15. Removal of the two remaining nodes produces the two triangular diagrams

The diagram in Fig. 8.11 is now brought into algebraic form and combined with the corresponding result in Fig. 8.15. We recall from Fig. 3.4 that, the left circle has the value $(2a+1)$ and the triangular diagram is $1/\sqrt{2a+1}$ and, similarly, for b.

$$\sum_x (2x+1)(-1)^{a+b+x} \begin{Bmatrix} c & a & a \\ x & b & b \end{Bmatrix} =$$

$$= \sum_x (2x+1)(-1)^{a+b+x} \begin{Bmatrix} a & b & x \\ b & a & c \end{Bmatrix} = \delta_{c,0}\sqrt{2a+1}\sqrt{2b+1}\,.$$

8.1.5 Sum with a 9j-Symbol

$$\sum_x (2x+1) \begin{Bmatrix} f & x & e \\ b & e & a \\ d & f & c \end{Bmatrix} = ?$$

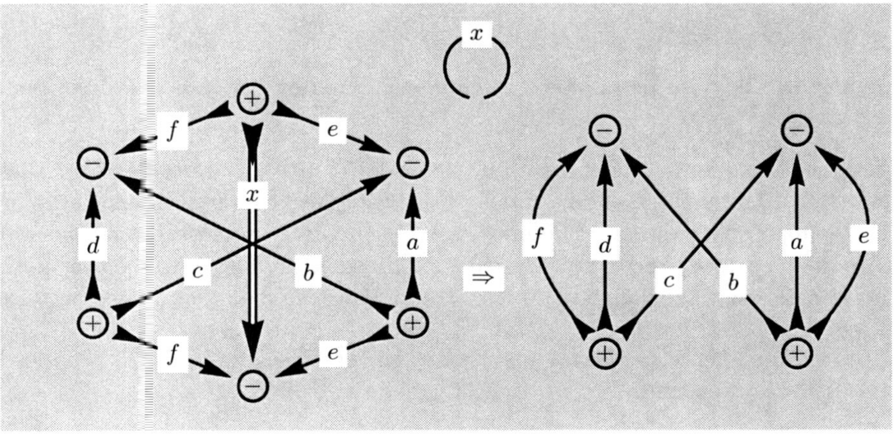

Fig. 8.16. The definition of the 9j-symbol is given in Fig. 6.2. Note that the same lines e and f go into and out from the nodes connected by the (double) x-line

The direction of the angular momentum lines and the node signs are already perfect for the summation procedure, for once. In the diagram on right-hand side of the figure the remaining parts of the e- and the f-line have been reconnected.

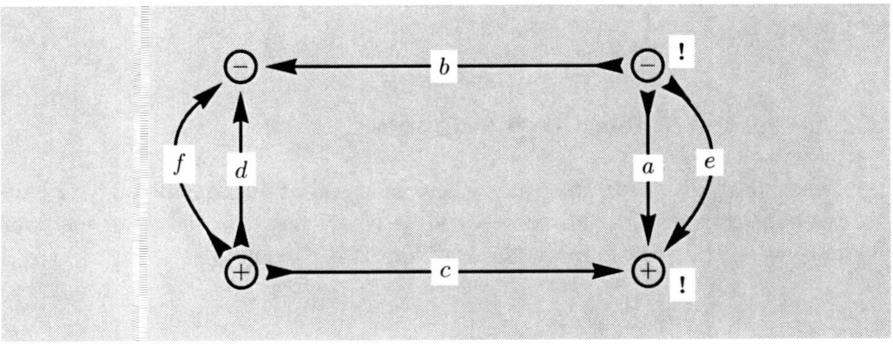

Fig. 8.17. The nodes on the right-hand side have been flipped around a horizontal axis and the changes in the cyclic order are reflected by '!' labelling the altered node signs

This diagram can be separated by analogy with Fig. 7.8 and the result gives two 3j-symbols and a quadratic diagram.

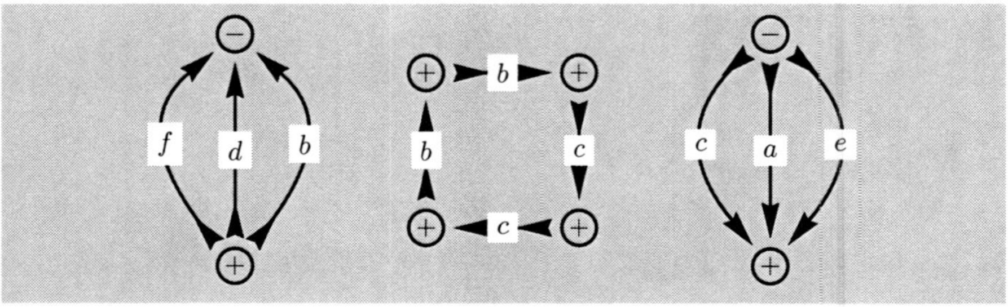

Fig. 8.18. The two parts of the diagram connected by just two lines have been separated

Because the quadrangular diagram, element (5) in Fig. 3.4, leads to a Kronecker $\delta_{b,c}$, the angular momentum c can be replaced by b without changing the value of the diagram. The node signs in the $3j$-symbol on the right-hand side do not correspond exactly to the definition, but it is easy to see that both node signs may be changed and this would bring no change in phase because $\Phi = (-1)^{2e+2a+2b} = 1$.

The sum over a $9j$-symbol analysed by the graphical method has produced the following relation:

$$\sum_x (2x+1) \begin{Bmatrix} f & x & e \\ b & e & a \\ d & f & c \end{Bmatrix} =$$

$$= \sum_x (2x+1) \begin{Bmatrix} a & b & e \\ c & d & f \\ e & f & x \end{Bmatrix} = \frac{\delta_{b,c}}{(2b+1)} \, 1_{\Delta\{a,b,e\}} \, 1_{\Delta\{b,d,f\}} \, .$$

The second $9j$-symbol is derived from the first by a cyclic permutation of rows and columns.

8.2 Sums with Two $3nj$-Symbols

In the context of closed diagrams there is no need to consider $3j$-symbols here because these will lead to cases already discussed in previous sections. Instead, we start with an example involving two $6j$-symbols.

8.2.1 Sum with Two $6j$-Symbols

The first example corresponds to an orthogonality relation for $6j$-symbols.

$$\sum_x (2x+1) \begin{Bmatrix} a & d & p \\ c & b & x \end{Bmatrix} \begin{Bmatrix} a & d & q \\ c & b & x \end{Bmatrix} = ?$$

The angular momentum x that is to be summed is present in both $3nj$-symbols and has common nodes with the same angular momentum lines. For the graphical summation to proceed, the two symbols have to be joined so that there is only one x-line in the diagram.

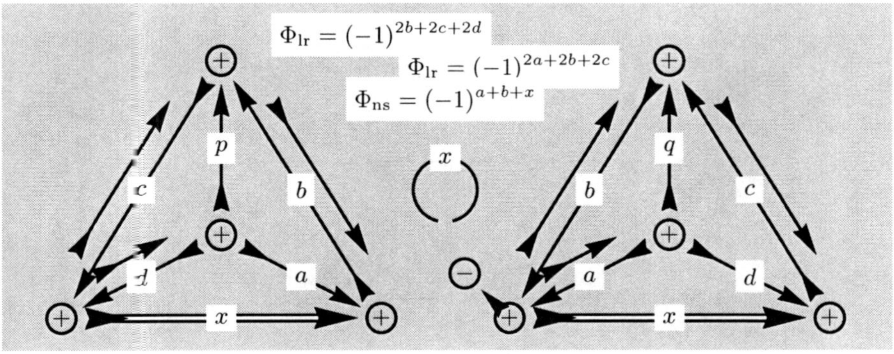

Fig. 8.19. The double line for the sum over angular momentum x appears in both $3j$-symbols and the factor $(2x + 1)$ is part of the sum. As preparation for a three line joining one node sign has to be changed. Several line directions are changed in anticipation of the requirements for the sum over x

Several phase factors arise from the transformations indicated in the figure. To simplify the phase we can drop the parts $(-1)^{4b} = 1$ and $(-1)^{4c} = 1$. The remaining phase is $\Phi = (-1)^{2a+2d}(-1)^{a+b+x}$.

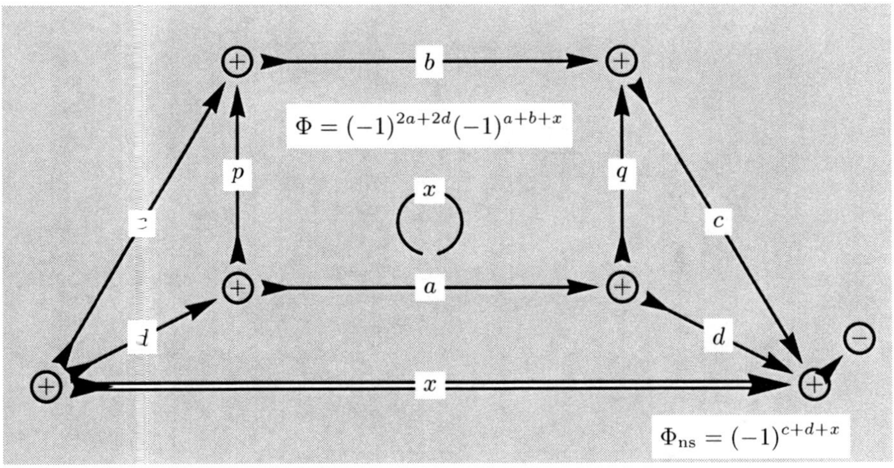

Fig. 8.20. Both $6j$-symbols have been joined by a three line connection. The line directions at the nodes connected by the x double line are appropriate for the sum over x. A further change of node sign is required for the summation procedure

From this figure we can understand the reason for the changes in line direction effected in Fig. 8.19. The line directions at the nodes connected by the double line x are appropriate for the summation. Note, however, that the two changes of node sign have introduced the phase $(-1)^x$ twice and this phase can, of course, no longer be present after the summation has been performed. The phase $(-1)^{2x} = (-1)^{-2x} = (-1)^{2a+2b}$ because we know that a, b and x satisfy a triangle condition and also that $a + b + x$ is an integer which guarantees $(-1)^{2a+2b+2x} = 1$.

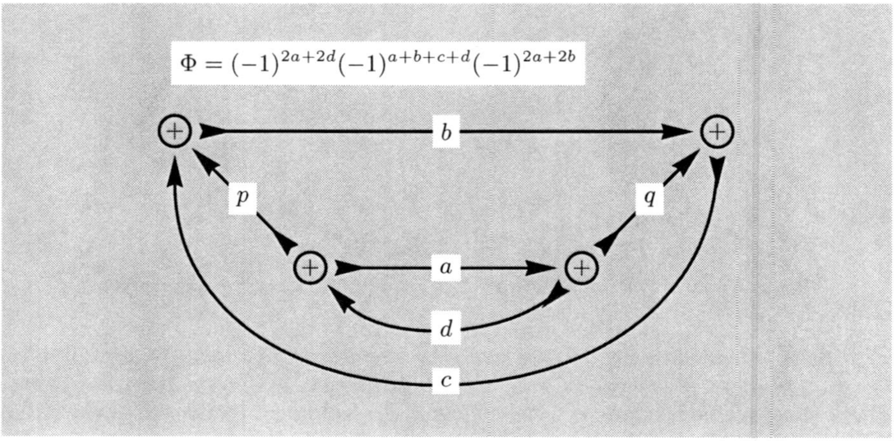

Fig. 8.21. As a result of the sum the c-and the d-line have been reconnected. Note that $(-1)^{2x}$ has been replaced by $(-1)^{2a+2b}$ in Φ

This diagram consists of two parts joined just by two lines and can be separated.

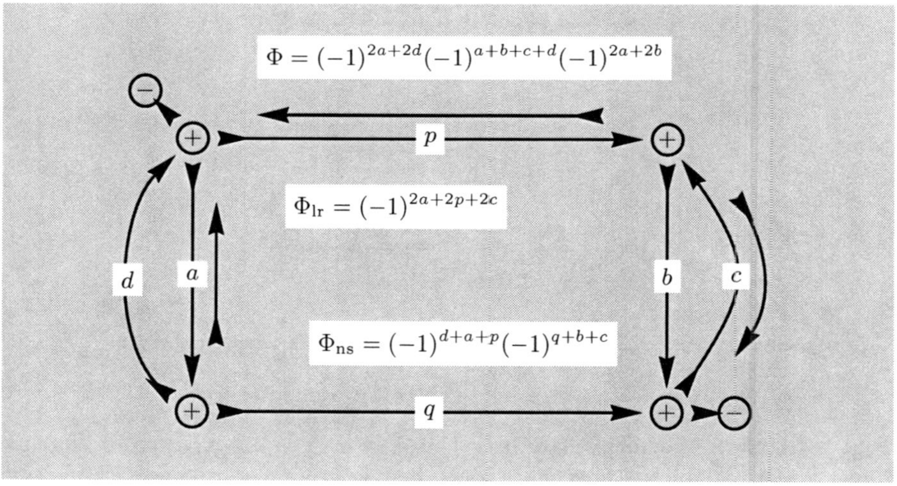

Fig. 8.22. The diagram has been rotated and rearranged to emphasize the two-line connection. The changes of line direction prepare the diagram for the following separation

The total phase can be simplified by removing $(-1)^{4a} = 1$ and all similar terms from Φ. The phase $\Phi = (-1)^{2p+q+p}$ remains. Separating the two diagrams leads to a simple result:

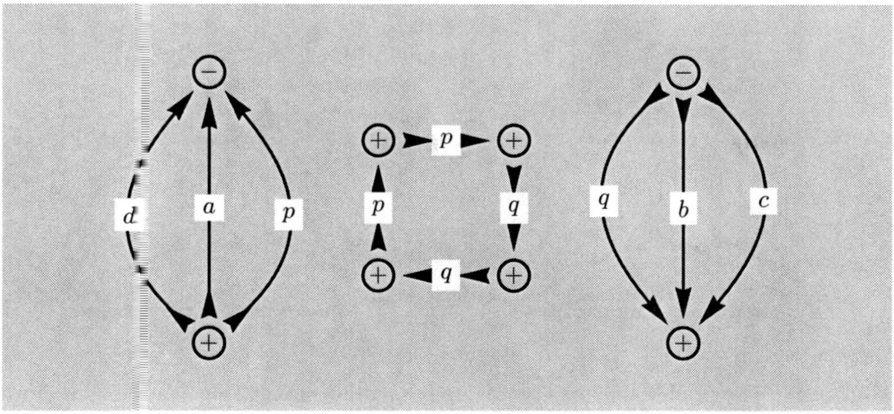

Fig. 8.23. The two parts of the diagram connected by just two lines have been separated. The quadratic diagram contains a Kronecker $\delta_{p,q}$ and the phase $\Phi = (-1)^{2p+q+p} \Rightarrow 1$ vanishes

The final result has collapsed into just two $3j$-symbols and a quadratic diagram. The relation derived by the graphical method follows from a translation of Fig. 8.23 inserted into the right-hand side of our starting equation.

$$\sum_x (2x+1) \begin{Bmatrix} a & d & p \\ c & b & x \end{Bmatrix} \begin{Bmatrix} a & d & q \\ c & b & x \end{Bmatrix} = \frac{\delta_{p,q}}{(2p+1)} 1_{\Delta\{a,d,p\}} 1_{\Delta\{b,c,p\}} \, .$$

Here, the first two columns in the first $6j$-symbol have been exchanged relative to the diagram in Fig. 8.19.

8.2.2 Alternating Sum with Two $6j$-Symbols

There is only one variation possible for the case of two $6j$-symbols, due to the stringent requirements for joining the diagrams and for the summation over the variable x. In this variation of Fig. 8.19 the c and the d angular momentum change place.

$$\sum_x (-1)^x (2x+1) \begin{Bmatrix} a & d & p \\ c & b & x \end{Bmatrix} \begin{Bmatrix} a & c & q \\ d & b & x \end{Bmatrix} = ?$$

In anticipation of the following transformations it is helpful to introduce the phase factor $(-1)^x$ into the first diagram to prevent problems at a later stage. We conclude from the presence of such a phase factor, with similar factors encountered already in Figs. 8.4 and 8.11, that the graphical summation procedure requires a cancellation of this phase factor by a suitable node sign change of the node to which the x-line is attached. This is because the final result cannot depend on the summation variable x and, therefore, we have to achieve $(-1)^{2x} \Rightarrow (-1)^{2a+2b}$ or a similar replacement in Fig. 8.24.

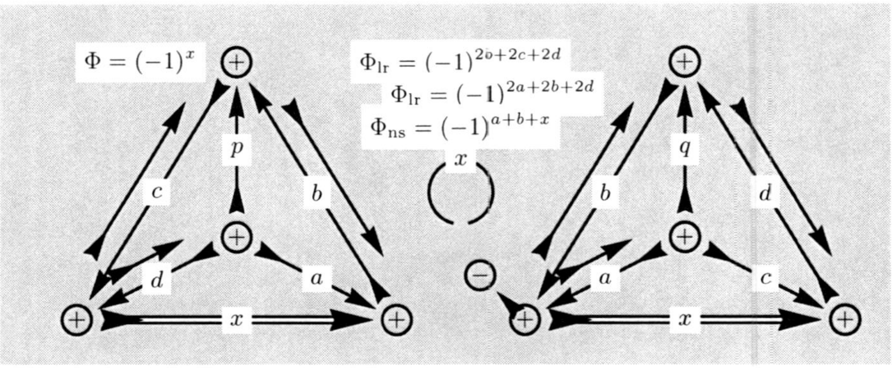

Fig. 8.24. The double line for the sum over angular momentum x appears in both $3j$-symbols and the factor $(2x + 1)$ is part of the sum. As preparation for a three line connection one node sign has to be changed. Several line directions are changed in anticipation of the requirements for the sum over x

Several phase factors arise from the transformations indicated on the figure. To simplify the phase we can drop the parts $(-1)^{4b} = 1$ and $(-1)^{4d} = 1$. The remaining phase is $\Phi = (-1)^{2a+2c}(-1)^{a+b+x}$.

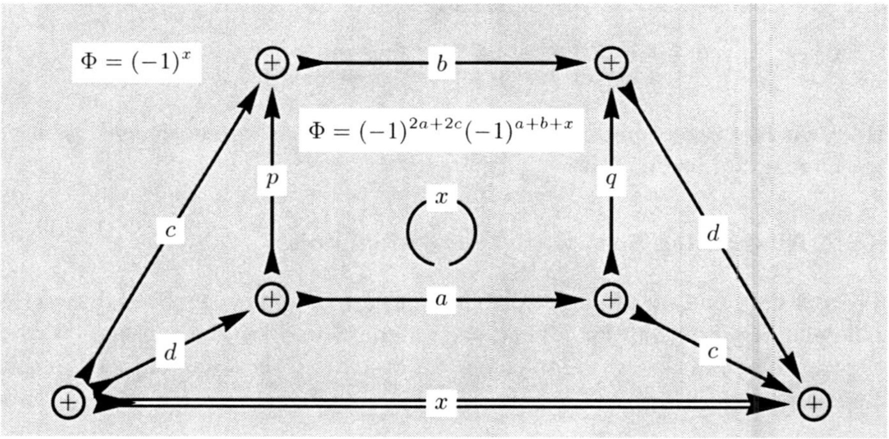

Fig. 8.25. Both $6j$-symbols have been joined by a three line connection. The line directions at the nodes connected by the x double line are appropriate for the sum over x

Note, however, that the original phase $(-1)^x$ and the phase introduced by the change of node sign lead to the phase $(-1)^{2x}$ and this phase should no longer be present after the summation has been performed. The phase $(-1)^{2x} = (-1)^{2a+2b}$ because we know $(-1)^{2a+2b+2x} = 1$. The line directions at the nodes connected by the double line x is appropriate for the summation but in this case the c- and the d-line will cross because the nodes connected by x have the same node sign.

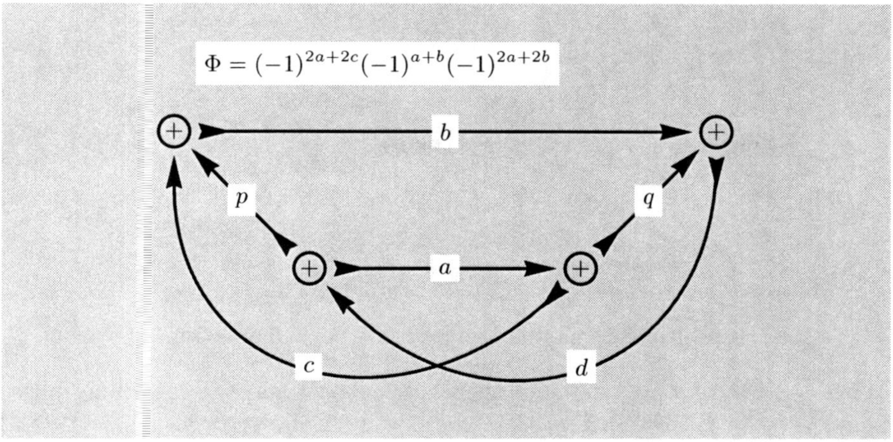

Fig. 8.26. As a result of the sum the c- and the d-line have been reconnected. Note that $(-1)^{2x}$ has been replaced by $(-1)^{2a+2b}$ in Φ

In this diagram we can straighten the angular momentum lines c and d and thereby alter the cyclic arrangement for all the nodes.

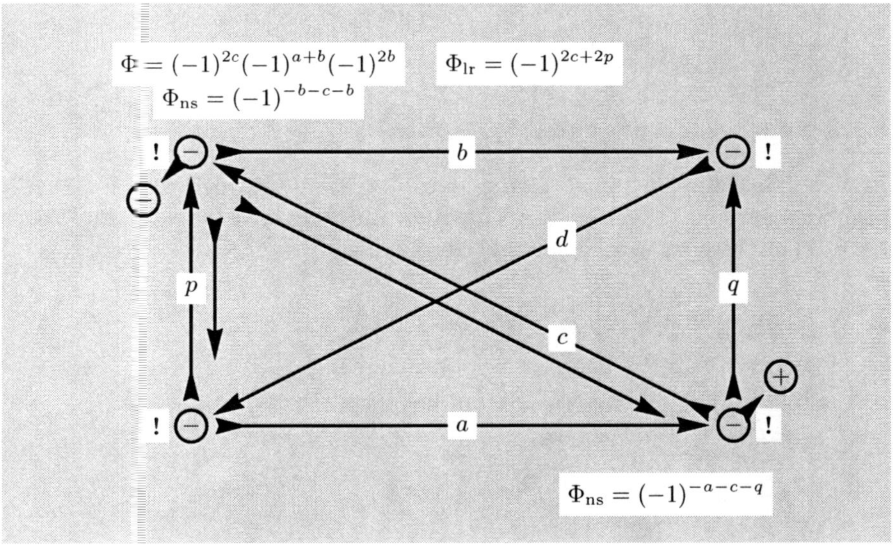

Fig. 8.27. The diagram has been rearranged and the cyclic order at all nodes has changed

The phase Φ has been simplified by removing $(-1)^{4a} = 1$. In anticipation of the goal, the signs of the upper left and lower right node are changed and lead to corresponding phase factors. For convenience, in both cases the exponent has been changed to negative values to simplify the final phase. The upper left node is now moved across the d-line.

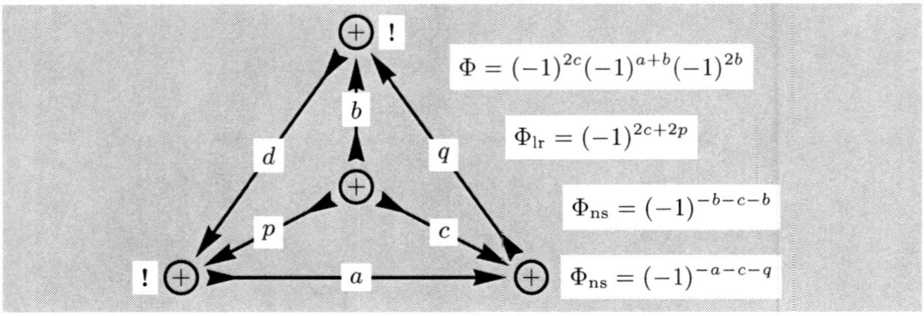

Fig. 8.28. The last transformation has produced the definition of a $6j$-symbol

With $(-1)^{2b+2c+2p} = 1$ some of the phases cancel and the remaining phase factor is $(-1)^{-p-q}$. From Fig. 8.24, where we note the extra phase $(-1)^x$, and from Fig. 8.28 we obtain the following result by translating the diagrams into their algebraic counterparts:

$$\sum_x (-1)^x(2x+1) \begin{Bmatrix} a & d & p \\ c & b & x \end{Bmatrix} \begin{Bmatrix} a & c & q \\ d & b & x \end{Bmatrix} = (-1)^{-p-q} \begin{Bmatrix} c & b & p \\ d & a & q \end{Bmatrix} .$$

The phase can be expressed in a different way:

$$(-1)^{-p-q} = (-1)^{2p+2q}(-1)^{p+q} = (-1)^{2c+2d}(-1)^{p+q} .$$

8.2.3 Sum with a $6j$- and a $9j$-Symbol

In this sub-section we analyse the sum of a $9j$- and a $6j$-symbol with the graphical method. The angular momentum variables have been chosen in such a way that the sum can be carried out graphically.

$$\sum_x (2x+1) \begin{Bmatrix} e & x & b \\ q & f & c \\ d & a & p \end{Bmatrix} \begin{Bmatrix} s & a & b \\ x & e & f \end{Bmatrix} = ?$$

In the following figure the $9j$-diagram has been rotated to the left.

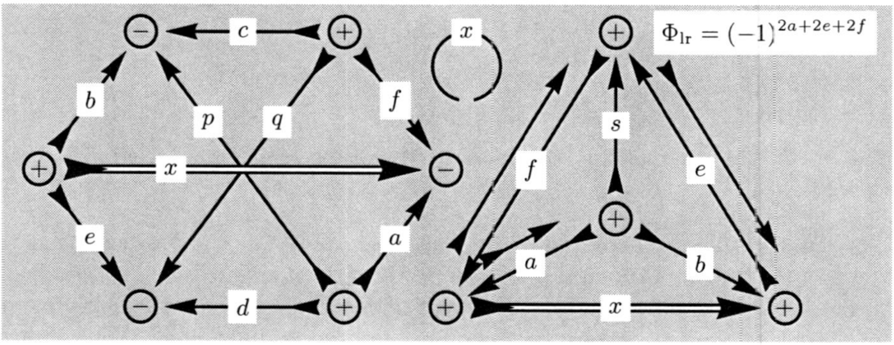

Fig. 8.29. The diagrams are chosen in accordance with the definition of the $9j$- and the $6j$-symbol

The central node in the $6j$-symbol is now moved across the x-line.

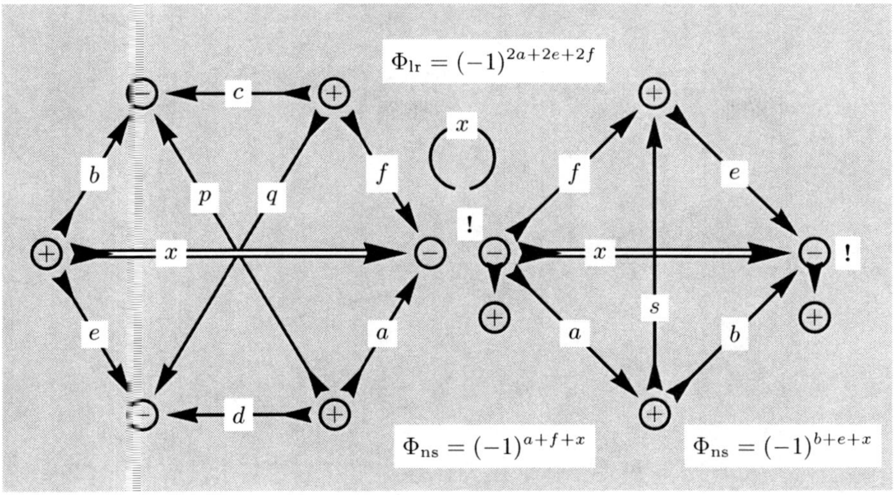

Fig. 8.30. The $9j$- and the $6j$-symbol can be joined at the nodes where the a, f and x-lines meet if the two nodes have different node signs

The two nodes connected by the double x-line in the $6j$-symbol have changed sign due to the deformation of the diagram, note the exclamation marks. The nodes are then changed once more directly to fulfil requirements for joining two diagrams and subsequent summation.

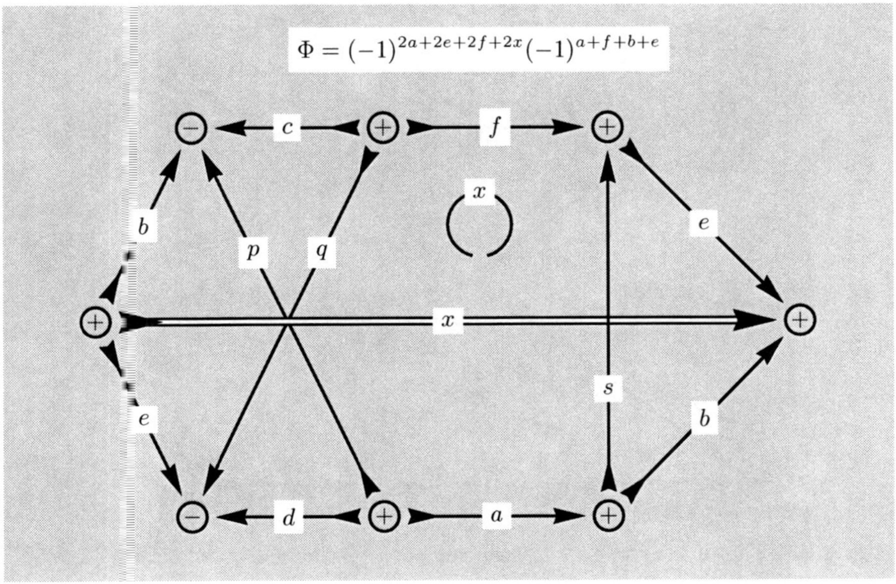

Fig. 8.31. The new diagram has 8 nodes and is therefore not a $3nj$-symbol. The line directions and node signs are appropriate for the summation over x

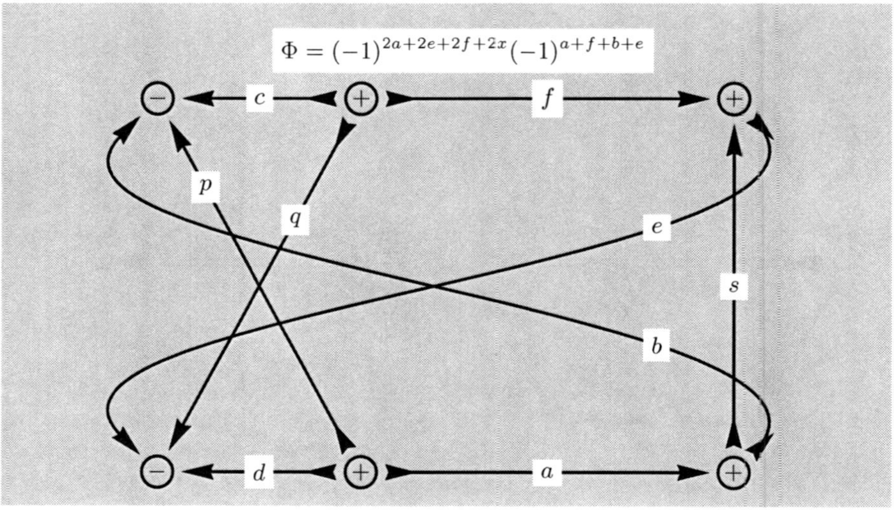

Fig. 8.32. The b- and the e-line have been reconnected as a result of the summation

Transforming the b- and e-lines into straight node connections alters the node signs at four nodes.

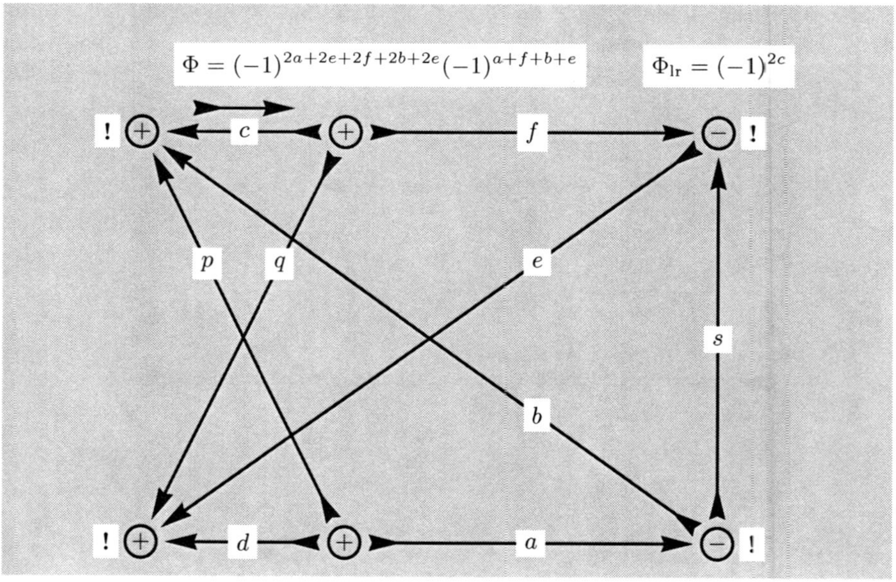

Fig. 8.33. The phase $(-1)^{2x}$ has been replaced by the equivalent term $(-1)^{2b+2e}$

At first glance the diagram in Fig. 8.33 seems rather confusing and it is not easy to see how it could be expressed by $3nj$-symbols. There are 6 nodes but the diagram can not be transformed into a 9j-symbol. An indication of a possible, useful transformation are the two triangles formed by a, b, p and e, f, q.

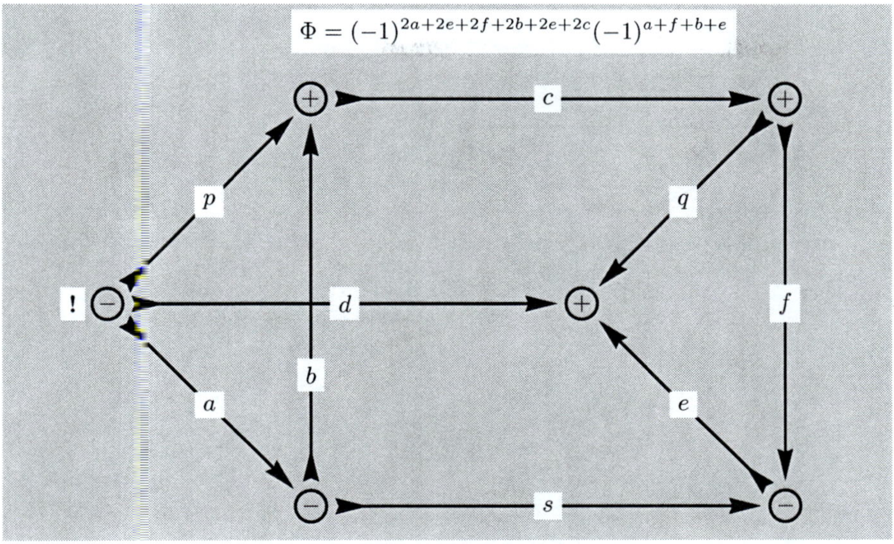

Fig. 8.34. The deformation of the diagram shows that two parts are connected by three angular momentum lines and can, thus, be separated

Following this deformation, we have reached a situation corresponding to Fig. 7.15 and can proceed to separate the two parts as shown in Fig. 7.21.

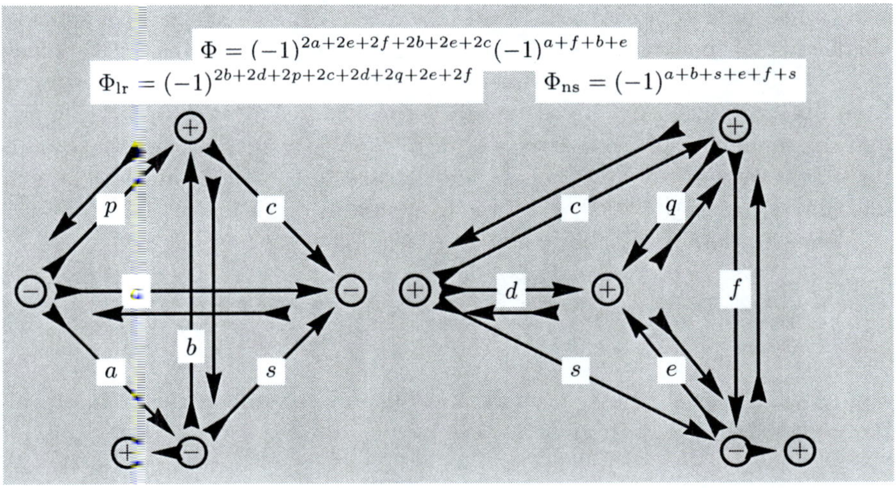

Fig. 8.35. The separation according to Fig. 7.21 results in two 6j-symbols

Changes in line direction and node sign changes have to be performed to achieve the diagrammatic form for the 6j-symbols.

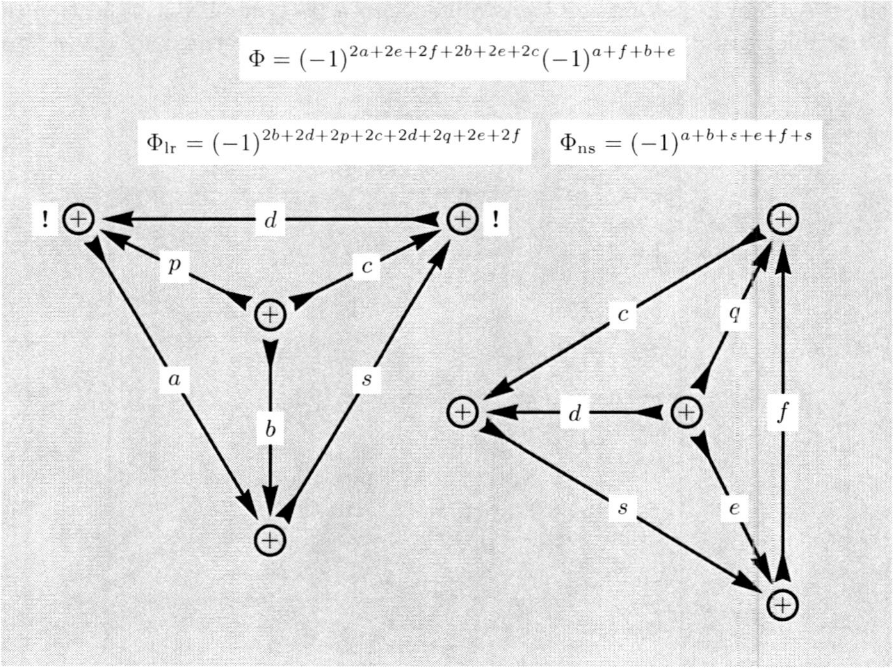

Fig. 8.36. The top-most node in the left diagram has been moved across the d-line and the nodes of the d-line have changed sign

It remains to simplify the phase. Due to direct node sign changes there are two terms, namely, $(-1)^{a+f+b+e}$ and $(-1)^{a+b+s+e+f+s}$ which have sums of single angular momentum variables in the exponent. Both terms together lead to $(-1)^{2a+2b+2e+2f+2s}$ and this phase is merged with the other phases coming from line reversals. Next we remove all terms like $(-1)^{4a} = 1$, $(-1)^{4b} = 1$ etc. and the remaining phase is $\Phi = (-1)^{2b+2p+2q+2f+2s}$. Further simplification is possible if we use $(-1)^{2b+2p+2c} = 1$ and replace $2b + 2p$ by $2c$ in the exponent and find with $(-1)^{2c+2q+2f} = 1$ that the resultant phase factor is $\Phi = (-1)^{2s}$.

Transcription of the final result gives the following relation

$$\sum_x (2x+1) \begin{Bmatrix} e & x & b \\ q & f & c \\ d & a & p \end{Bmatrix} \begin{Bmatrix} s & a & b \\ x & e & f \end{Bmatrix} = (-1)^{2s} \begin{Bmatrix} b & c & p \\ d & a & s \end{Bmatrix} \begin{Bmatrix} d & e & q \\ f & c & s \end{Bmatrix} ,$$

and by using the symmetry properties of $6j$- and $9j$-symbols this relation can be brought into the form given in [12].

$$\sum_x (2x+1) \begin{Bmatrix} a & f & x \\ d & q & e \\ p & c & b \end{Bmatrix} \begin{Bmatrix} x & e & f \\ s & a & b \end{Bmatrix} = (-1)^{2s} \begin{Bmatrix} a & b & s \\ c & d & p \end{Bmatrix} \begin{Bmatrix} c & d & s \\ e & f & q \end{Bmatrix} .$$

8.2.4 Sum with Two 9j-Symbols

In this sub-section we consider a sum that involves two 9j-symbols. We will find that the sum can be performed but the resulting diagram will no longer be separable. Thus, the result will be interpreted as a new entity defining a 3nj-symbol of higher order.

$$\sum_x (2x+1) \begin{Bmatrix} a & b & p \\ c & d & q \\ r & s & x \end{Bmatrix} \begin{Bmatrix} e & f & p \\ g & h & q \\ r & s & x \end{Bmatrix} = ?$$

As in most of our examples the variable names follow those chosen in [12] to facilitate comparison.

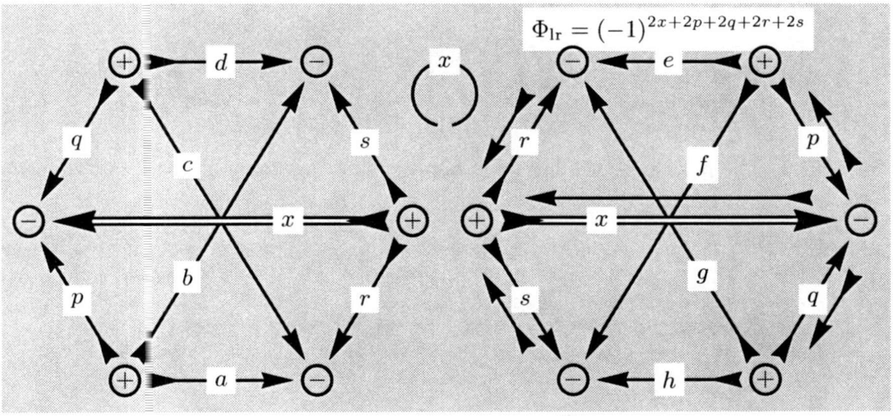

Fig. 8.37. The diagrams are chosen in accordance with the definition for the 9j-symbols but the diagrams have been rotated

In anticipation of the need for appropriate line directions, some angular momentum lines are reversed.

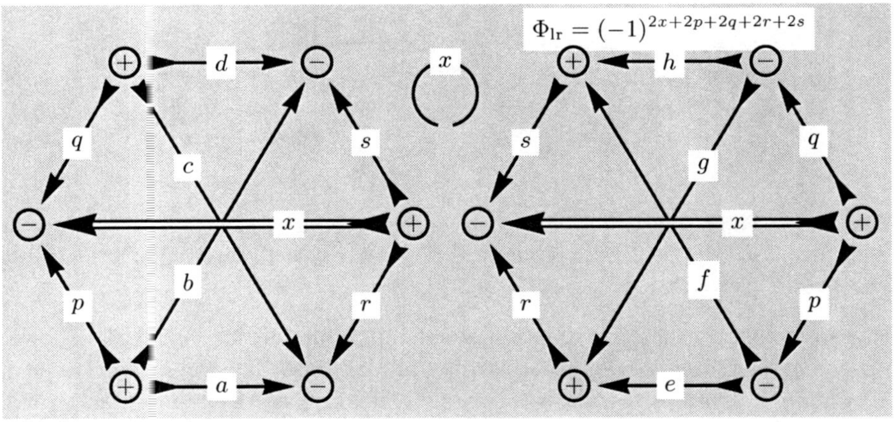

Fig. 8.38. The diagram on the right-hand side has been rotated around a horizontal line which gives a mirror image through which all nodes change sign

The two diagrams can now be joined at the two nodes in the centre of Fig. 8.38, both connected to lines r, s and x.

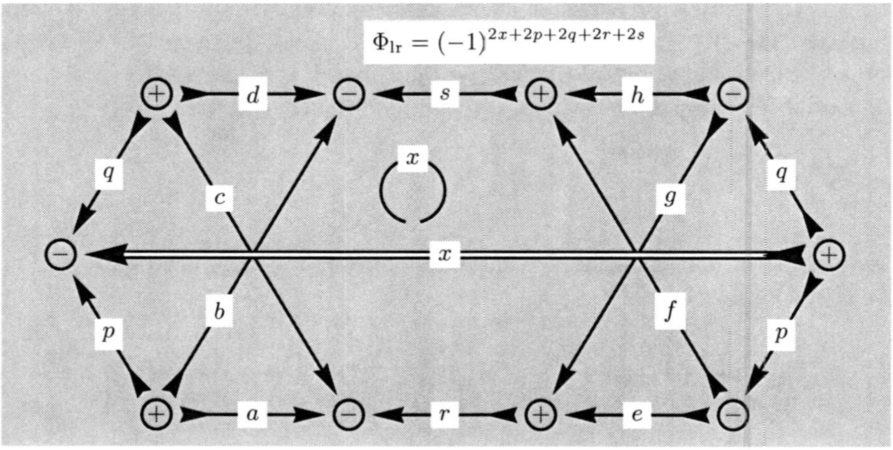

Fig. 8.39. The p- and q-lines have the appropriate directions to carry out the sum over x

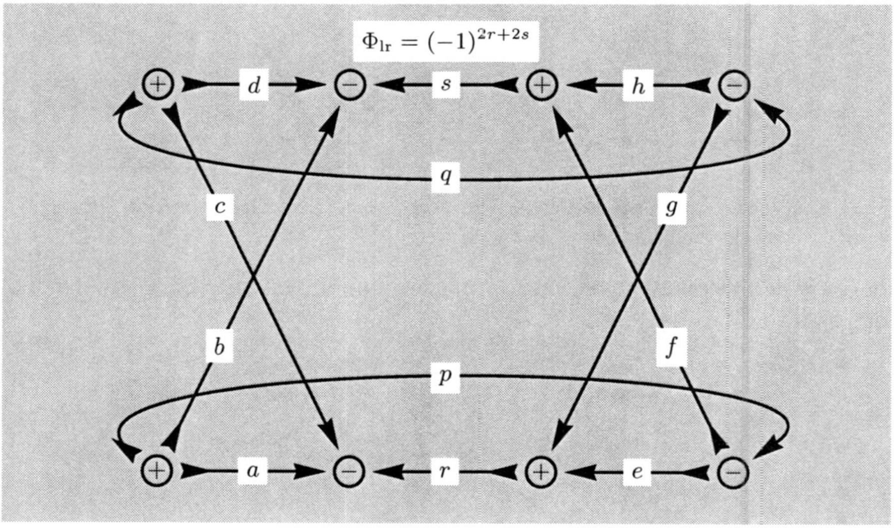

Fig. 8.40. The p- and q-lines have been reconnected as a result of the sum over x and the phase $(-1)^{2x+2p+2q} = 1$ can be dropped because all three angular momenta belong to the same node

Inspection shows that this diagram is not separable into $3nj$-symbols of lower order. In fact, one could use this diagram as basis for the definition of a $12j$-symbol. There are, however, two distinct kinds of $12j$-symbol, as we will discover. We next bring Fig. 8.40 into a more attractive and useful form.

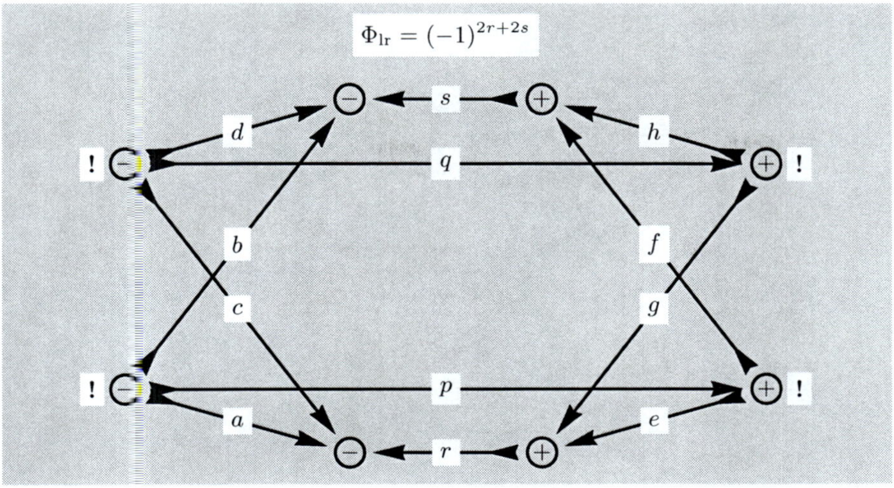

Fig. 8.41. The diagram requires several transformations to reach the definition of a 12j-symbol

The diagram is now rotated around a vertical axis and, as a consequence, all nodes change sign.

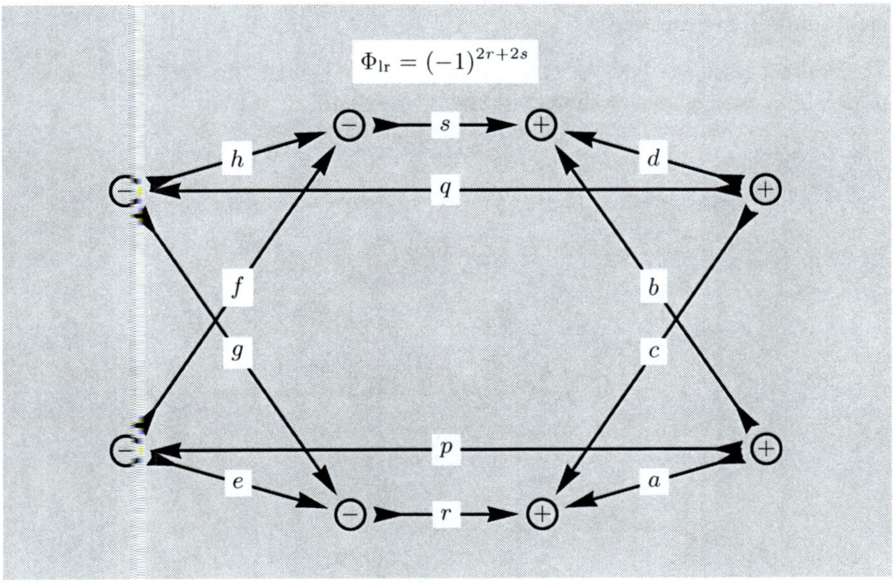

Fig. 8.42. The diagram is a mirror transformation, exchanging left and right, of the diagram in Fig. 8.41

In the following transformation the e-line is moved past the g-line and, similarly, the d-line is moved past the b-line. The two nodes affected by this change of cyclic order change their sign.

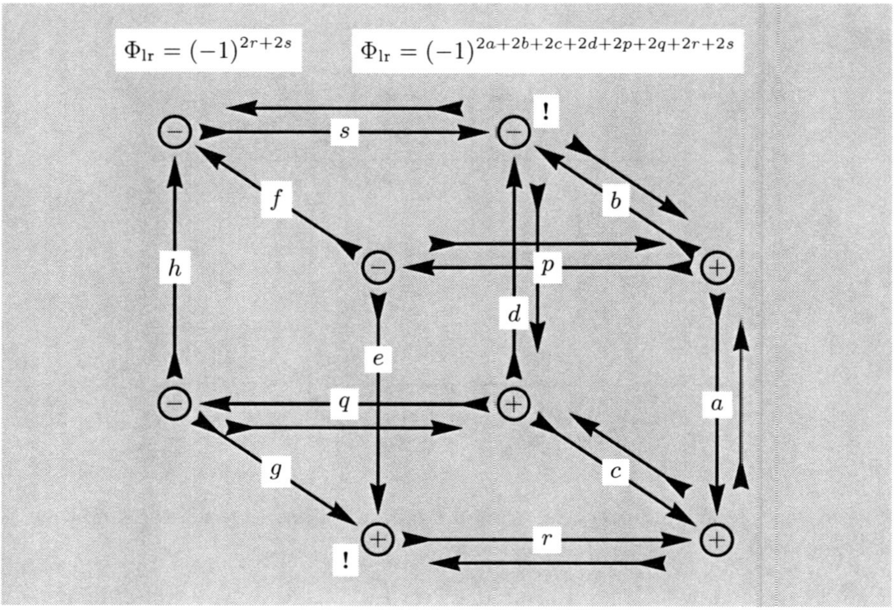

Fig. 8.43. In preparation for the definition of the 12j-symbol several angular momentum lines are reversed

To facilitate the analysis of the various steps taken in the derivation, all the node sign changes are gathered in the corresponding entries.

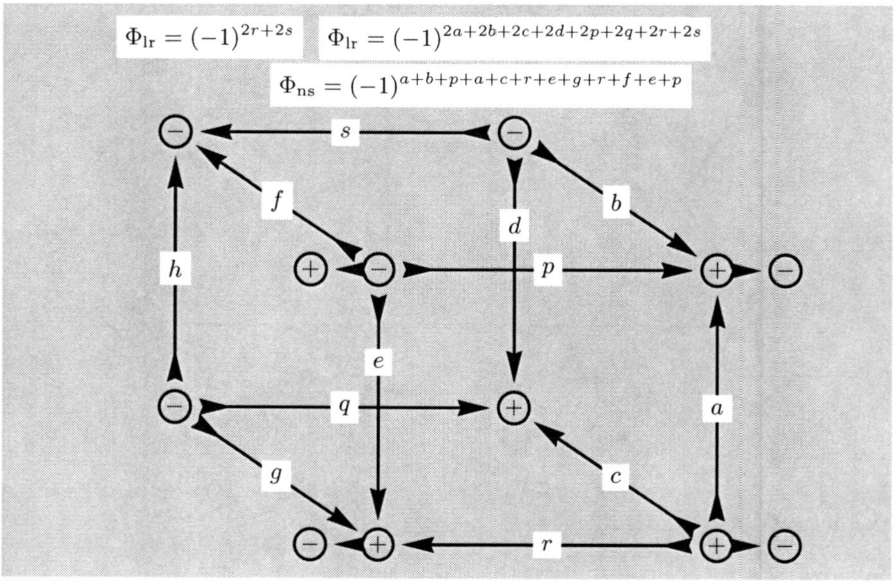

Fig. 8.44. The node signs of four nodes are changed directly in order to conform with the definition of the 12j-symbol

Although the eye tends to view these diagrams as a three-dimensional structure, it should be emphasized that the diagrams occupy just two dimensions. One more rotation is needed to conform with the definition of a 12j-symbol.

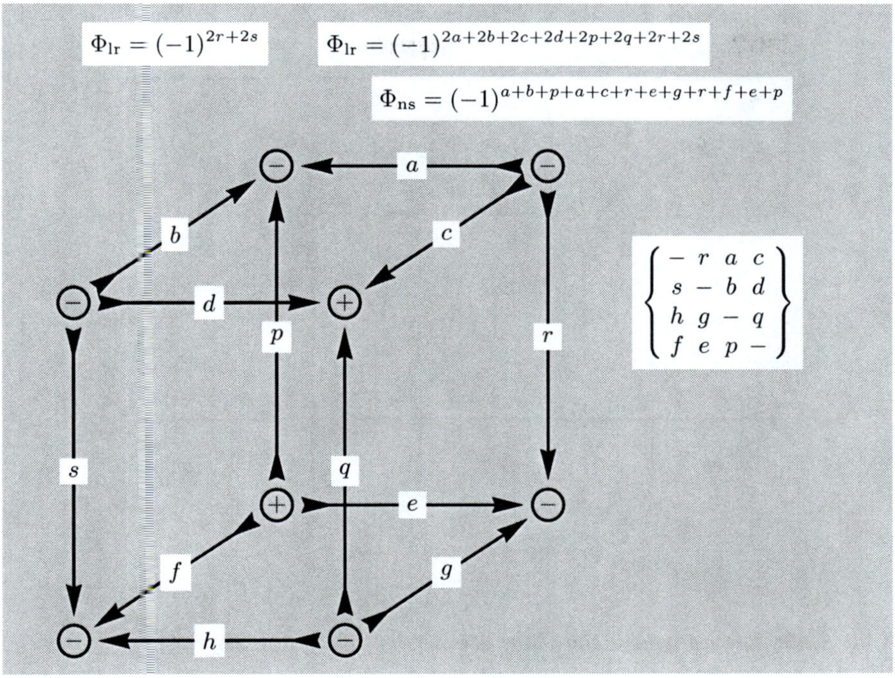

Fig. 8.45. This diagram is the result of an counter-clockwise rotation by 90 degrees

In this figure the diagram is shown together with the algebraic expression representing the diagram. For the final result it is necessary to analyse the phase factors. From the two terms $\Phi_{\rm lr}$ we remove $(-1)^{4r} = 1$ and $(-1)^{4s} = 1$ and find with $(-1)^{2a+2b+2p} = 1$ and $(-1)^{2c+2d+2q} = 1$ that the phase reduces to 1. It remains to consider the phase factor $\Phi_{\rm ns}$. Collecting equal values one has $\Phi_{\rm ns} = (-1)^{2a+2e+2p+2r+b+c+f+g}$ and in the exponent we replace $2p \to 2e+2f$ and $2r \to 2a+2c$. Eliminating fourfold values one finds the remaining phase to be $\Phi = (-1)^{b+3c+3f+g}$ and this is equivalent to $\Phi = (-1)^{b-c-f+g}$.

$$\sum_x (2x+1) \left\{\begin{matrix} a & b & p \\ c & d & q \\ r & s & x \end{matrix}\right\} \left\{\begin{matrix} e & f & p \\ g & h & q \\ r & s & x \end{matrix}\right\} = (-1)^{b-c-f+g} \left\{\begin{matrix} - & r & a & c \\ s & - & b & d \\ h & g & - & q \\ f & e & p & - \end{matrix}\right\} .$$

As mentioned before, our graphical analysis of a sum of two 9j-symbols has not led to a separation into 3nj-symbols of lower order but rather to the definition of a new symbol, namely a 12j-symbol. In fact, the theory of angular momentum shows [10] that there are two kinds of 12j-symbols and the diagram in Fig. 8.45 is one of the two representations for a $12j$(II)-symbol of the *second kind*. We will encounter the $12j$(I)-symbol of the first kind in Sect. 8.3.2.

The above diagram is equivalent to a second form, displayed in Fig. 8.47, which can be reached easily by transforming the diagram in Fig. 8.45.

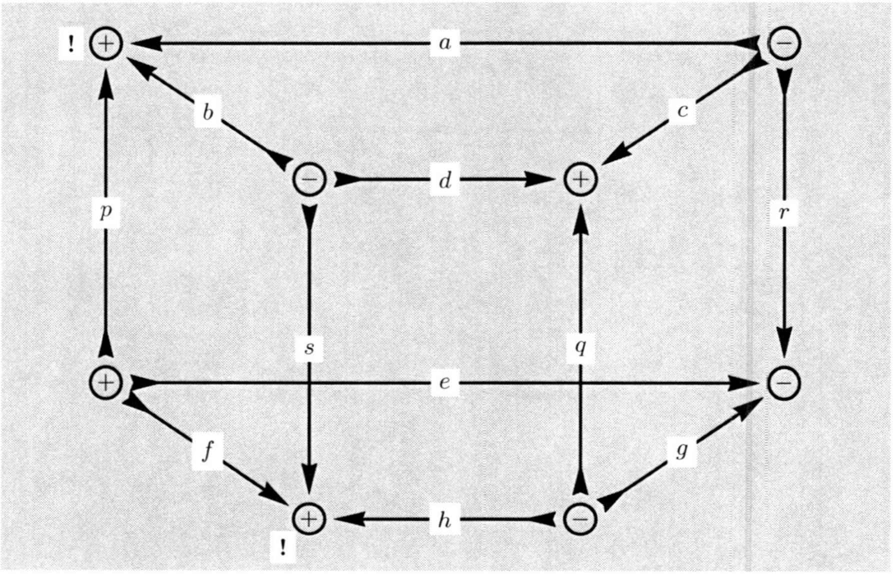

Fig. 8.46. The a-line and the e-line is extended relative to Fig. 8.45

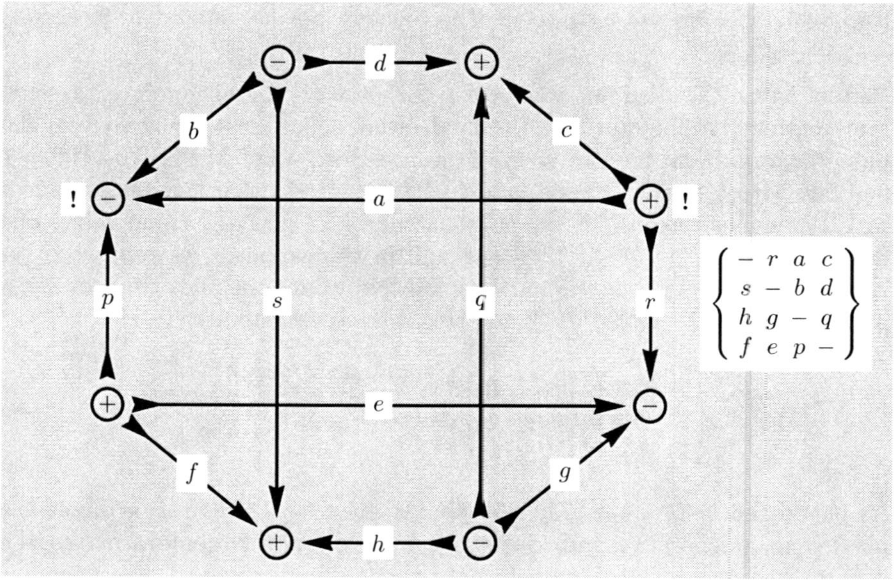

Fig. 8.47. The d-line is moved across the a-line and this produces a characteristic octagon shape

This is the equivalent graphical representation of the 12j-symbol (second kind) as defined in [12] and denoted as 12j(II)-symbol.

8.3 Sums with Three 3nj-Symbols

We now examine relations that involve three 3nj-symbols. although we will not venture beyond this level, Varshalovich et al. [12] consider single and multiple sums involving up to four 3nj-symbols.

8.3.1 The 9j-Symbol Expressed Through 6j-Symbols

In this sub-section the starting point is not, as before, an algebraic expression but the diagram of a 9j-symbol. As an example of the application of the three line separation, derived in Sect. 7.8, we consider the separation of a 9j-symbol into smaller units.

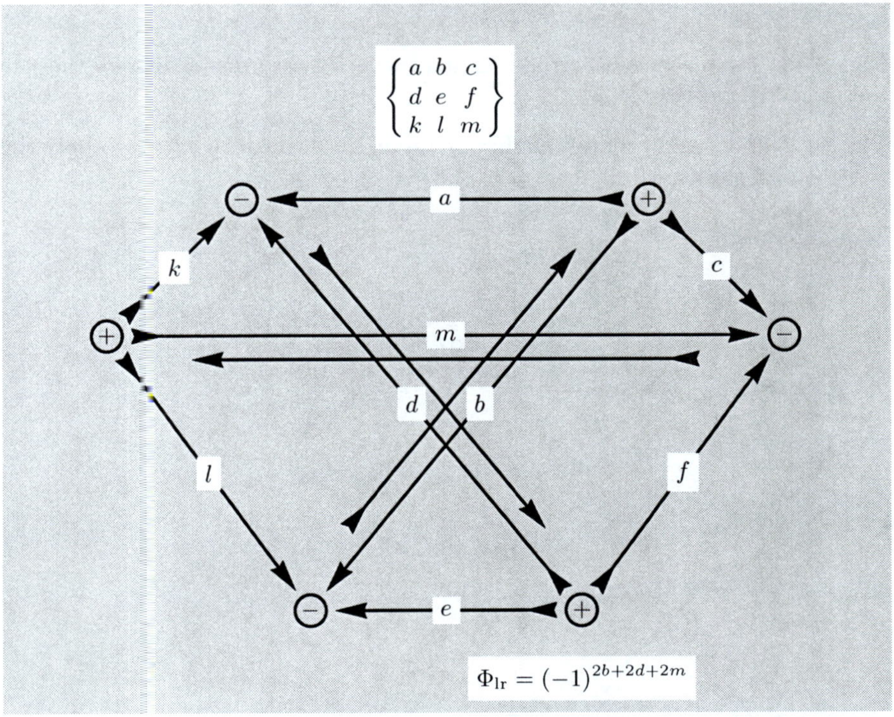

Fig. 8.48. The diagram shows a distorted but instructive graphical representation of the 9j-symbol. Three line reversals are indicated together with the resulting phase

In the following step the a- and the m-line are reduced to one line through line contraction, i.e. introduction of an additional sum over angular momentum x and its projection ξ.

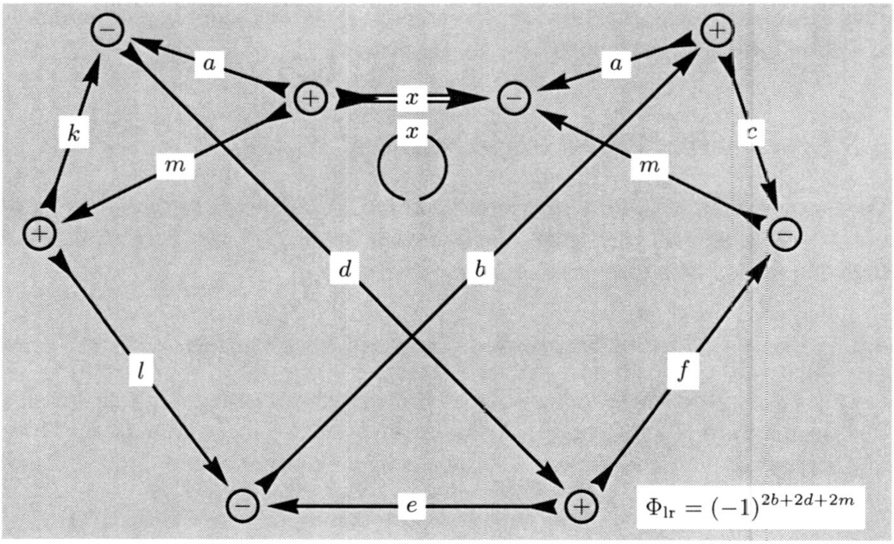

Fig. 8.49. Two new nodes introduced and connected by a double line for the sum over x and its projection

The x-, b- and f-lines represent three parallel lines to which we now apply the three line separation rule from Sect. 7.8.

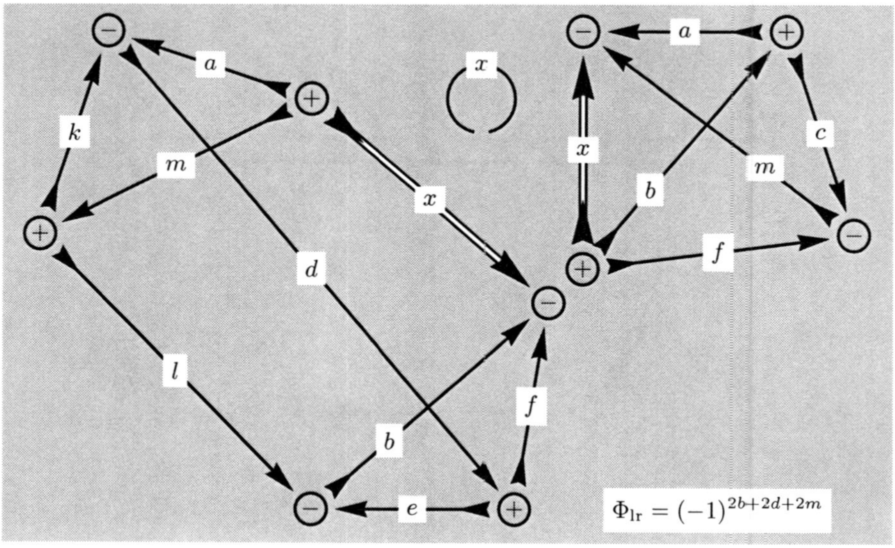

Fig. 8.50. Two new nodes have been introduced to allow a three line separation in the right-hand part of the diagram

The same procedure is now applied to the three parallel lines l, d and x and leads to another separation.

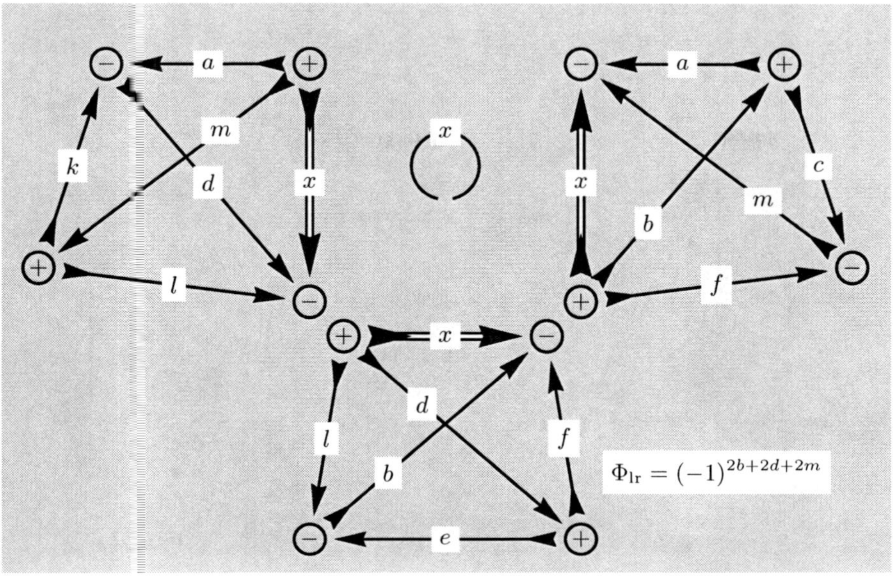

Fig. 8.51. Two new nodes have been introduced in preparation of a three line separation in the left-hand part of the diagram

As a result of our manipulations, the $9j$-symbol has been separated into three $6j$-symbols which, however, have not yet assumed their proper form.

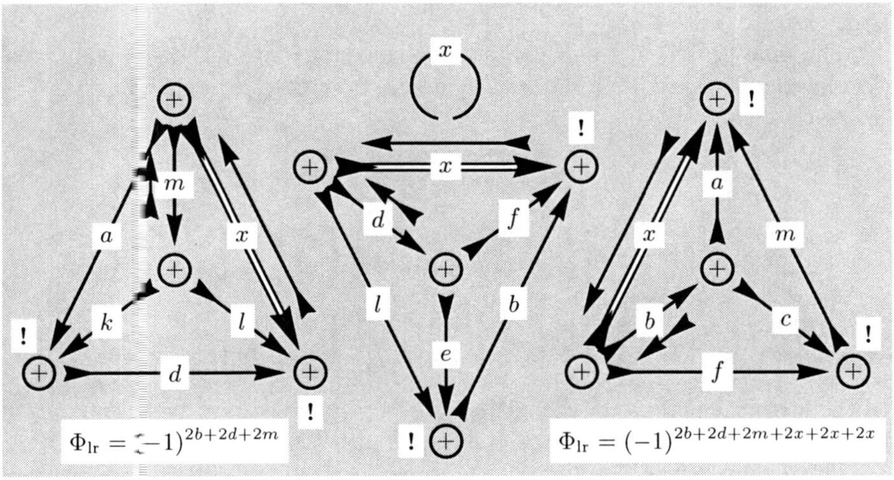

Fig. 8.52. In these diagrams nodes have been moved across the d-line (left), the b-line (middle) and across the m-line (right). Through this manipulation two nodes in each $6j$-symbol change from $-$ to $+$

Because $(-1)^{4b+4d+4m+4x+2x} = (-1)^{2x}$ the remaining phase factor is very simple and all $6j$-symbols appear in their basic form.

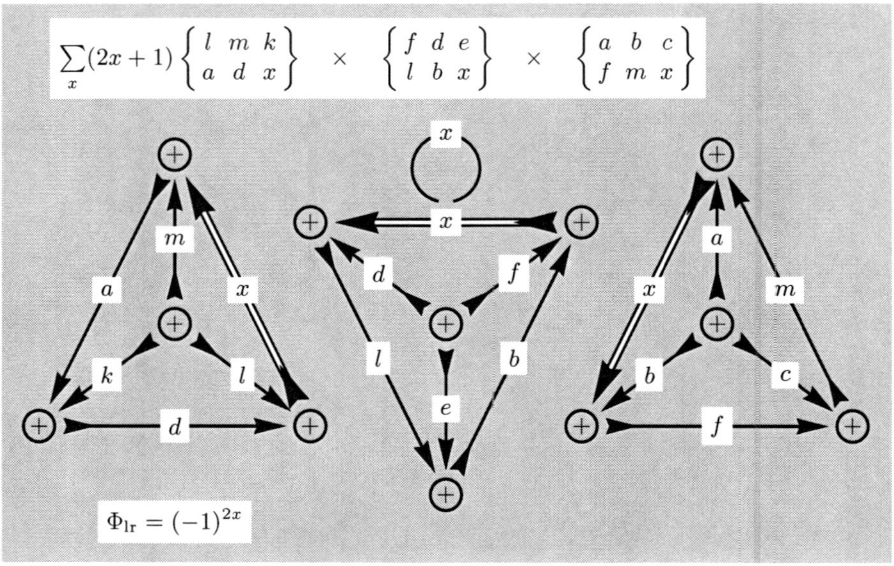

Fig. 8.53. The double lines indicate the sum over x and every $6j$-symbol contains this summation variable

The $6j$-symbols have the correct form, where three angular momentum lines go out from the central node and all the nodes at the corners of the triangle are connected by lines in counter-clockwise direction. All nodes have a positive node sign.

The final graphical result can now be converted into an algebraic expression for the $9j$-symbol represented by a sum over three $6j$-symbols.

$$\begin{Bmatrix} a & b & c \\ d & e & f \\ k & l & m \end{Bmatrix} =$$
$$= \sum_x (-1)^{2x}(2x+1) \begin{Bmatrix} l & m & k \\ a & d & x \end{Bmatrix} \begin{Bmatrix} f & d & e \\ l & b & x \end{Bmatrix} \begin{Bmatrix} a & b & c \\ f & m & x \end{Bmatrix}$$
$$= \sum_x (-1)^{2x}(2x+1) \begin{Bmatrix} a & d & k \\ l & m & x \end{Bmatrix} \begin{Bmatrix} b & e & l \\ d & x & f \end{Bmatrix} \begin{Bmatrix} c & f & m \\ x & a & b \end{Bmatrix} .$$

In the second equality the elements of the $6j$-symbols have been rearranged to conform with the result (3.1) in [9], and, also (19) on p. 466 in [12].

This example has the virtue of a repeat performance of the prescription how to proceed with a sum over an angular momentum which is present in several diagrams, e.g. in Fig. 8.53. If the double-lined arrow has different lines at both nodes it is necessary to seek out and join corresponding nodes until there is only one double-lined arrow left, with necessarily identical lines at both ends. At that point the sum can be performed and we have essentially retraced our steps and returned to Fig. 8.48

8.3.2 Sum with One 9j- and Two 6j-Symbols

$$\sum_x (2x-1)\begin{Bmatrix} e & f & x \\ a & b & p \\ c & d & q \end{Bmatrix}\begin{Bmatrix} f & e & x \\ l & k & h \end{Bmatrix}\begin{Bmatrix} p & q & x \\ k & l & g \end{Bmatrix} = ?$$

The algebraic quantities are first translated and arranged in such a way as to facilitate the joining of the diagrams.

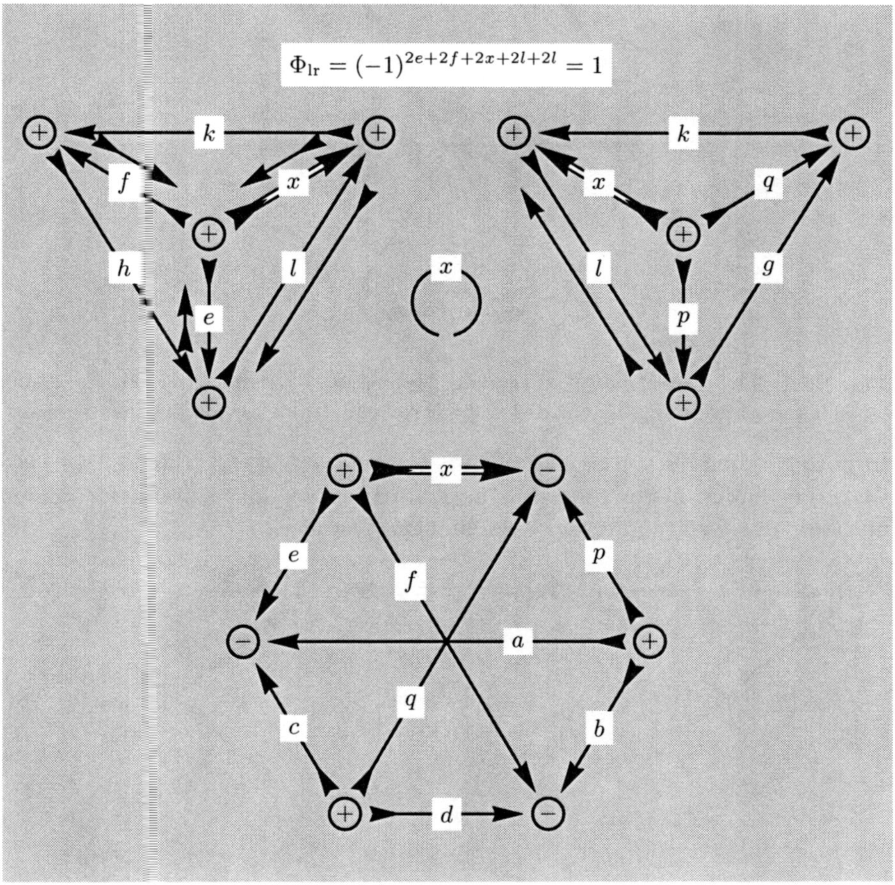

Fig. 8.54. The sum over x combines two $6j$- and one $9j$-symbol. Note that the phase resulting from line reversals amounts to 1 because e, f and x belong to the same node. The two reversals of l-lines adds the phase $(-1)^{4l} = 1$

The central node of the left $6j$-symbol has a counterpart in the $9j$-symbol, the node with the angular momenta e, f and x. By analogy, the right $6j$-symbol has a central node with p, q and x that will join the corresponding node from the $9j$-symbol.

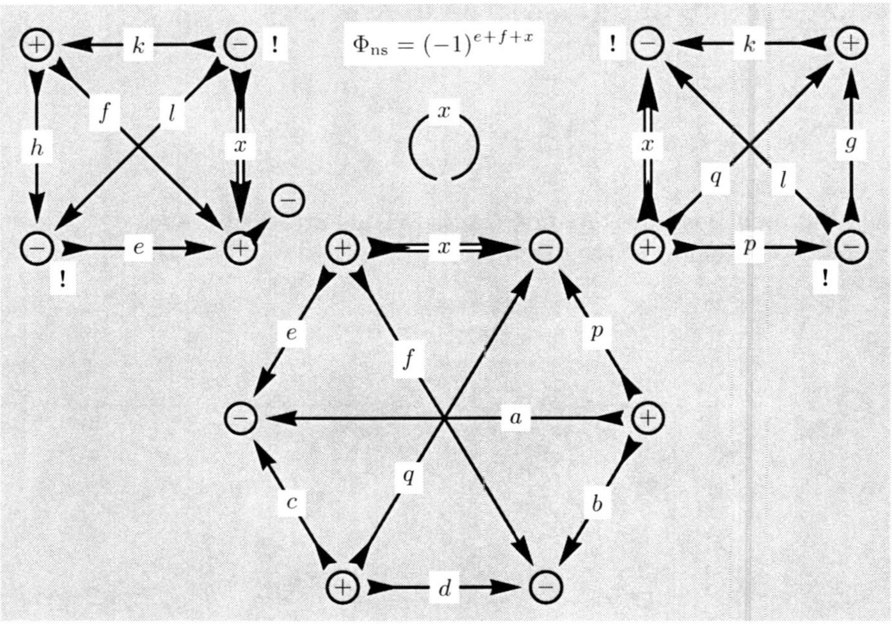

Fig. 8.55. The central nodes of the two $6j$-symbols have moved across the l-line. One change of node sign is required to fulfil the conditions for joining the diagrams

In both $6j$-symbols the central node is moved outside the triangle. We note that, the angular momentum lines have already directions suitable for joining but one node sign requires a direct change of node sign.

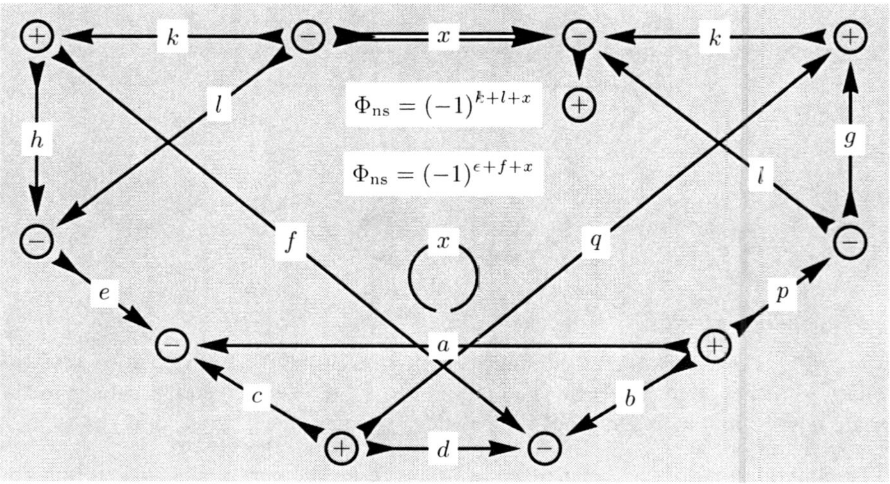

Fig. 8.56. The diagrams have been joined at two nodes with different node signs. The x-line appears only once in the diagram and a further change of node sign is required for the application of the summation procedure

A further change of node sign for one of the nodes connected by the x-line achieves the appropriate conditions for the sum over x.

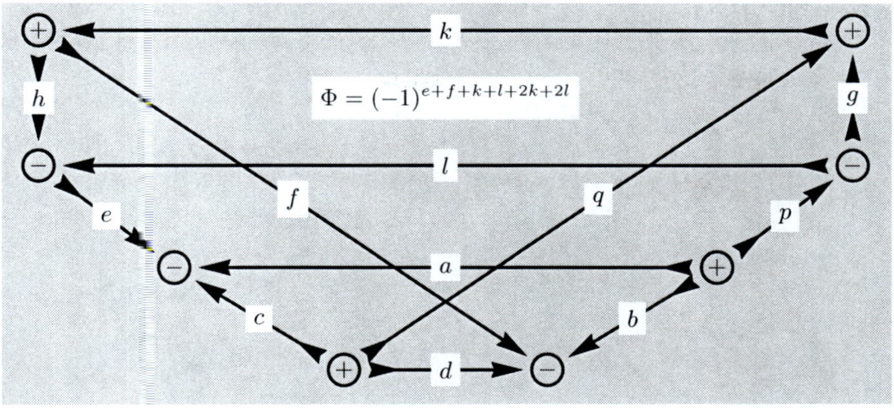

Fig. 8.57. The k- and the l-line have been reconnected by the summation over x. In the phase $(-1)^{2x}$ is replaced by $(-1)^{2k+2l}$

Performing the sum over x has created a diagram with eight nodes which can be brought into an octagon shape familiar from Fig. 8.47.

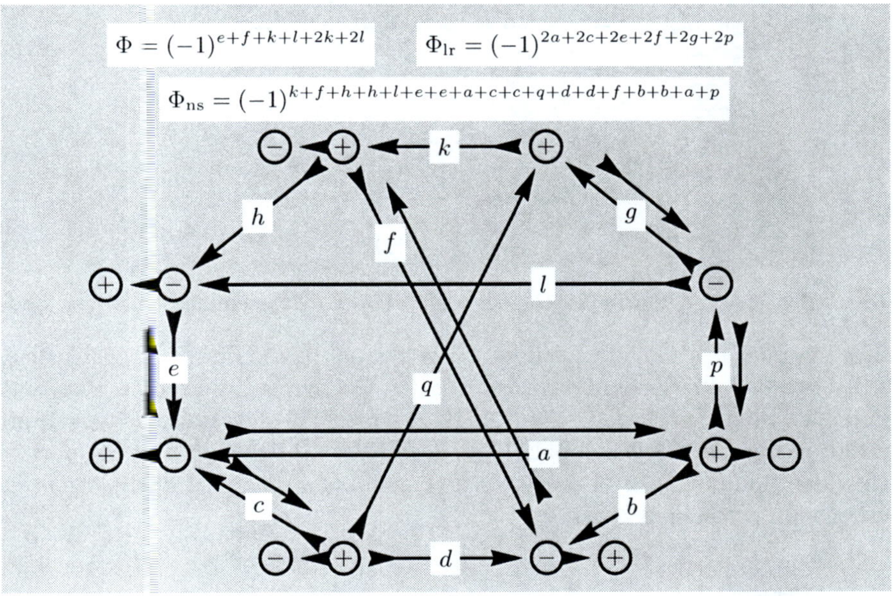

Fig. 8.58. Several line directions and node signs are changed to approach the definition of a $3nj$-symbol

As first step in the simplification of the phase we remove all angular momenta that appear four times in the exponent. The remaining parts provide $\Phi = (-1)^{e+f+p+q}$ and $\Phi = (-1)^{2b+2d+2g+2h+2p}$. In the latter phase we set

$(-1)^{2g} \Rightarrow (-1)^{2k+2p}$ and combine $(-1)^{2b+2d} = (-1)^{2f}$ with $(-1)^{2h+2k} = (-1)^{2f}$. With these manipulations we have $(-1)^{e+f+3p+3q}$ which is equivalent to $(-1)^{e+f-p-q}$.

The changes in line direction and node signs have created the graphical form of a new $12j$-symbol.

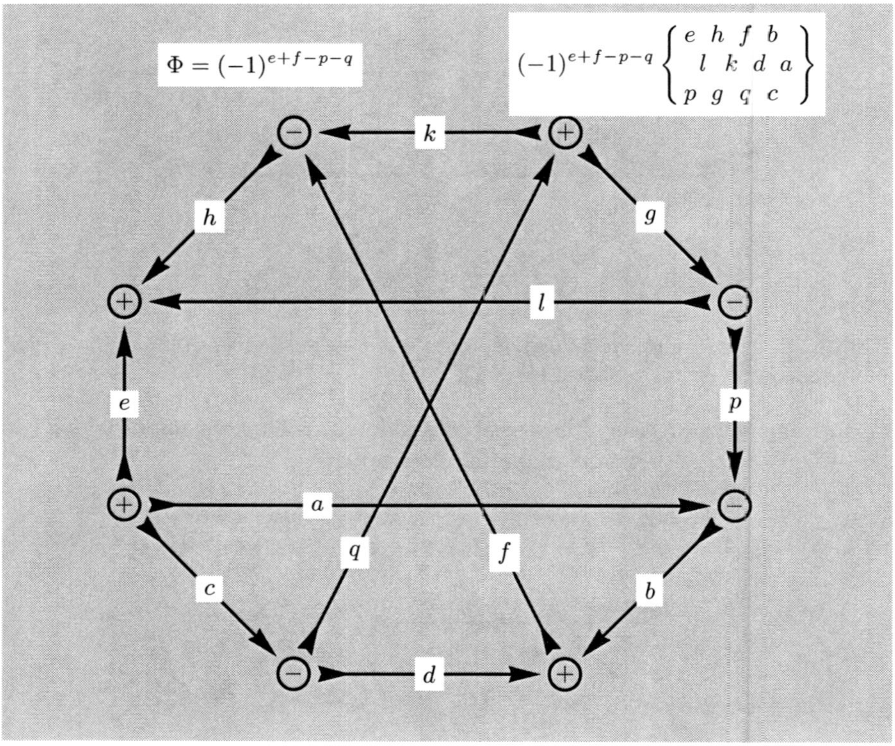

Fig. 8.59. The diagram represents the definition of a $12j$-symbol of the first kind

The diagram in this figure defines a $12j$-symbol of the first kind [12] which will be called $12j$(I)-symbol in contrast to the symbol defined in Figs. 8.45 and 8.47 which is referred to as $12j$(II). Figure 8.59 shows also the algebraic notation for the $12j$(I)-symbol. The phase factor $(-1)^{e+f-p-q}$ is not part of the definition of the $12j$(I)-symbol, of course, but it is a part of the algebraic expression considered here.

$$\sum_x (2x+1) \begin{Bmatrix} e & f & x \\ a & b & p \\ c & d & q \end{Bmatrix} \begin{Bmatrix} f & e & x \\ l & k & h \end{Bmatrix} \begin{Bmatrix} p & q & x \\ k & l & g \end{Bmatrix} =$$

$$= (-1)^{e+f-p-q} \begin{Bmatrix} e & h & f & b \\ & l & k & d & a \\ p & g & q & c \end{Bmatrix} .$$

This result can be made to agree with the formula given in [12] if the symmetry properties of the $12j$(I)-symbol given there are observed.

In the previous case of a $12j$(II) we have seen that there are two equivalent graphical representations and the same is true for the $12j$(I)-symbol here. The necessary transformations will be performed in the following figures.

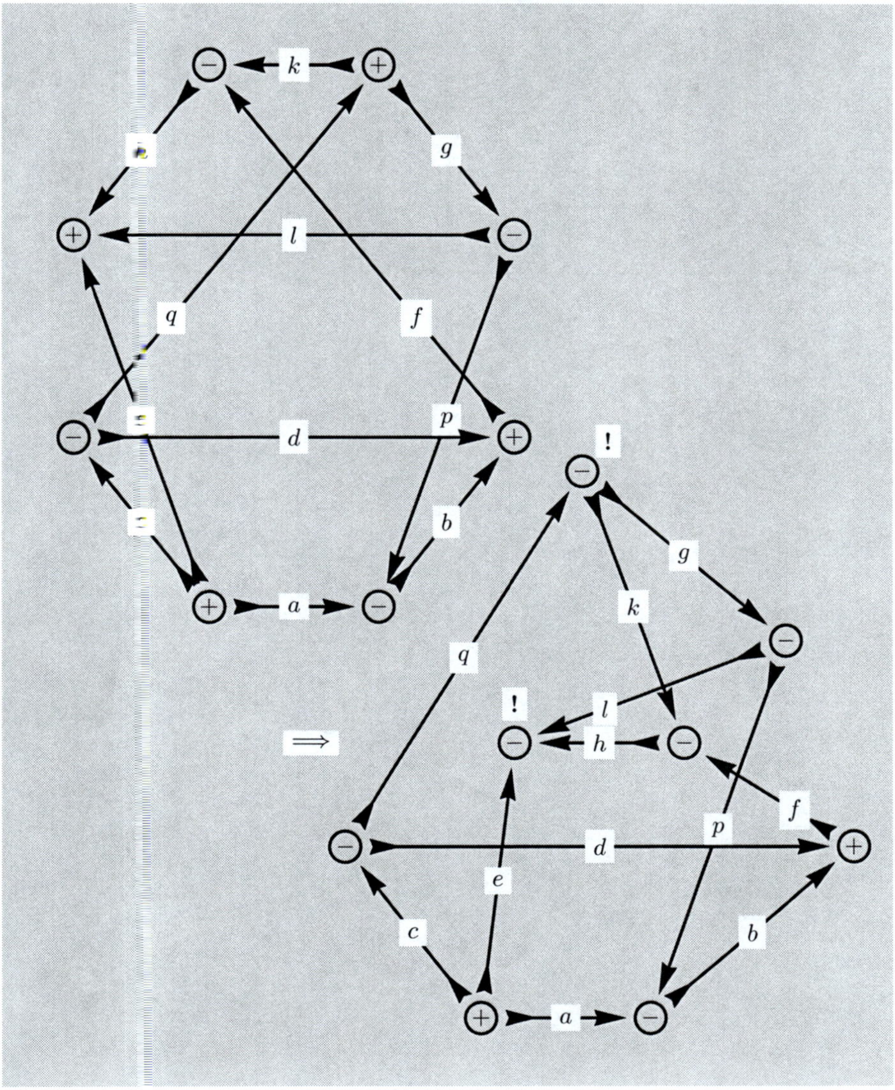

Fig. 8.60. In the first step the a-line is decreased in length and moved below the d-line which is extended. No node sign is changed by this deformation

Figure 8.60 displays a second step in which the h-line is moved across the q-line and two nodes change sign.

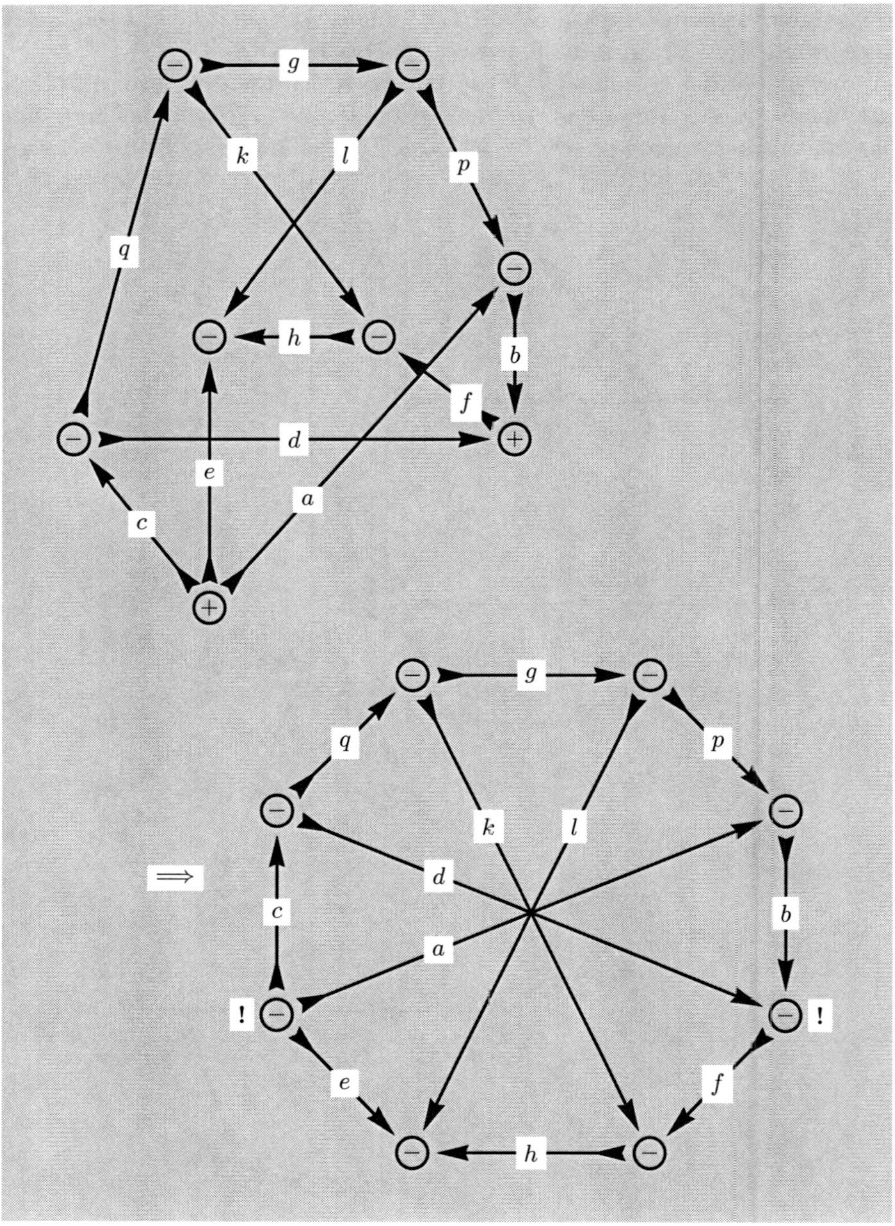

Fig. 8.61. In the next step the node with the lines a, b and p is moved across the d- and the f-line

In the lower part of the figure the h-line is brought down to the bottom and two nodes change sign in the process. The last step has created a diagram that is very close to the desired form.

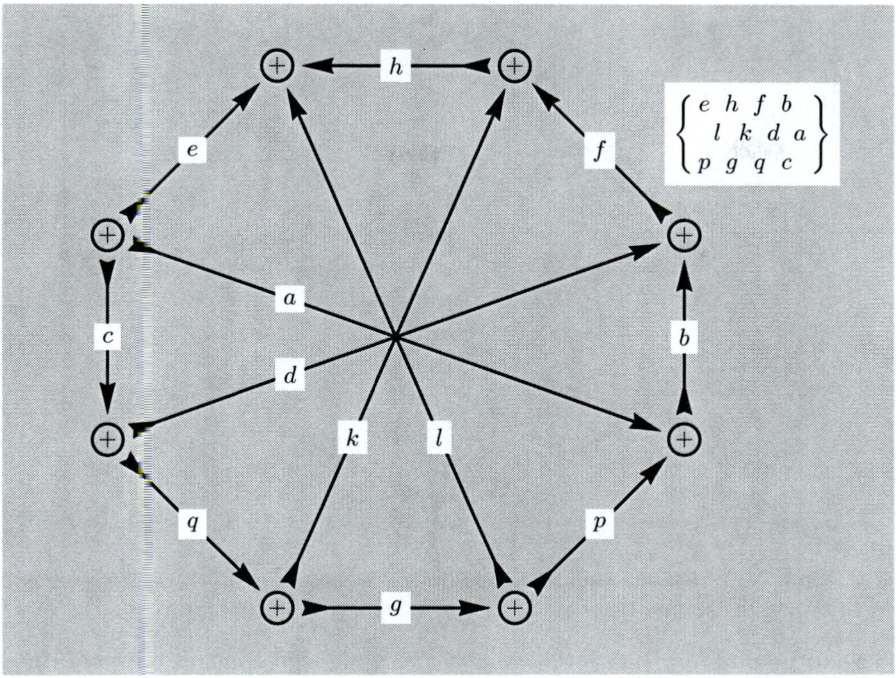

Fig. 8.62. In the final transformation, the whole diagram is turned about a horizontal axis

The diagram in Fig. 8.62 is a mirror image of the preceding one and it is obvious that the cyclic order of all nodes must change through this transformation. The graphical form for the $12j$(I)-symbol agrees perfectly with the definition used in [12].

8.3.3 Alternating Sum with One $9j$- and Two $6j$-Symbols

In analogy to previous expressions, here we consider a product of two $6j$-symbols and a $9j$-symbol summed over x but with alternating signs.

$$\sum_x (2x+1)(-1)^x \begin{Bmatrix} e & f & x \\ a & b & p \\ c & d & q \end{Bmatrix} \begin{Bmatrix} f & e & x \\ l & k & h \end{Bmatrix} \begin{Bmatrix} p & q & x \\ l & k & g \end{Bmatrix} =$$

$$= \sum_x (2x+1)(-1)^x \begin{Bmatrix} h & l & e \\ x & f & k \end{Bmatrix} \begin{Bmatrix} e & f & x \\ a & b & p \\ c & d & q \end{Bmatrix} \begin{Bmatrix} k & p & g \\ q & l & x \end{Bmatrix} = ?$$

In the second line of the formula the $6j$-symbols have been rearranged for easier comparison with the following figure. We recall that, the bottom row of each $6j$-symbol defines the boundary and and the x-lines are now no longer inside.

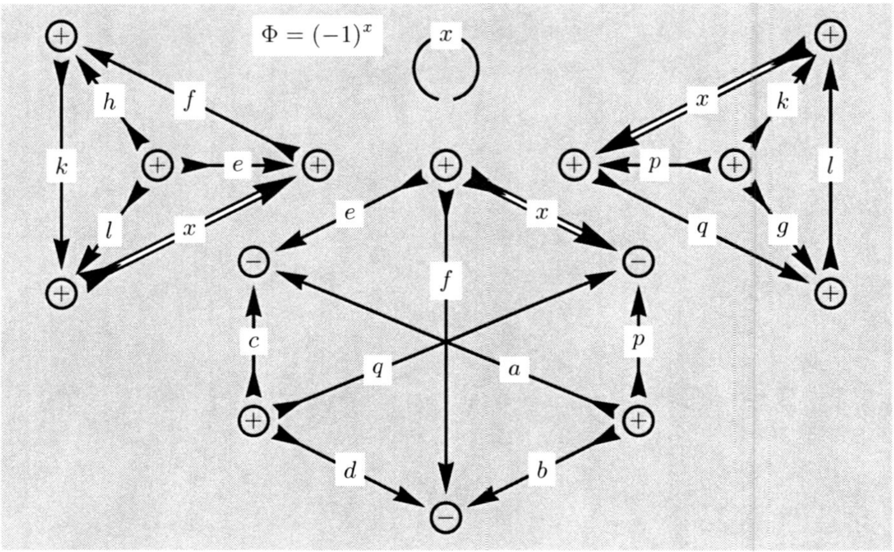

Fig. 8.63. Note that the angular momentum x is present in all three diagrams and in this form the summation cannot be carried out in graphical form

Several preparatory line reversals and two node sign changes are performed in the next figure, and concomitant phases taken into account.

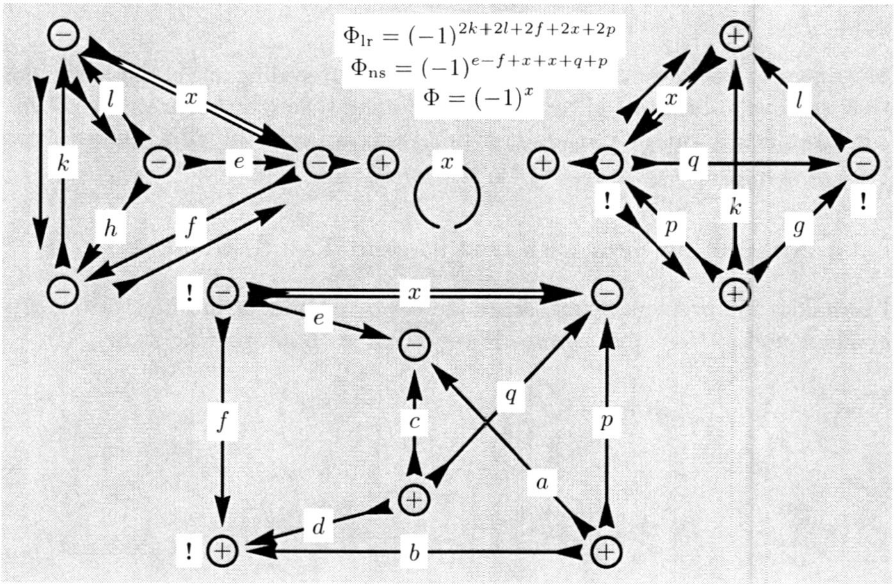

Fig. 8.64. Because the left $6j$-symbol is flipped around the direction of e all its nodes have changed sign. In the right $6j$-symbol the central node has been moved across the q-line. In the bottom diagram both nodes connected by the c-line have been moved across the f-line

After line reversals have taken effect it will be possible to join all three diagrams.

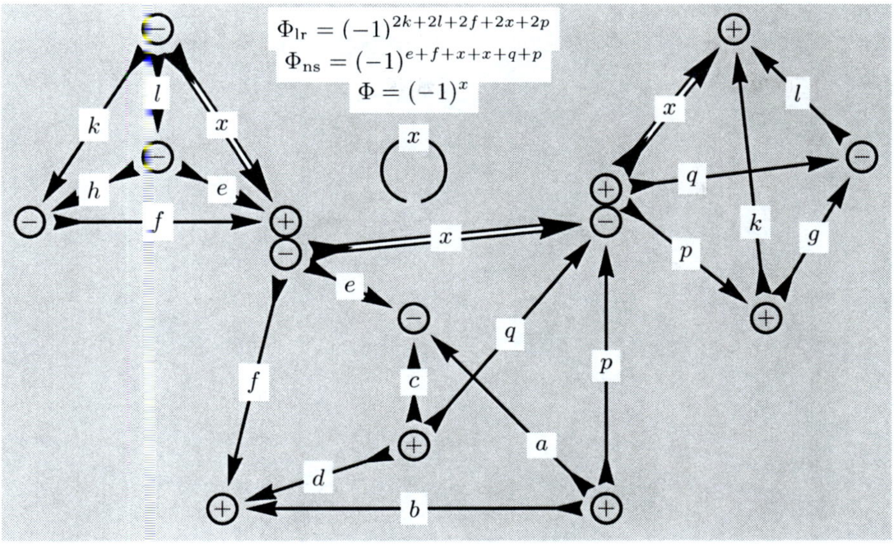

Fig. 8.65. The adjacent nodes with different signs indicate the way to join the diagrams. As required, lines entering and leaving the nodes have commen angular momenta and have the same directions

By joining the three diagrams we are left with only one (double) x-line which connects to k- and l-lines at both ends.

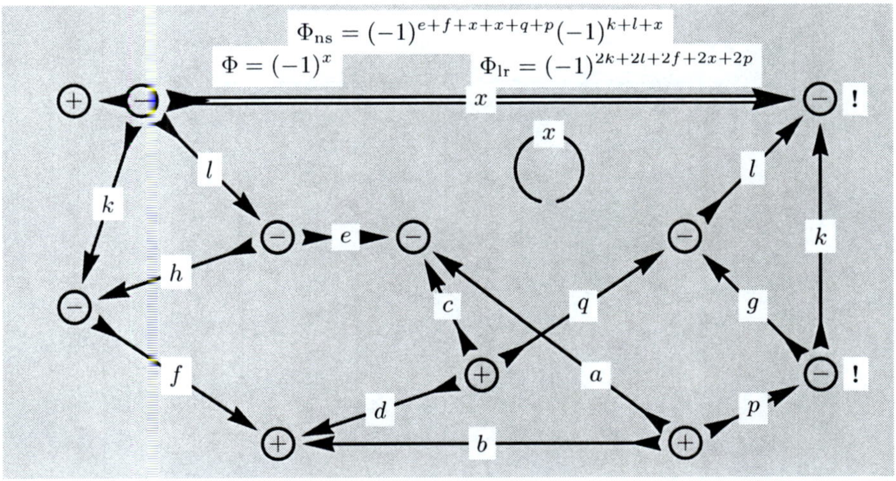

Fig. 8.66. For the sum over x the top left node needs a sign change, which introduces an additional phase. Note the node sign changes at the right of the diagram

The sum removes the x-line and the circular diagram. However, there are several occurrences of x in the phase factor. If the sum is possible, the phase may no longer depend on x. Inspection of the phase shows that a single x appears four times and $(-1)^{4x} = 1$. The remaining dependence on x vanishes since $(-1)^{2k+2l+2x} = 1$.

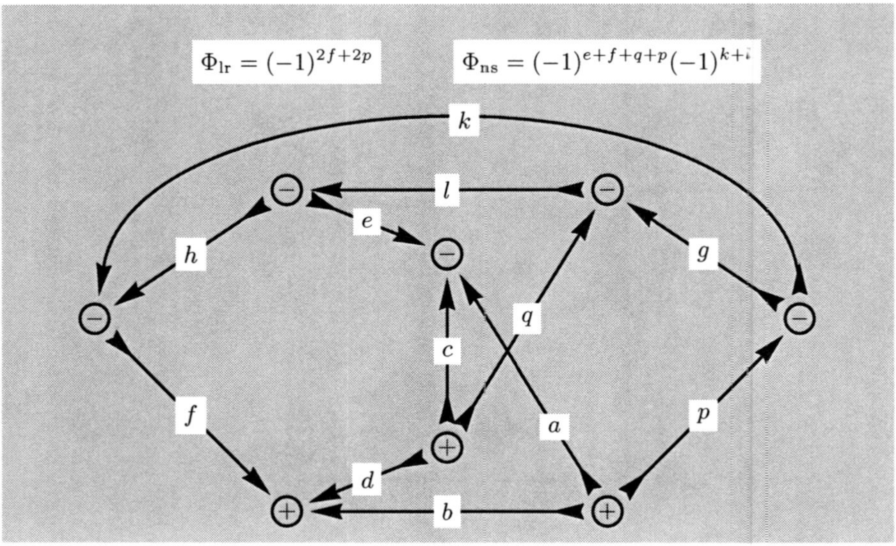

Fig. 8.67. Note the simplified phase

The resulting diagram will be subjected to several further transformations.

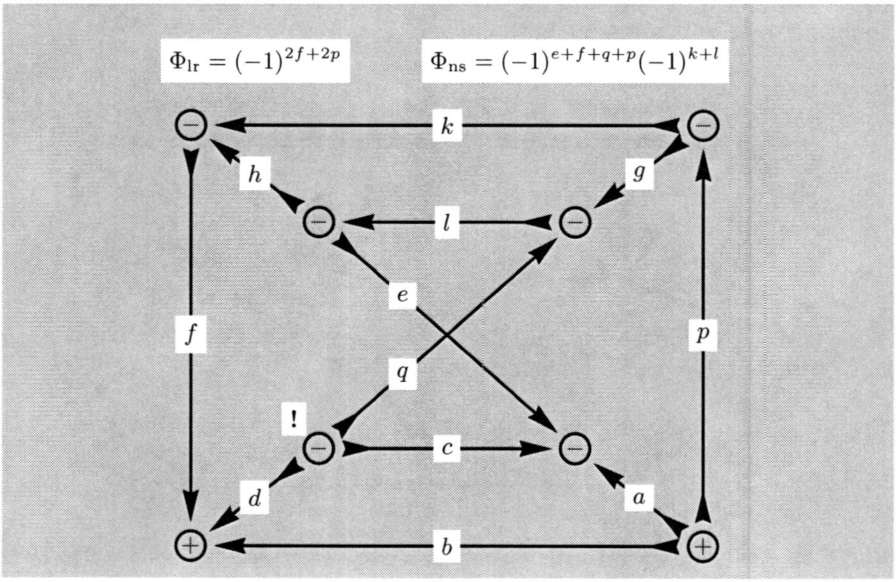

Fig. 8.68. The c-line is moved through the q-line and the common node changes sign because the cyclic arrangement has changed

In the next step, the l- and the c-line are moved up and below the square, respectively.

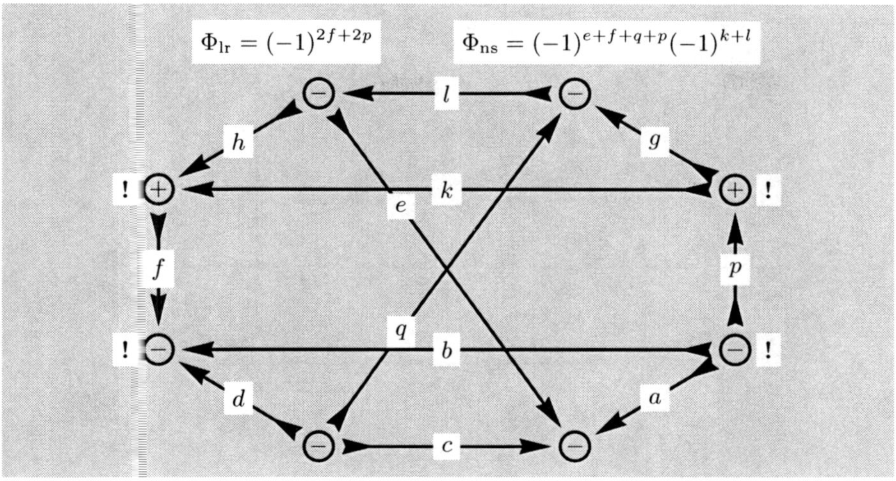

Fig. 8.69. The transformation leads to node sign changes on the left and the right of the diagram

The transformations have produced a diagram that is closely related to the diagram in Fig. 8.59. Several directions and node signs are, however, different from the definition of a $12j$(I)-symbol.

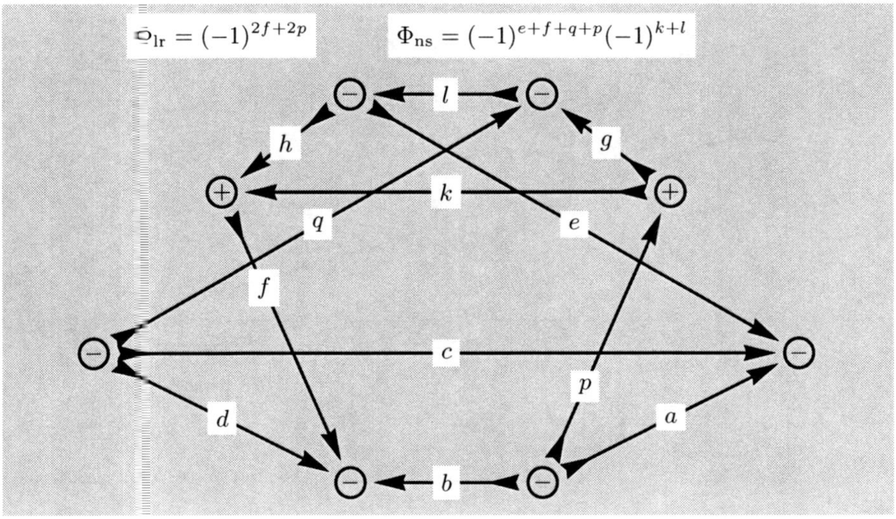

Fig. 8.70. The c-line is extended and moved upward without node sign change in the bottom nodes

We try further transformations to reach a diagram related to the second form of the $12j$(II)-symbol as given in Fig. 8.62.

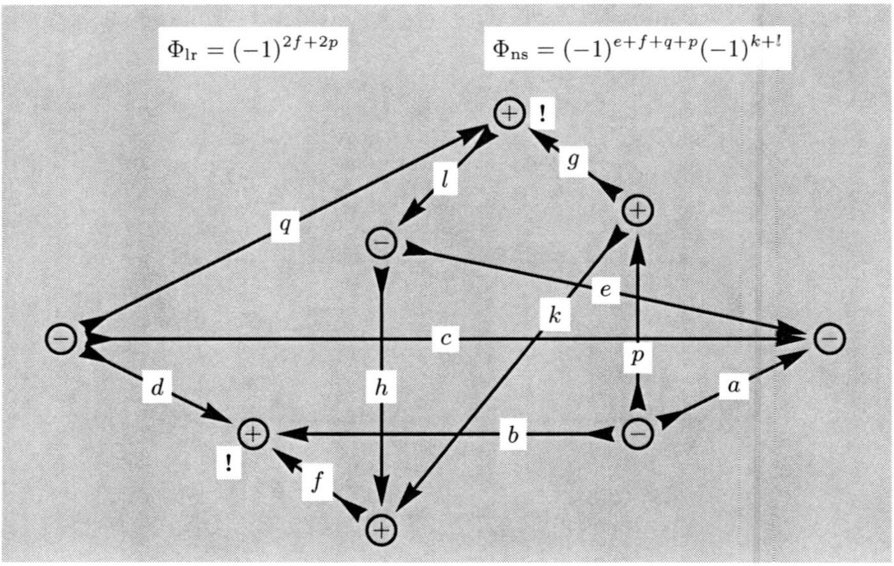

Fig. 8.71. The main change is the movement of the h-line across the q-line with node sign changes at the nodes connected to the l- and f-line

In the next step we move the node connected to the e, h and l-line.

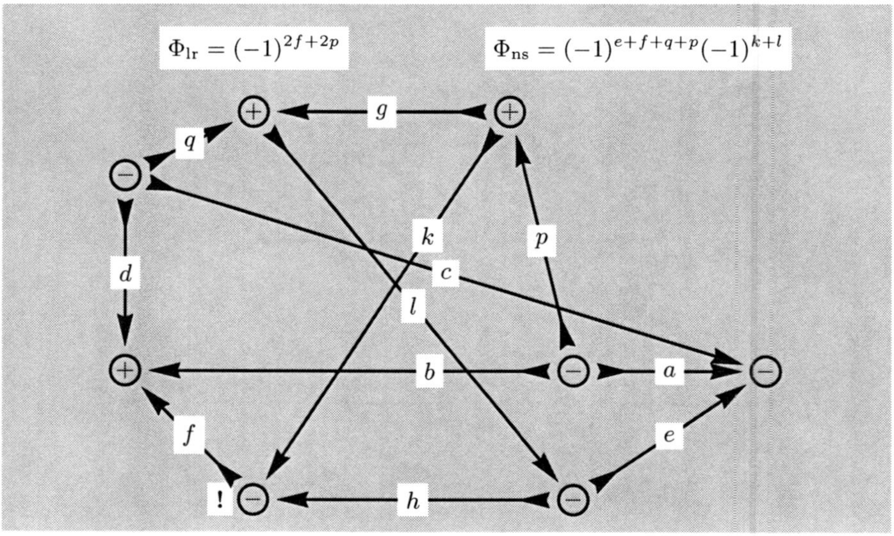

Fig. 8.72. Note that the node sign at the extreme right node does not change because the cyclic arrangement is not altered

The last transformation brings the diagram into a form that closely resembles a $12j$(I)-symbol and the necessary changes en route to the definitive representation are implemented in Figs. 8.73 and 8.74.

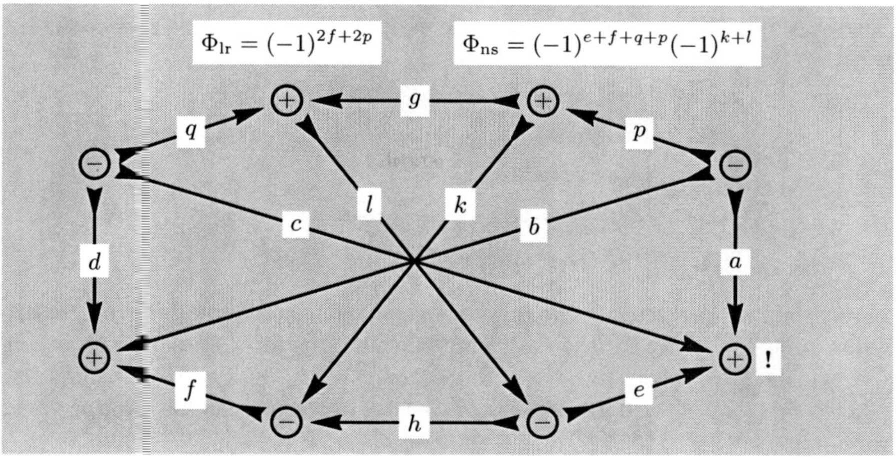

Fig. 8.73. For the new diagram only one node sign has changed

It is not difficult to show that $(-1)^{2f+2p}$ is equivalent to $(-1)^{2g+2h}$.

In order to recover the proper form of the $12j$(I)-symbol we return to Fig. 8.73 and implement the necessary line reversals and node sign changes.

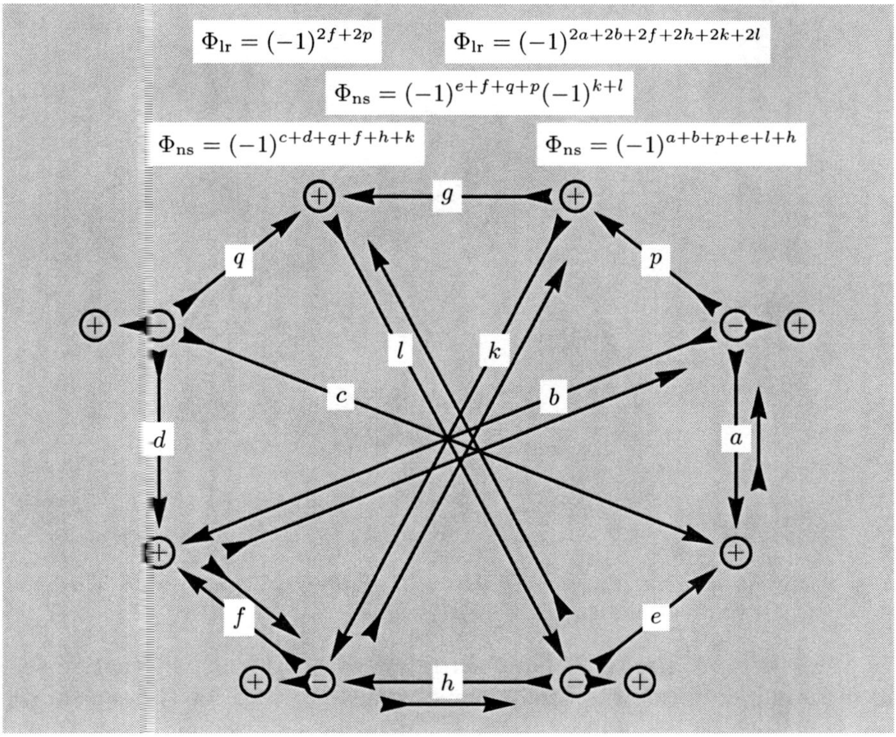

Fig. 8.74. Taking Fig. 8.62 as reference for a $12j$(I)-symbol the line reversals and node sign changes are indicated

The accumulated phase factor may be simplified and we first collect all the terms

$$\begin{aligned}\Phi &= (-1)^{2f+2p}(-1)^{e+f+q+p+k+l} \\ &\quad\times(-1)^{2a+2b+2f+2h+2k+2l}(-1)^{c+d+q+f+h+k+a+b+p+e+l+h} \\ &= (-1)^{2f+2p}(-1)^{2e+2f+2q+2p+2k+2l} \\ &\quad\times(-1)^{2a+2b+2f+2h+2k+2l}(-1)^{2h}(-1)^{c+d+a+b} \\ &= (-1)^{2a+2b+2e+2f+2q}(-1)^{a+b+c+d} .\end{aligned}$$

In passing to the second equality we have paired all matching single angular momenta and in the last equality all angular momenta appearing four times have been removed. If we use $(-1)^{2e} = (-1)^{2a+2c}$ and also $(-1)^{2f} = (-1)^{2b+2d}$, both $4a$ and $4b$ in the exponent can be dropped and we are left with $(-1)^{2c+2d+2q} = 1$ and the phase $\Phi = (-1)^{a+b+c+d}$.

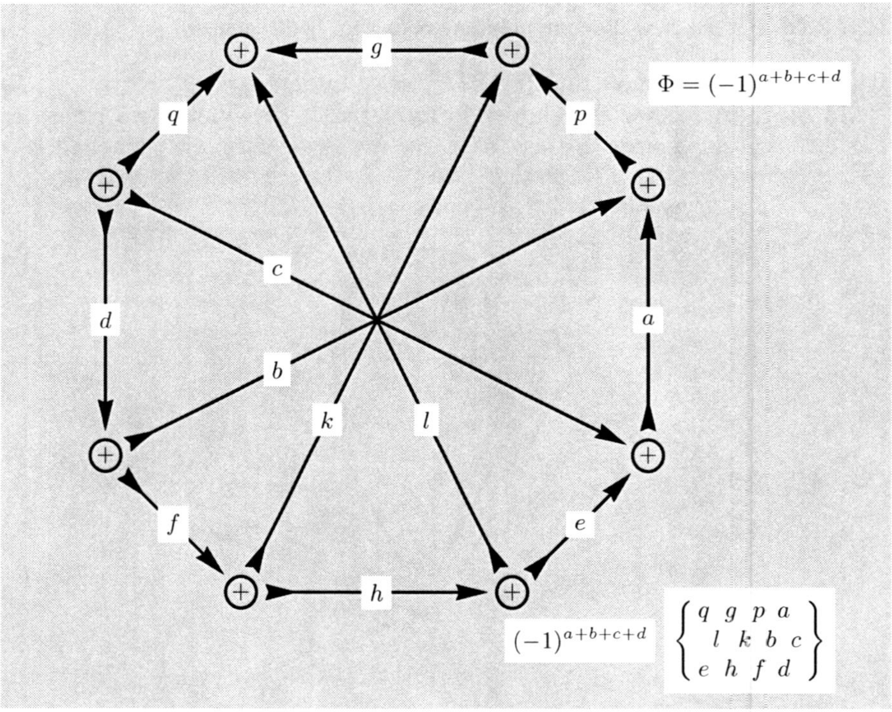

Fig. 8.75. The diagram has assumed the correct form of a $12j$(I)-symbol. The phase Φ is not part of the definition of the $12j$(I)-symbol

In a way it is, at first sight, quite surprising to find for the alternating sum also a result in form of a $12j$(I)-symbol similar to that for the sum in the preceding section. In addition to the $(-1)^x$ in the sum the positions of k and l are exchanged in one of the two $6j$-symbols compared to Sect. 8.3.2; thus we are analysing sums of different terms. The result for the sum is

$$\sum_x (2x-1)(-1)^x \begin{Bmatrix} e & f & x \\ a & b & p \\ c & d & q \end{Bmatrix} \begin{Bmatrix} f & e & x \\ l & k & h \end{Bmatrix} \begin{Bmatrix} p & q & x \\ l & k & g \end{Bmatrix} =$$

$$= (-1)^{a+b+c+d} \begin{Bmatrix} q & g & p & a \\ & l & k & b & c \\ e & h & f & d \end{Bmatrix} = (-1)^{a+b+c+d} \begin{Bmatrix} e & h & f & d \\ & l & k & b & c \\ q & g & p & a \end{Bmatrix} .$$

We note that the graphical method has not led us to an expression in simpler terms but has rather shown the sum to be equivalent to the definition of a $12j$(I)-symbol. To reach the final equality above, one of the symmetry relations given in [12] for the $12j$(I)-symbol is used.

It is worth noting that, the derived result is not listed in the tables of Varshalovich et al. [12] and this demonstrates the power of the graphical method to go beyond known relations. On close inspection, however, we find that our result may be transformed into the formula given in Sect. 8.3.2. This requires two steps. Firstly we rename, and effectively exchange, some variables, namely $a \leftrightarrow c$, $b \leftrightarrow d$ and $p \leftrightarrow q$, and this has no effect on the equality. Secondly, we call upon a symmetry property of the $9j$-symbol,

$$\begin{Bmatrix} e & f & x \\ c & d & q \\ a & b & p \end{Bmatrix} = (-1)^{a+b+c+d+e+f+p+q+x} \begin{Bmatrix} e & f & x \\ a & b & p \\ c & d & q \end{Bmatrix} ,$$

and undo the effect of renaming the variables in the $9j$-symbol. Note that the last $6j$-symbol and the $12j$-symbol retain the renamed variables.

$$\sum_x (2x+1)(-1)^x (-1)^{a+b+c+d+e+f+p+q+x}$$

$$\times \begin{Bmatrix} e & f & x \\ a & b & p \\ c & d & q \end{Bmatrix} \begin{Bmatrix} f & e & x \\ l & k & h \end{Bmatrix} \begin{Bmatrix} q & p & x \\ l & k & g \end{Bmatrix}$$

$$= (-1)^{a+b+c+d} \begin{Bmatrix} e & h & f & b \\ & l & k & d & a \\ p & g & q & c \end{Bmatrix} .$$

With $(-1)^{2x} = (-1)^{-2x} = (-1)^{-2e-2f}$ and a partial cancellation of phase factors on both sides of the equality we transfer the remaining phase factor $(-1)^{-e-f+p+q} = (-1)^{e+f-p-q}$ to the right-hand side. Because the $6j$-symbol is symmetric under exchange of columns we regain the same relation as given on p. 106 in Sect. 8.3.2, thereby strengthening our confidence in the graphical methods applied in both sections.

9

Open Diagrams

Simple examples of open diagrams with external lines have been encountered in our discussion of orthogonality relations for $3jm$-symbols. One might say that, open diagrams are more general than closed diagrams of the preceding section. Furthermore, open diagrams will be nearer to the calculation of matrix elements of spherical tensor operators, which contain a $3jm$-symbol according to the Wigner–Eckart Theorem, and a reduced matrix element with one or more $3nj$-symbols. In the end, a realistic calculation can involve products of open and closed diagrams and in Chap. 10 an example is analysed. From Chap. 7, on general rules, we recall that separations are possible if all the external lines stay with one of the separated parts.

As before, we will seek a close correspondence with the formulae found in Varshalovich et al. [12]. An exception is the new formula derived in in Sect. 9.6 which deals with a product of two $3jm$-symbols, while all other sections analyse sums over products.

The term 'recoupling' is used frequently to describe the fact that angular momenta present in initial $3jm$-symbols appear recombined differently in the final expression.

9.1 Sum with Two $3jm$-Symbols

Let us consider the following sum with two $3jm$-symbols

$$\sum_{\kappa}(-1)^{q-\kappa}\begin{pmatrix} a & b & q \\ \alpha & \beta & -\kappa \end{pmatrix}\begin{pmatrix} q & d & c \\ \kappa & \delta & \gamma \end{pmatrix} = ?$$

This expression has no result in the sense of a reduction to simpler terms. It can, however, be transformed into an equivalent, albeit more complicated, expression and this will be achieved with the graphical method.

The end result is insight to angular momentum coupling schemes represented by the $6j$-symbol. In this sense, the sum is recoupled to a sum over angular momentum and projection, x and ξ.

E. Balcar, S.W. Lovesey, *Introduction to the Graphical Theory of Angular Momentum*, STMP 234, 119–157, DOI 10.1007/978-3-642-03118-2_9,

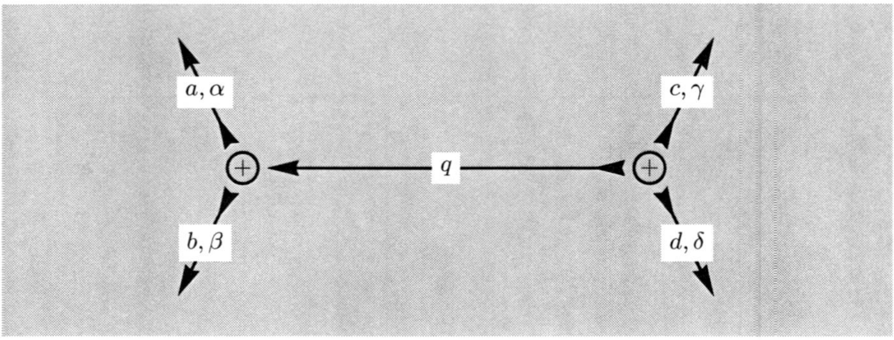

Fig. 9.1. A product of two $3jm$-symbols and the line joining the two nodes represents the sum over the projection κ of angular momentum q

The diagram has four external lines and it is not obvious which further transformations are possible. In the rearranged diagram below we see that the introduction of a new variable x and a corresponding sum may be used to reduce the number of lines.

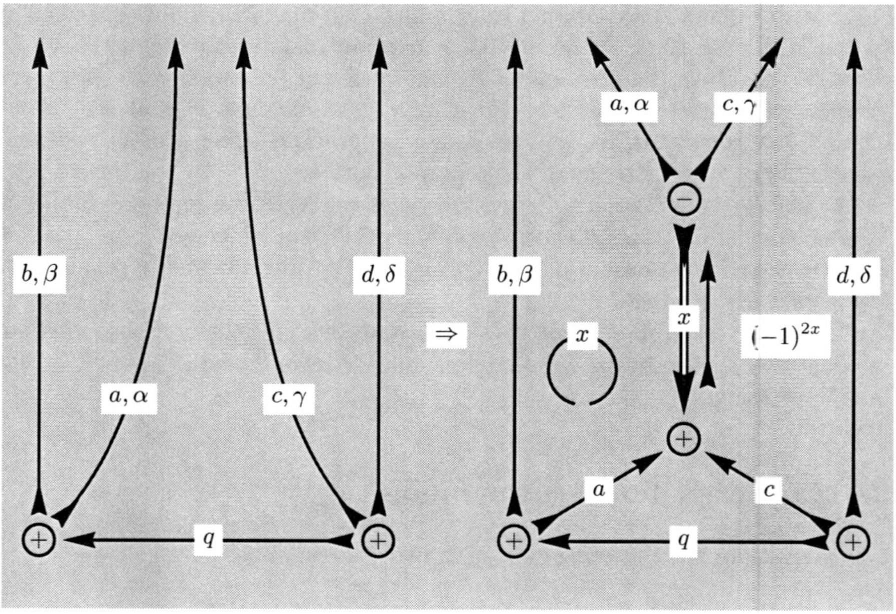

Fig. 9.2. The diagram has been rearranged and on the right-hand side the sum over x is introduced

In preparation for a three-line separation the direction of the angular momentum line x is reversed and the resulting phase factor is $\Phi_{\mathrm{lr}} = (-1)^{2x}$. Because the origin of the phase is obvious by now, only the factor $(-1)^{2x}$ is given in the figure and $\Phi_{\mathrm{lr}} =$ is left out.

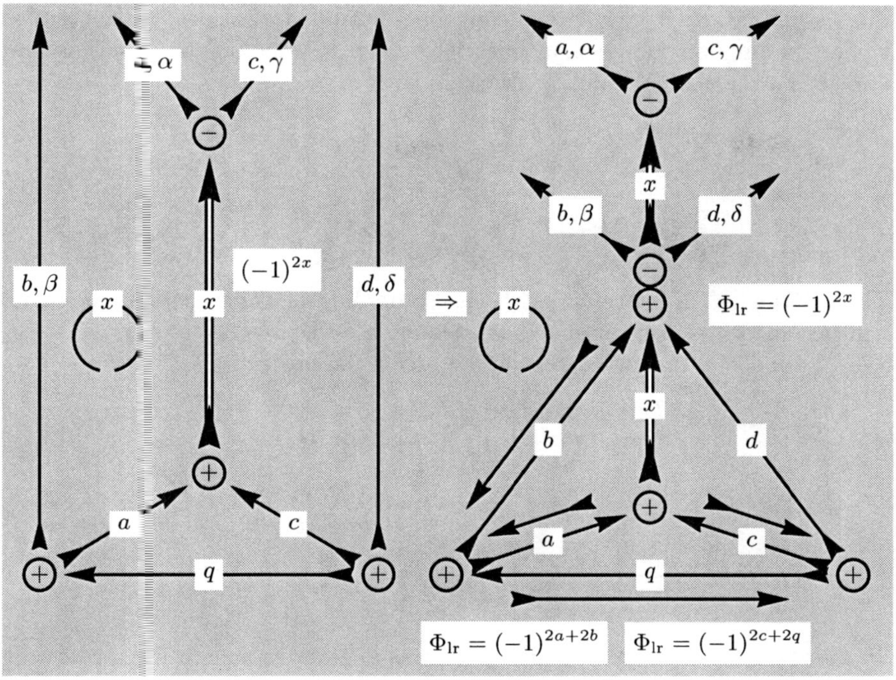

Fig. 9.3. The diagram has been extended to create room for the separation of three lines which is shown on the right-hand side of the figure

Line reversals in the diagram on the right-hand side are necessary to recover the definition of a $6j$-symbol. The total phase factor simplifies because $(-1)^{2a+2c+2x} = 1$ and, furthermore, with $(-1)^{2b+2q} = (-1)^{2a}$ the remaining phase is just $(-1)^{2a}$.

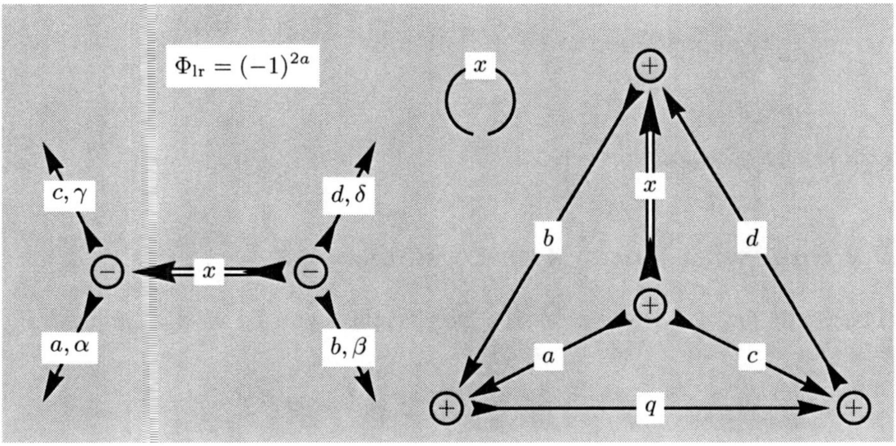

Fig. 9.4. The diagram contains a sum over a product of two $3jm$-symbols and a $6j$-symbol together with the usual factor $(2x + 1)$

The manipulations have led to an equivalent formula for the original expression where the angular momentum lines have been reordered such that a is now joined in a node with c and, similarly, b with d.

$$\sum_{\kappa}(-1)^{q-\kappa}\begin{pmatrix} a & b & q \\ \alpha & \beta & -\kappa \end{pmatrix}\begin{pmatrix} q & d & c \\ \kappa & \delta & \gamma \end{pmatrix}$$
$$= (-1)^{2a}\sum_{x,\xi}(2x+1)(-1)^{x-\xi}\begin{pmatrix} a & c & x \\ \alpha & \gamma & -\xi \end{pmatrix}\begin{pmatrix} x & d & b \\ \xi & \delta & \beta \end{pmatrix}\begin{Bmatrix} a & c & x \\ d & b & q \end{Bmatrix} .$$

The relation of this formula with the recoupling of three angular momenta will become more transparent if we rename $a \to j_3$, $b \to j_2$, $d \to j_1$, $q \to j_{23}$, $x \to j_{12}$ and $c \to J$ and, correspondingly, for the projections.

$$\sum_{m_{23}}(-1)^{j_{23}-m_{23}}\begin{pmatrix} j_3 & j_2 & j_{23} \\ m_3 & m_2 & -m_{23} \end{pmatrix}\begin{pmatrix} j_{23} & j_1 & J \\ m_{23} & m_1 & M \end{pmatrix}$$
$$= (-1)^{2j_3}\sum_{j_{12},m_{12}}(2j_{12}+1)(-1)^{j_{12}-m_{12}}\begin{pmatrix} j_3 & J & j_{12} \\ m_3 & M & -m_{12} \end{pmatrix}\begin{pmatrix} j_{12} & j_1 & j_2 \\ m_{12} & m_1 & m_2 \end{pmatrix}$$
$$\times\begin{Bmatrix} j_3 & J & j_{12} \\ j_1 & j_2 & j_{23} \end{Bmatrix} .$$

To see the relation with recoupling of three angular momenta in more detail, it is necessary to convert the $3jm$-symbols into Clebsch-Gordan coefficients needed for the construction of states,

$$|(j_1J_2)j_{12}j_3, JM\rangle \quad \text{and} \quad |j_1(j_2J_3)j_{23}, JM\rangle ,$$

coupled from product states $|j_1m_1, j_2m_2, j_3m_3\rangle$. The relation between the two couplings is given by Judd [7] as,

$$|j_1(j_2J_3)j_{23}, JM\rangle$$
$$= \sum_{j_{12}}[(2j_{12}+1)(2j_{23}+1)]^{1/2}(-1)^{j_1+j_2+j_3+J}\begin{Bmatrix} j_3 & J & j_{12} \\ j_1 & j_2 & j_{23} \end{Bmatrix}$$
$$\times|(j_1J_2)j_{12}j_3, JM\rangle ,$$

also briefly indicated in Sect. 2.2.

9.2 Sum with Three $3jm$-Symbols

The result found for a sum of two $3jm$-symbols can be used to analyse the case of a sum over three $3jm$-symbols.

$$\sum_{\kappa,\delta,\beta}(-1)^{q-\kappa}\begin{pmatrix} a & b & q \\ \alpha & \beta & -\kappa \end{pmatrix}\begin{pmatrix} q & d & c \\ \kappa & \delta & \gamma \end{pmatrix}(-1)^{d-\delta+b-\beta}\begin{pmatrix} d & b & f \\ -\delta & -\beta & \varphi \end{pmatrix} = ?$$

The diagram that corresponds to this algebraic expression will have three external and three internal lines.

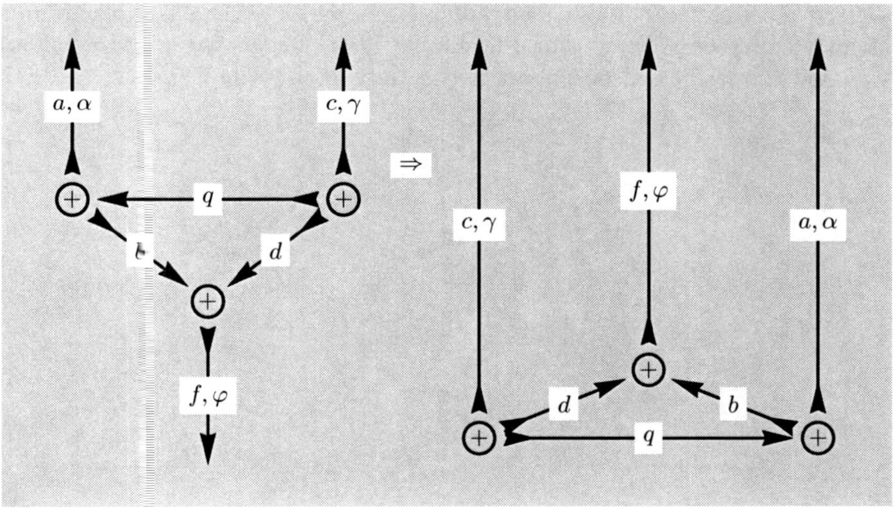

Fig. 9.5. A product of three $3jm$-symbols and the lines joining the three nodes represent the sums over the projections κ, δ and β

The three external lines allow a three-line separation but instead of following the usual procedure, we turn to the result given in Fig. 9.4, where we introduce the additional $3jm$-symbol and the sums over β and δ:

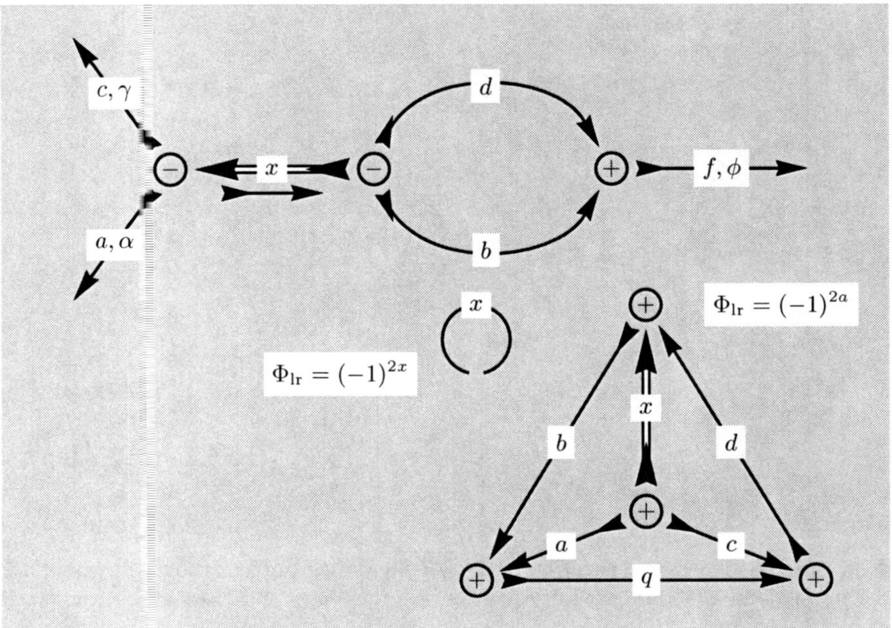

Fig. 9.6. The diagram, derived from Fig. 9.4, contains sums over a product of three $3jm$-symbols and a $6j$-symbol together with the usual factor $(2x + 1)$

The reversal of the x-line has introduced the phase $(-1)^{2x}$. This diagram is slightly more complicated than Fig. 4.9 but it can be treated in an analogous way and the result will be similar to the diagram given in Fig. 4.14.

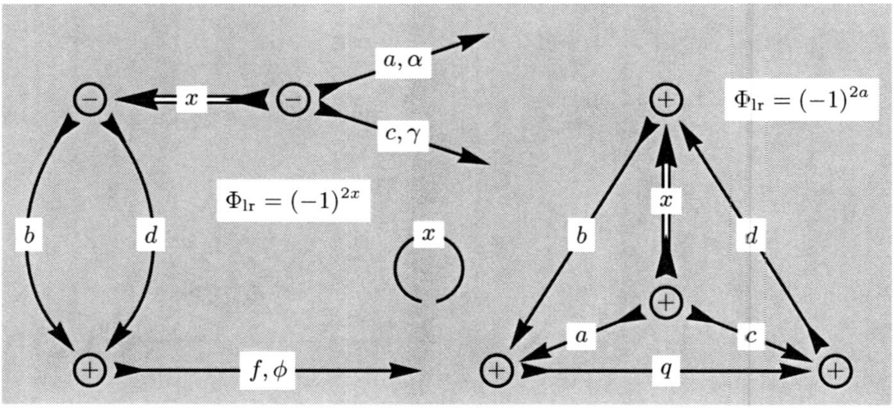

Fig. 9.7. The diagram has been rearranged to indicate the possibility of a two-line separation. The direction of the x-line is reversed to achieve the required line directions

The left-hand part of the diagram has been rearranged to facilitate the application of the two-line separation rule described in Sect. 7.5.

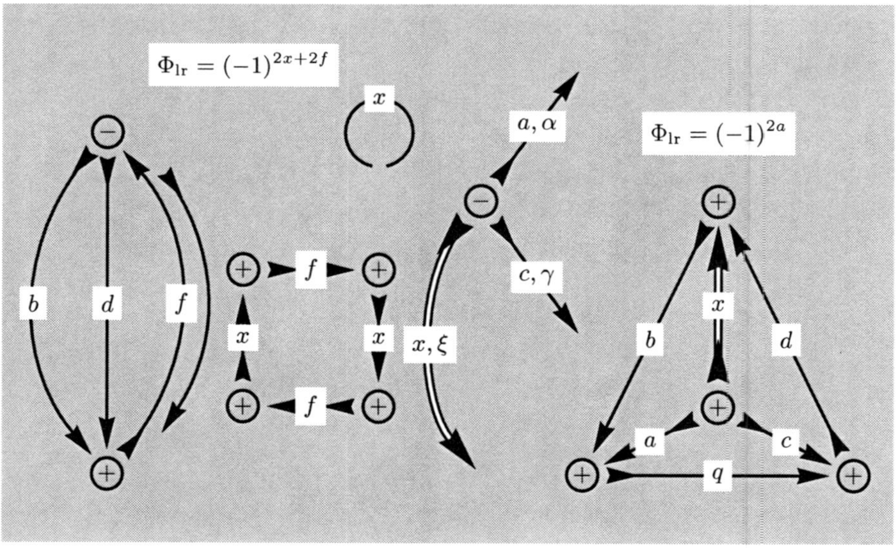

Fig. 9.8. The two-line separation of the left-hand diagram leads to a $3j$-symbol and a $3jm$-symbol. The quadrangular diagram contains $\delta_{x,f}$ and the sum over x, ξ

The phase factor from the reversal of the f-line cancels $(-1)^{2x}$ because x is replaced by f. The factors from the quadrangular diagram and the circle cancel with $x = f$.

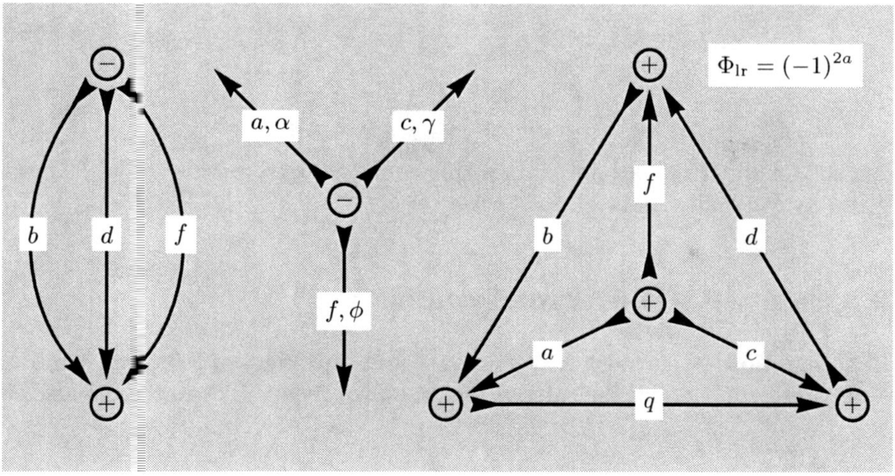

Fig. 9.9. The Kronecker $\delta_{x,f}\delta_{\xi,\varphi}$ make the sum over x, ξ straightforward and x, ξ is replaced by f, φ

The translation of the diagram into an algebraic expression produces the following result for the sum over three 3jm-symbols.

$$\sum_{\kappa,\delta,\beta}(-1)^{q-\kappa}\begin{pmatrix} a & b & q \\ \alpha & \beta & -\kappa \end{pmatrix}\begin{pmatrix} q & d & c \\ \kappa & \delta & \gamma \end{pmatrix}(-1)^{d-\delta+b-\beta}\begin{pmatrix} d & b & f \\ -\delta & -\beta & \varphi \end{pmatrix}$$
$$= (-1)^{2a}\begin{pmatrix} a & c & f \\ \alpha & \gamma & \varphi \end{pmatrix}\begin{Bmatrix} a & c & f \\ d & b & q \end{Bmatrix} .$$

The 3j-symbol is not required here because the triangle condition $\triangle\{b, d, f\}$ is also embodied in the 6j-symbol.

In this form the equation is not easily comparable with the equivalent formula given in [12]. In order to facilitate comparison we need to change the expression and concentrate first on the left-hand side. For each 3jm-symbol we perform an odd permutation and change the summation index $\delta \to -\delta$

$$\sum_{\kappa,\delta,\beta}(-1)^{q-\kappa}(-1)^{d+\delta+b-\beta}\begin{pmatrix} b & a & q \\ \beta & \alpha & -\kappa \end{pmatrix}\begin{pmatrix} q & c & d \\ \kappa & \gamma & -\delta \end{pmatrix}\begin{pmatrix} d & f & b \\ \delta & \varphi & -\beta \end{pmatrix}$$
$$\times (-1)^{a+b+2q+c+2d+f+b} .$$

The change of sign for the summation index δ introduces a phase factor because $(-1)^{d+\delta} = (-1)^{2d}(-1)^{-d+\delta}$ and this is equivalent to $(-1)^{2d}(-1)^{d-\delta}$.

In the total phase on the left-hand side $(-1)^{2d}(-1)^{a+c+f}(-1)^{2b+2d+2q}$ we have $(-1)^{4d} = 1$ and with $(-1)^{2b+2q} \Rightarrow (-1)^{2a}$ there is a phase $(-1)^{2a}$ on both sides of the equation and may be cancelled. The remaining phase factor $(-1)^{a+c+f}$ can be brought to the right-hand side and used to change the sign of the projections in the 3jm-symbol.

$$\sum_{\kappa,\delta,\beta}(-1)^{q-\kappa}(-1)^{d-\delta+b-\beta}\begin{pmatrix} b & a & q \\ \beta & \alpha & -\kappa \end{pmatrix}\begin{pmatrix} q & c & d \\ \kappa & \gamma & -\delta \end{pmatrix}\begin{pmatrix} d & f & b \\ \delta & \varphi & -\beta \end{pmatrix}$$
$$= \begin{pmatrix} a & c & f \\ -\alpha & -\gamma & -\varphi \end{pmatrix}\begin{Bmatrix} a & c & f \\ d & b & q \end{Bmatrix} .$$

Except for differing variable names this equation corresponds to the formula given by Varshalovich et al. in [12].

9.3 Sum with Four $3jm$-Symbols I

In this section we consider a fivefold sum over a product of four $3jm$-symbols and transform it under the rules of the graphical theory. The initial expression is

$$\sum_{\kappa,\rho,\sigma,\psi,\tau}(-1)^{q-\kappa+r-\rho+s-\sigma+p-\psi+t-\tau}\begin{pmatrix} p & a & q \\ \psi & -\alpha & \kappa \end{pmatrix}\begin{pmatrix} q & r & t \\ -\kappa & \rho & \tau \end{pmatrix}$$
$$\times\begin{pmatrix} r & b & s \\ -\rho & \beta & \sigma \end{pmatrix}\begin{pmatrix} s & p & t \\ -\sigma & -\psi & -\tau \end{pmatrix} = \ ?$$

To facilitate the transcription into graphical symbols the phases are now associated with each corresponding $3jm$-symbol.

$$(-1)^{a-\alpha}\sum_{\substack{\kappa,\rho,\sigma,\\ \psi,\tau}}(-1)^{a-\alpha}\begin{pmatrix} p & a & q \\ \psi & -\alpha & \kappa \end{pmatrix}(-1)^{q-\kappa}\begin{pmatrix} q & r & t \\ -\kappa & \rho & \tau \end{pmatrix}$$
$$\times(-1)^{r-\rho}\begin{pmatrix} r & b & s \\ -\rho & \beta & \sigma \end{pmatrix}(-1)^{s-\sigma+p-\psi+t-\tau}\begin{pmatrix} s & p & t \\ -\sigma & -\psi & -\tau \end{pmatrix} = \ ?$$

Note that the phase $(-1)^{a-\alpha}$ has been introduced twice into the sum in order to provide the first $3jm$-symbol with the required phase factor.

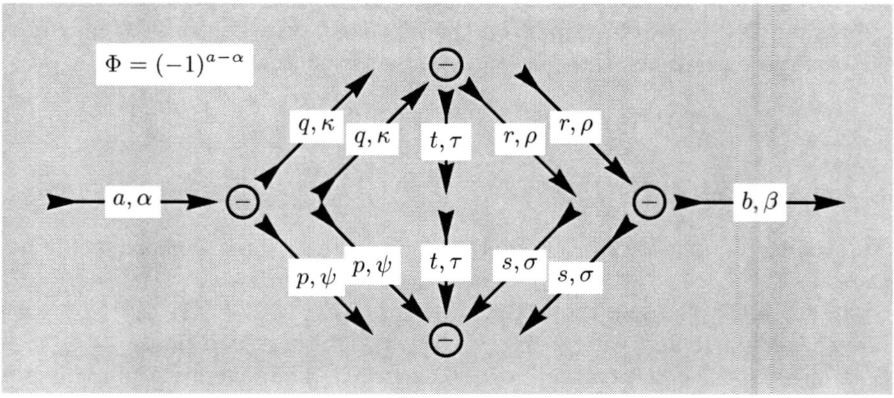

Fig. 9.10. The four symbols are arranged as to indicate the future node connections for the sums over projections

The five sums over angular momentum projections allow us to connect the appropriate nodes.

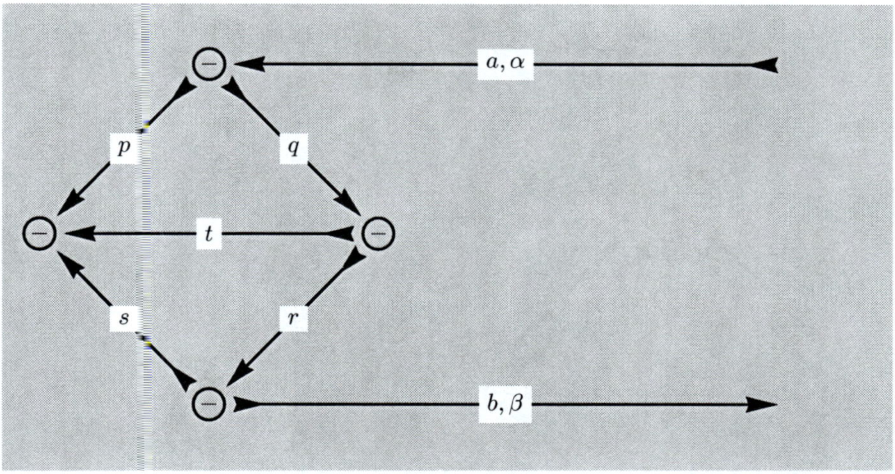

Fig. 9.11. After a rotation two external lines pointing to the right remain when the five summations are taken into account. It is now possible to perform a two-line separation as in Sect. 7.5 or for external lines in Sect. 4.3

This figure is found to be an analogue of Fig. 4.10, thus leading to a corresponding result.

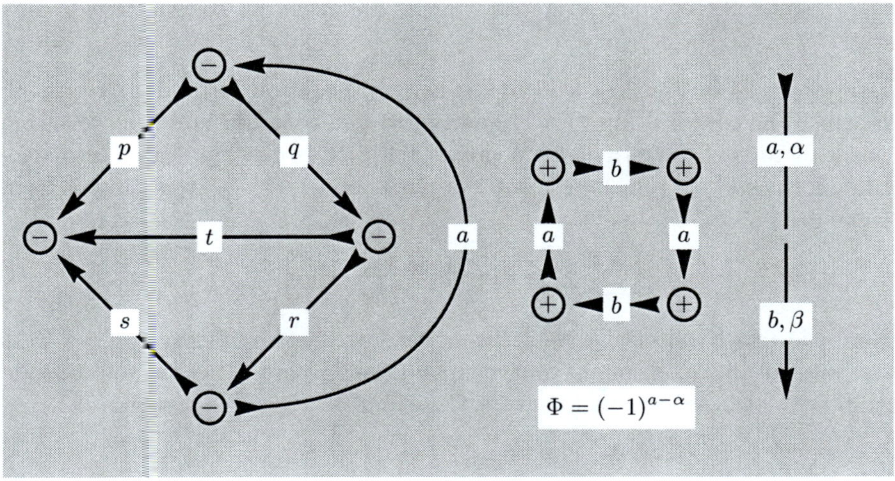

Fig. 9.12. A two-line separation has been carried out as in 4.10 and following figures

The diagram has been separated into a free line, a factor as described in Fig. 3.4 and a diagram with four nodes which can be converted into a $6j$-symbol.

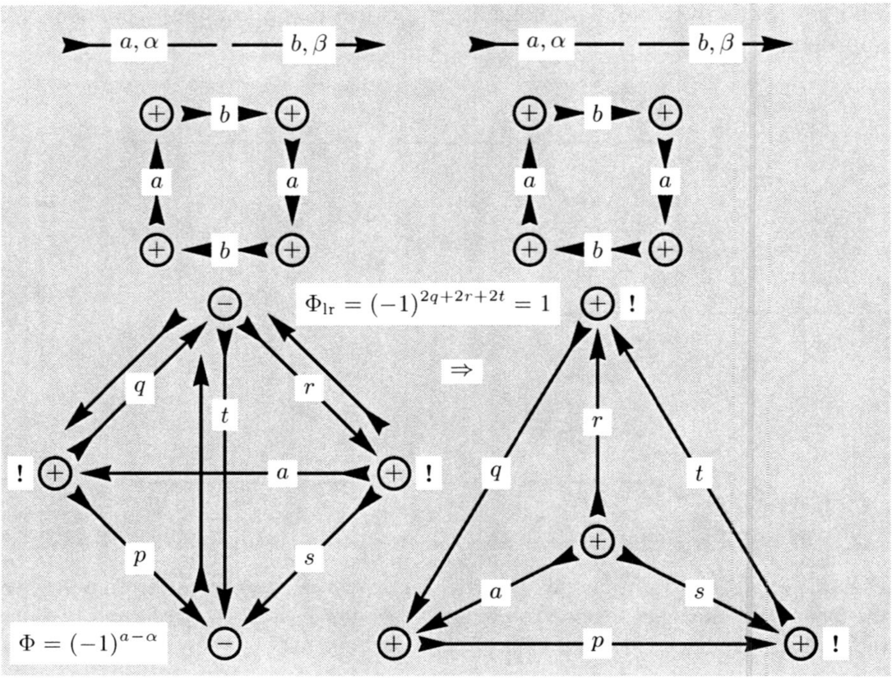

Fig. 9.13. In the left part the a-line has been moved into the centre of the diagram thereby changing the sign of two nodes. Passing to the right-hand side we move the node connected to a, r and s into the triangle and again two nodes change from $-$ to $+$

Line reversals in Fig. 9.13 lead to the correct graphical form of a $6j$-symbol. Because the reversed lines are connected to one common node the resulting phase factor is 1. The right-hand side of Fig. 9.13 is now translated into algebraic form and we remember that the phase $\Phi = (-1)^{a-\alpha}$ was present from the outset.

$$(-1)^{a-\alpha}\frac{\delta_{a,b}\delta_{\alpha,\beta}}{(2a+1)}\begin{Bmatrix} s & r & a \\ q & p & t \end{Bmatrix} = (-1)^{a-\alpha}\frac{\delta_{a,b}\delta_{\alpha,\beta}}{(2a+1)}\begin{Bmatrix} q & p & a \\ s & r & t \end{Bmatrix} .$$

Here the two Kronecker-δ result from the free line, the factor $1/(2a+1)$ is the value of the quadrangular diagram and the $6j$-symbol has the appropriate graphical form. The value found for the sum over four $3jm$-symbols is

$$\sum_{\kappa,\rho,\sigma,\psi,\tau} (-1)^{q-\kappa+r-\rho+s-\sigma+p-\psi+t-\tau}\begin{pmatrix} p & a & q \\ \psi & -\alpha & \kappa \end{pmatrix}\begin{pmatrix} q & r & t \\ -\kappa & \rho & \tau \end{pmatrix}$$
$$\times\begin{pmatrix} r & b & s \\ -\rho & \beta & \sigma \end{pmatrix}\begin{pmatrix} s & p & t \\ -\sigma & -\psi & -\tau \end{pmatrix} = (-1)^{a-\alpha}\frac{\delta_{a,b}\delta_{\alpha,\beta}}{(2a+1)}\begin{Bmatrix} q & p & a \\ s & r & t \end{Bmatrix} .$$

This result is not altogether surprising because we have seen in Sect. 5.1 that the $6j$-symbol is defined from a sum over four $3jm$-symbols.

9.4 Sum with Four $3jm$-Symbols II

We analyse a fourfold sum over a product of four $3jm$-symbols and use graphical theory to consider possible transformations. Unlike the previous product of four $3jm$-symbols which was summed over 5 projections, the new sum covers just 4 projections and the starting diagram is rather different. Thus, we also look forward to a different result. The initial expression is

$$\sum_{\kappa,\rho,\sigma,\psi} (-1)^{q-\kappa} \begin{pmatrix} p & a & q \\ \psi & \alpha & -\kappa \end{pmatrix} (-1)^{r-\rho} \begin{pmatrix} q & b & r \\ \kappa & \beta & -\rho \end{pmatrix} \times$$
$$\times (-1)^{s-\sigma} \begin{pmatrix} r & c & s \\ \rho & \gamma & -\sigma \end{pmatrix} (-1)^{p-\psi} \begin{pmatrix} s & d & p \\ \sigma & \delta & -\psi \end{pmatrix} = \ ?$$

All $3jm$-symbols are already in the form that is immediately transcribable into diagrams. In contrast to the previous section with a sum over five projections the present formula contains only a sum over four projections and, therefore, a larger number of external lines.

It is at this point not obvious, but we will find more than one way to re-express this formula. The following figure shows a diagram with four external lines and the graphical treatment of these external lines aims at a reduction of the number of lines such that a three line separation becomes possible. The required step is the introduction of an extra variable x and its projection ξ together with the corresponding sum.

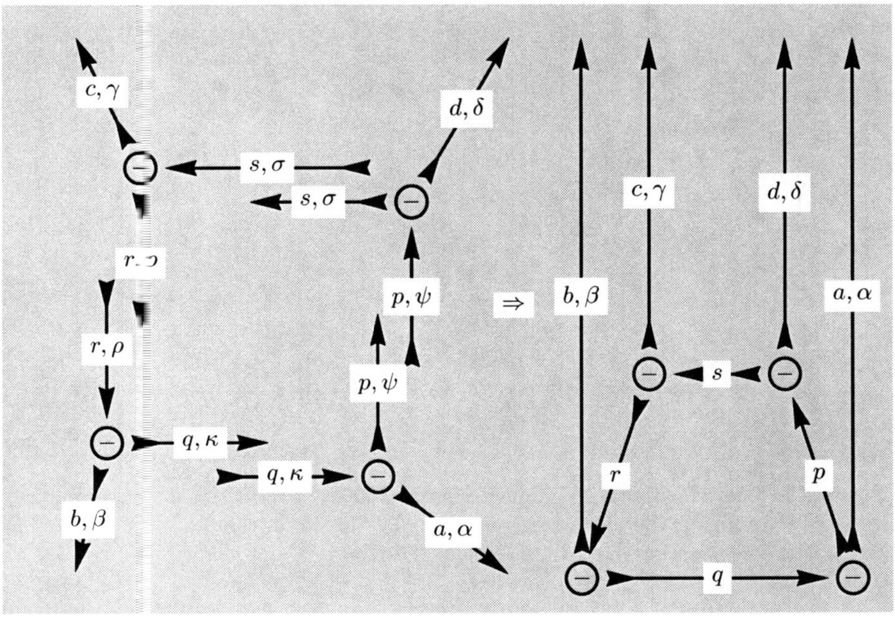

Fig. 9.14. The four symbols are arranged as to indicate the future node connections for the sums over projections. In the right-hand part of the figure the four sums over κ, ρ, σ and ψ are expressed by internal lines connecting the four nodes

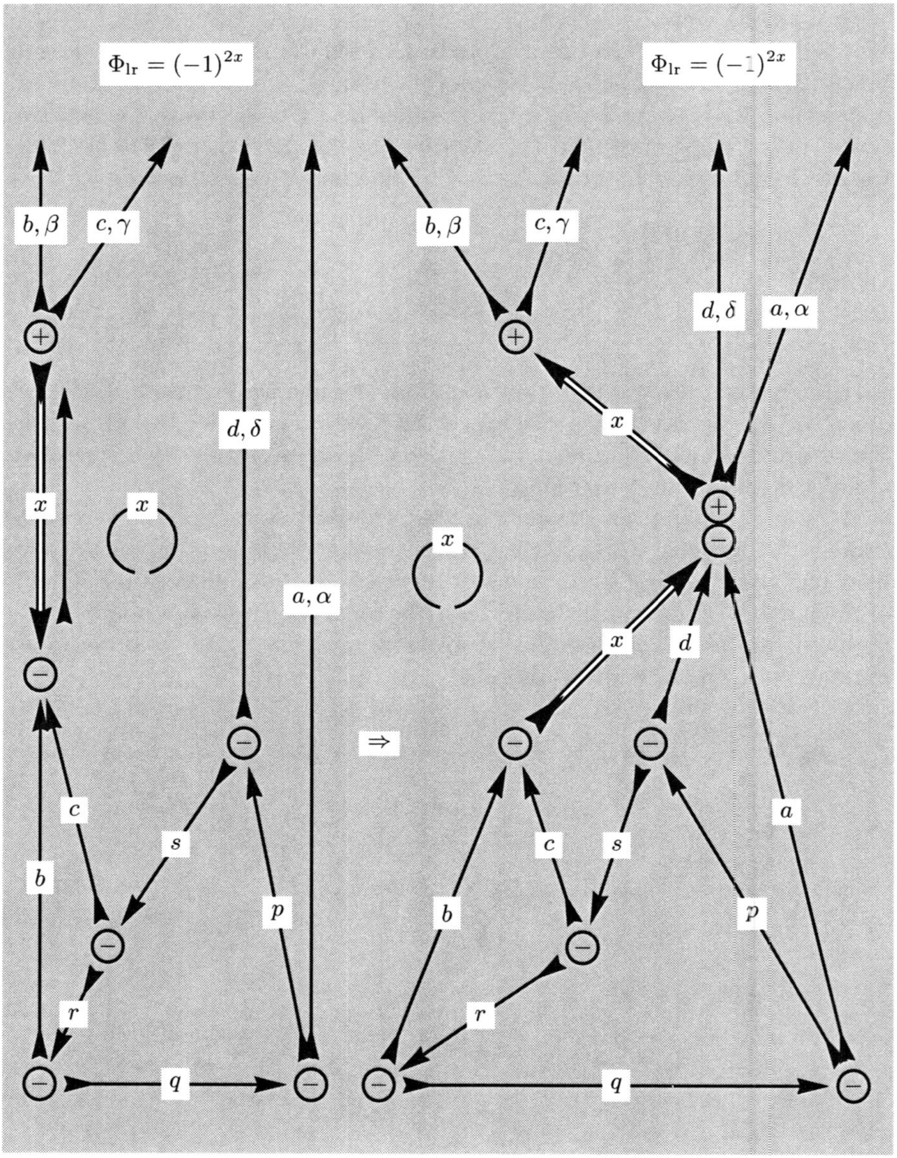

Fig. 9.15. The left part of this figure demonstrates the reduction of the number of lines by introducing the extra sum over x and ξ including the factor $(2x + 1)$. The reversal of direction for the x-line leads to the phase $(-1)^{2x}$

The x, d and a-lines have the same direction in the right-hand side diagram and allow a three line separation into a part with all the external lines and a second part without external lines.

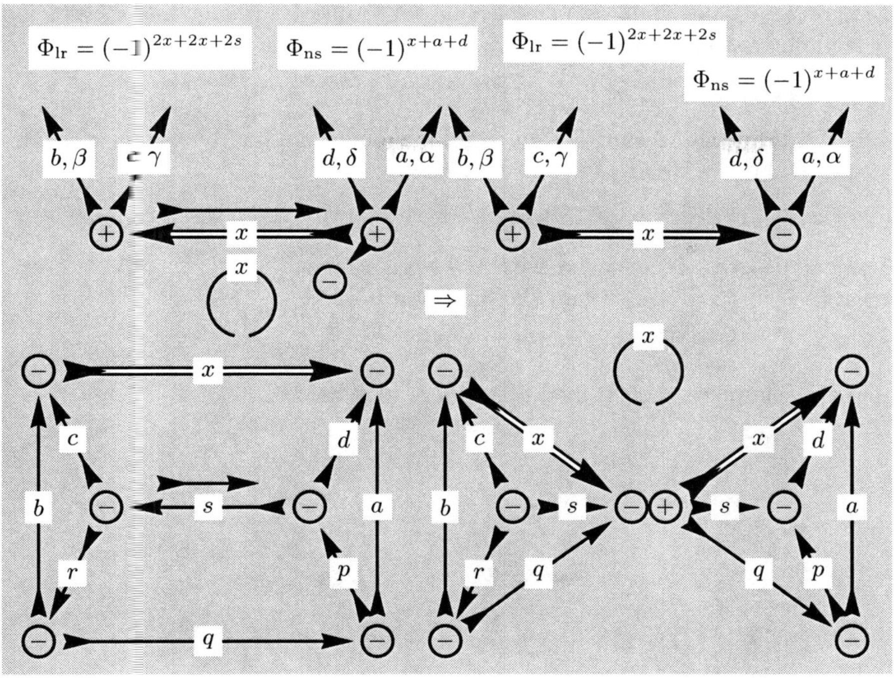

Fig. 9.16. The left part of this figure shows the result of the separation. The diagram without external lines is further separable. The diagrams are modified to influence the form of the final result

In the transition to the following figure the c- and the d-line have been turned inside out such that the corner node and the central node change place. Note that these transformations change all node signs.

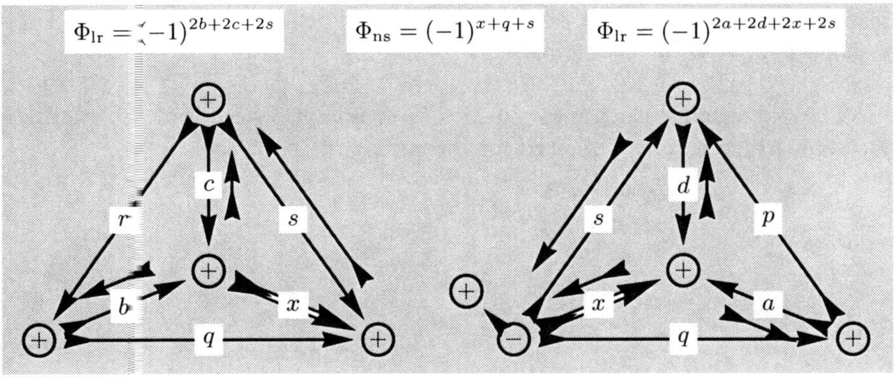

Fig. 9.17. This figure deals only with the transformations required for the two $6j$-symbols found in the right-hand side of Fig. 9.16. Several line reversals and one node sign change are performed to obtain the diagrams appropriate for the two $6j$-symbols

Line reversals and node sign changes have produced several phase factors and it is important to reduce the phases from Figs. 9.16 and 9.17 as far as possible.

$$\Phi = (-1)^{2x+2x+2s+x+a+d}(-1)^{2b+2c+2s}(-1)^{2a+2d+2x+2s+x+q+s} .$$

Every fourfold appearance of an angular momentum can be removed, for instance $(-1)^{4x} = 1$ and $(-1)^{4s} = 1$, to give

$$\Phi = (-1)^{2s+a+d}(-1)^{2b+2c}(-1)^{2a+2d+q+s} .$$

Further progress is achieved with the replacement $(-1)^{2b+2c} = (-1)^{2x}$ and then $(-1)^{2s+2x} = (-1)^{2q}$ and the final phase is

$$\Phi = (-1)^{3a+3d+3q+s} = (-1)^{s-a-d-q} ,$$

and the complete transformed diagram is seen below.

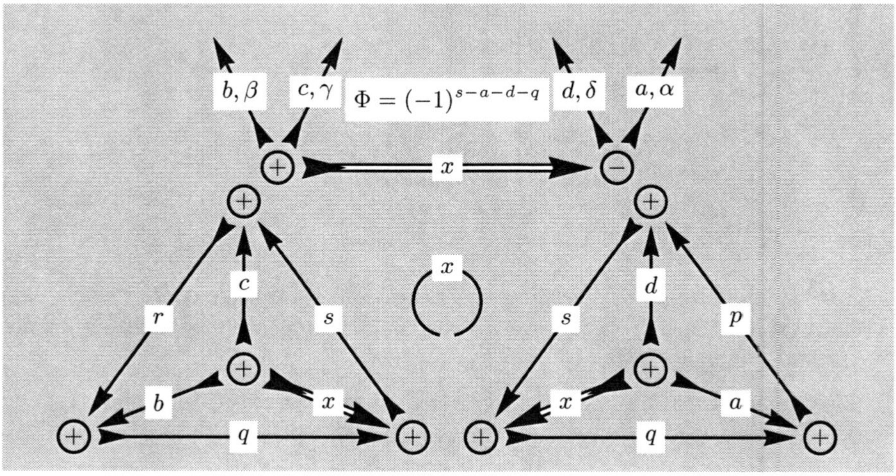

Fig. 9.18. The results from Fig. 9.16 and 9.17 are collected here and the two $6j$-symbols have now the required form

These diagrams are translated into their analytic form and the resulting formula, which can be exploited for recoupling, is

$$\begin{aligned}
&\sum_{\kappa,\rho,\sigma,\psi} (-1)^{q-\kappa} \begin{pmatrix} p & a & q \\ \psi & \alpha & -\kappa \end{pmatrix} (-1)^{r-\rho} \begin{pmatrix} q & b & r \\ \kappa & \beta & -\rho \end{pmatrix} \\
&\qquad \times (-1)^{s-\sigma} \begin{pmatrix} r & c & s \\ \rho & \gamma & -\sigma \end{pmatrix} (-1)^{p-\psi} \begin{pmatrix} s & d & p \\ \sigma & \delta & -\psi \end{pmatrix} \\
&= (-1)^{s-a-d-q} \sum_{x,\xi} (2x+1)(-1)^{x-\xi} \begin{pmatrix} a & x & d \\ \alpha & -\xi & \delta \end{pmatrix} \begin{pmatrix} b & x & c \\ \beta & \xi & \gamma \end{pmatrix} \\
&\qquad \times \begin{Bmatrix} a & x & d \\ s & p & q \end{Bmatrix} \begin{Bmatrix} b & x & c \\ s & r & q \end{Bmatrix} .
\end{aligned}$$

A Variation of this Formula

Instead of leading to a product of two 6j-symbols, a variation in the graphical procedure can produce a 9j-symbol. To this end, the diagram on the right-hand side of Fig. 9.14 is slightly modified by a rotation of the s-line such that the two nodes connected by this line change place.

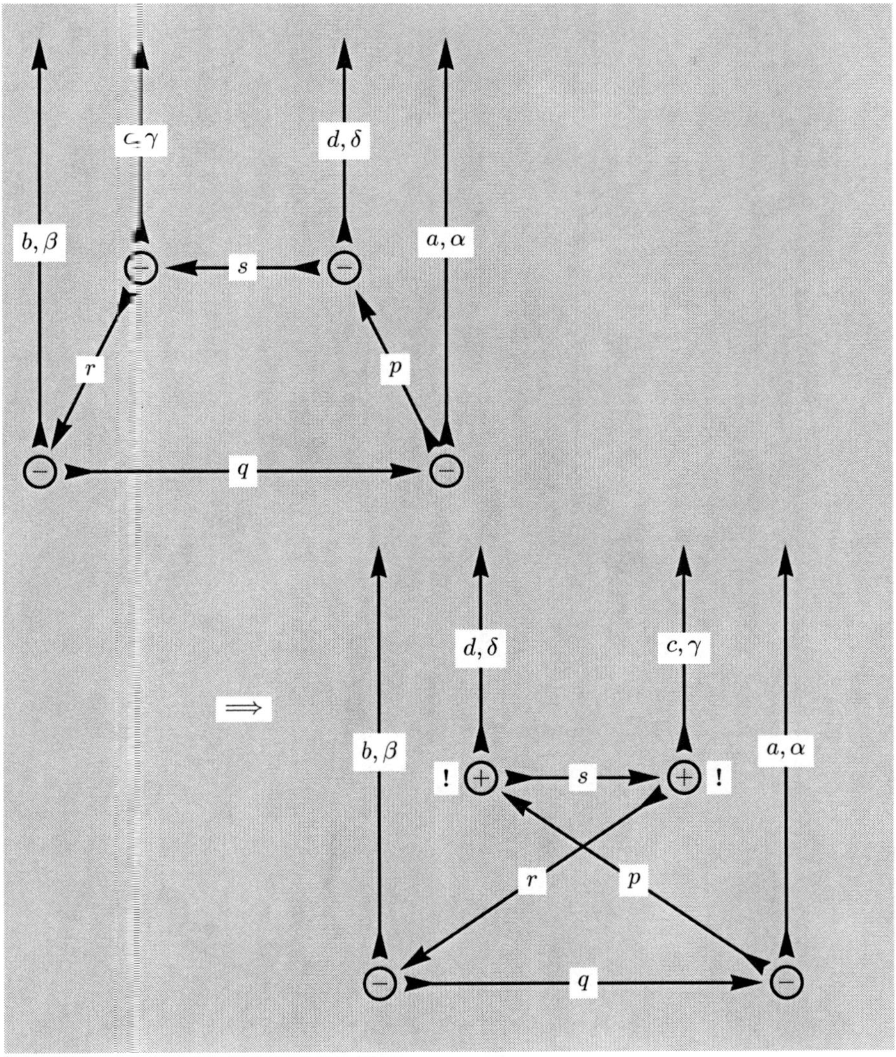

Fig. 9.19. Note that the two nodes connected by the s-line suffer a change of sign because in passing from top to bottom diagram the cyclic orientation of the lines changes at both nodes

A sum over x, ξ is introduced here and a three line separation is performed.

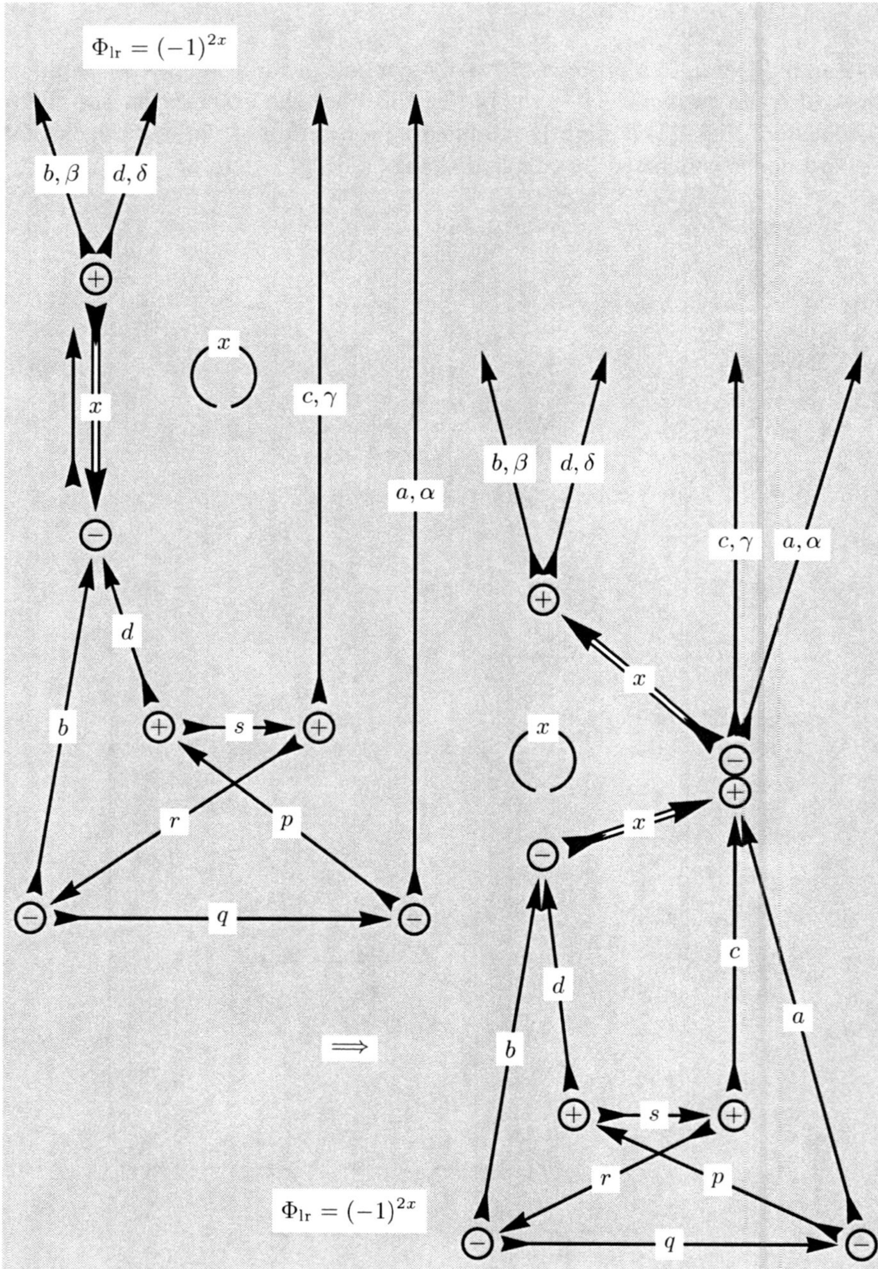

Fig. 9.20. This figure is a modified form of Fig. 9.15 and equivalent transformations are performed here

Compared to Fig. 9.15, different signs are chosen for the new nodes.

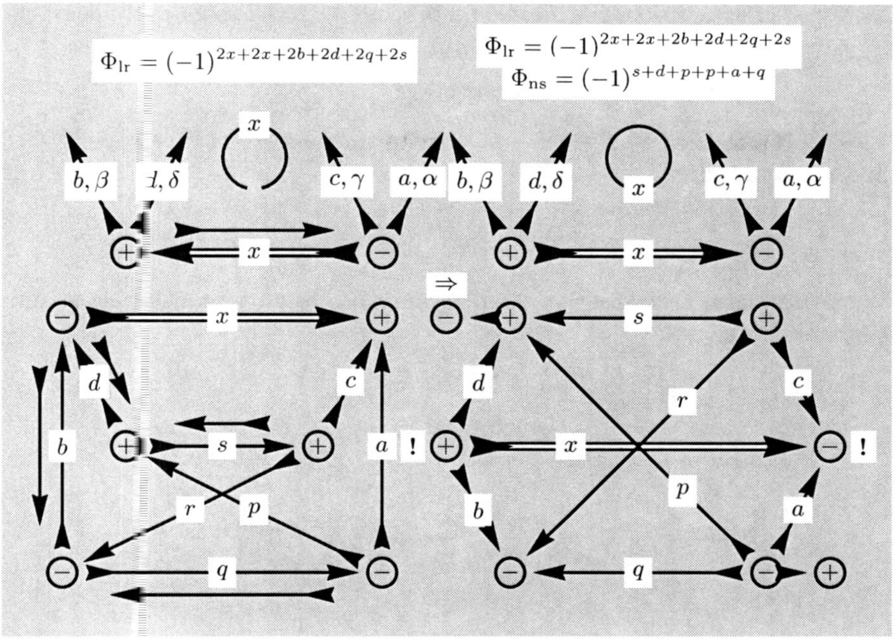

Fig. 9.21. The left part of this figure shows the result of the separation. The diagram without external lines is no longer separable and can be transformed into a 9j-symbol. Note the node sign changes effected by moving the s-line above the x-line

Two sign changes at the left top and bottom right nodes enter the additional phase $(-1)^{s+d+p+p+a+q}$ and with these alterations the 9j-symbol achieves the appropriate form.

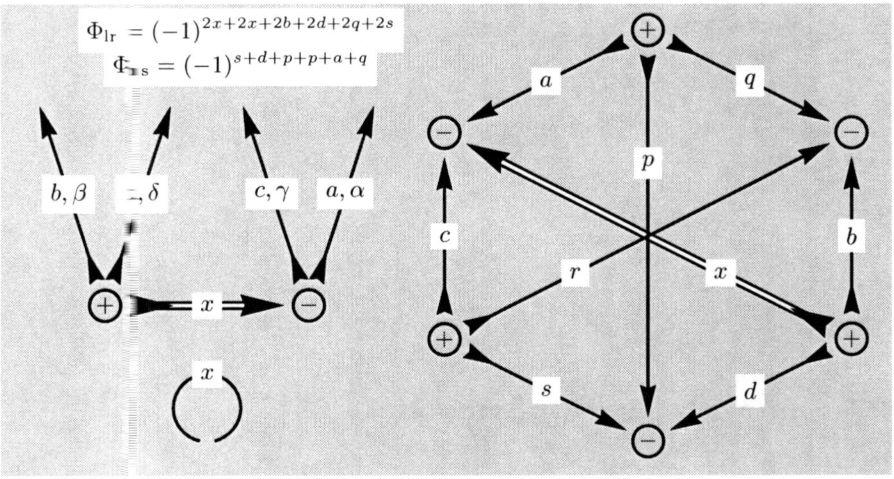

Fig. 9.22. Note that the diagram for the 9j-symbol has been rotated

As last step we consider the phase and proceed with the usual simplifications.

$$\begin{aligned}\Phi &= (-1)^{2x+2x+2b+2d+2q+2s}(-1)^{s+d+p+p+a+q} \\ &= (-1)^{2b+2d+2p+2q+2s}(-1)^{a+d+q+s} \,.\end{aligned}$$

To simplify further set $(-1)^{2p} \to (-1)^{2a+2q}$ and use $(-1)^{2b+2q} = (-1)^{2r}$ to find the total phase

$$\begin{aligned}\Phi &= (-1)^{2r+2s+2d+2a+2q}(-1)^{s+d+a+q} = (-1)^{2r+3s+3d+3a+3q} \\ &= (-1)^{2r-s-d-a-q} \,.\end{aligned}$$

With this phase we transcribe the diagrammatic result into algebraic quantities and obtain

$$\begin{aligned}&\sum_{\kappa,\rho,\sigma,\psi} (-1)^{q-\kappa} \begin{pmatrix} p & a & q \\ \psi & \alpha & -\kappa \end{pmatrix} (-1)^{r-\rho} \begin{pmatrix} q & b & r \\ \kappa & \beta & -\rho \end{pmatrix} \\ &\qquad \times (-1)^{s-\sigma} \begin{pmatrix} r & c & s \\ \rho & \gamma & -\sigma \end{pmatrix} (-1)^{p-\psi} \begin{pmatrix} s & d & p \\ \sigma & \delta & -\psi \end{pmatrix} \\ &= (-1)^{2r-s-d-a-q} \sum_{x,\xi} (2x+1)(-1)^{x-\xi} \begin{pmatrix} a & x & c \\ \alpha & -\xi & \gamma \end{pmatrix} \begin{pmatrix} b & x & d \\ \beta & \xi & \delta \end{pmatrix} \\ &\qquad \times \begin{Bmatrix} a & p & q \\ x & d & b \\ c & s & r \end{Bmatrix} .\end{aligned}$$

Both results derived in this section are in perfect agreement with the relations given in [12].

This equation can, of course, be combined with the earlier result from p. 132 to derive a new relation in which combinations of $3jm$- and $3nj$-symbols are treated. Since $(-1)^{2r-2s} = (-1)^{2r+2s} = (-1)^{2c}$ we find

$$\begin{aligned}&\sum_{x,\xi} (2x+1)(-1)^{x-\xi} \begin{pmatrix} a & x & d \\ \alpha & -\xi & \delta \end{pmatrix} \begin{pmatrix} b & x & c \\ \beta & \xi & \gamma \end{pmatrix} \\ &\qquad \times \begin{Bmatrix} a & x & d \\ s & p & q \end{Bmatrix} \begin{Bmatrix} b & x & c \\ s & r & q \end{Bmatrix} \\ &= (-1)^{2c} \sum_{x,\xi} (2x+1)(-1)^{x-\xi} \begin{pmatrix} a & x & c \\ \alpha & -\xi & \gamma \end{pmatrix} \begin{pmatrix} b & x & d \\ \beta & \xi & \delta \end{pmatrix} \\ &\qquad \times \begin{Bmatrix} a & p & q \\ x & d & b \\ c & s & r \end{Bmatrix} .\end{aligned}$$

9.5 Sum with Six $3jm$-Symbols

To continue our analysis of open diagrams we consider a sum over six $3jm$-symbols. As in the previous section we will find optional variations that are readily derived by the graphical method.

$$\sum_{\substack{\psi,\kappa,\mu,\rho\\ \sigma,\mu,\lambda}} (-1)^{q-\kappa}\begin{pmatrix} p & a & q \\ \psi & \alpha & -\kappa \end{pmatrix}(-1)^{v-\mu+g-\lambda}\begin{pmatrix} q & v & g \\ \kappa & -\mu & -\lambda \end{pmatrix}$$
$$\times(-1)^{p-\psi}\begin{pmatrix} g & u & p \\ \lambda & \nu & -\psi \end{pmatrix}(-1)^{r-\rho}\begin{pmatrix} v & b & r \\ \mu & \beta & -\rho \end{pmatrix}$$
$$\times(-1)^{s-\sigma}\begin{pmatrix} r & c & s \\ \rho & \gamma & -\sigma \end{pmatrix}(-1)^{u-\nu}\begin{pmatrix} s & d & u \\ \sigma & \delta & -\nu \end{pmatrix} = \quad ?$$

From the algebraic expressions we see that five $3jm$-symbols have one ingoing and two outgoing lines and the remaining node has two ingoing and one outgoing line.

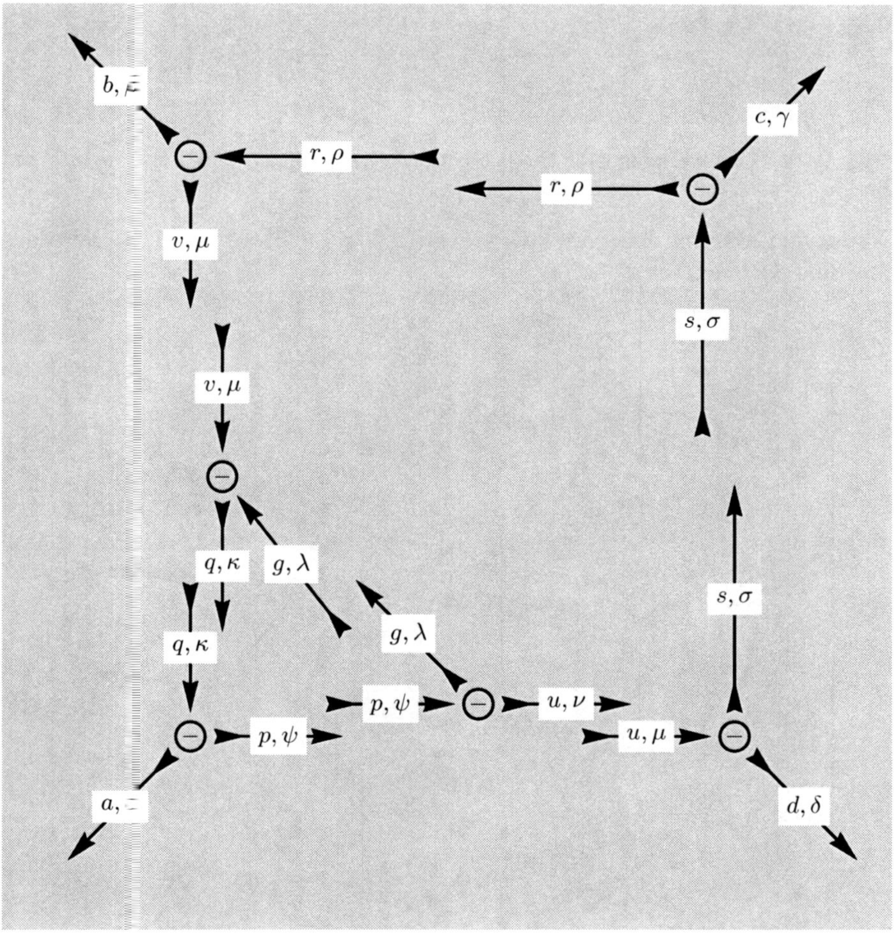

Fig. 9.23. The six symbols are arranged as to indicate the future node connections for the sum over projections.

The resulting diagram in the following figure has four external lines.

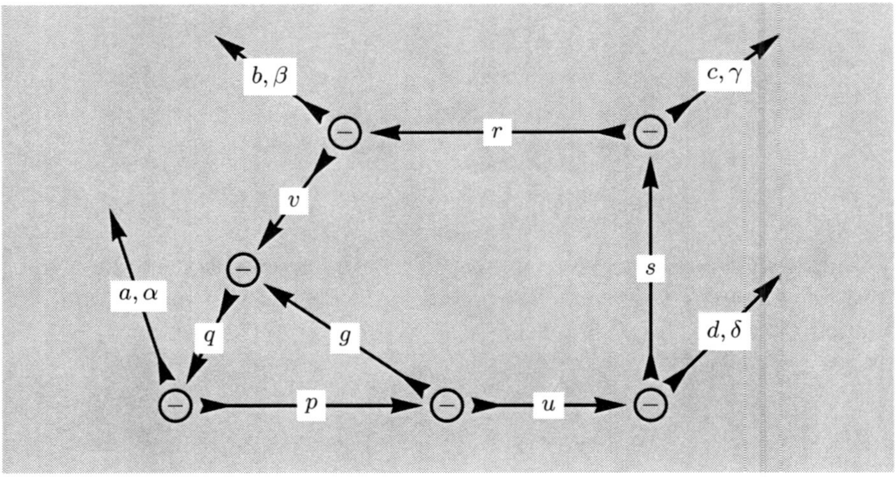

Fig. 9.24. The six sums over the projections $\psi, \kappa, \mu, \rho, \sigma, \mu$ and λ are represented by lines connecting the nodes

Nodes and external lines are now rearranged to facilitate further separations and line contractions.

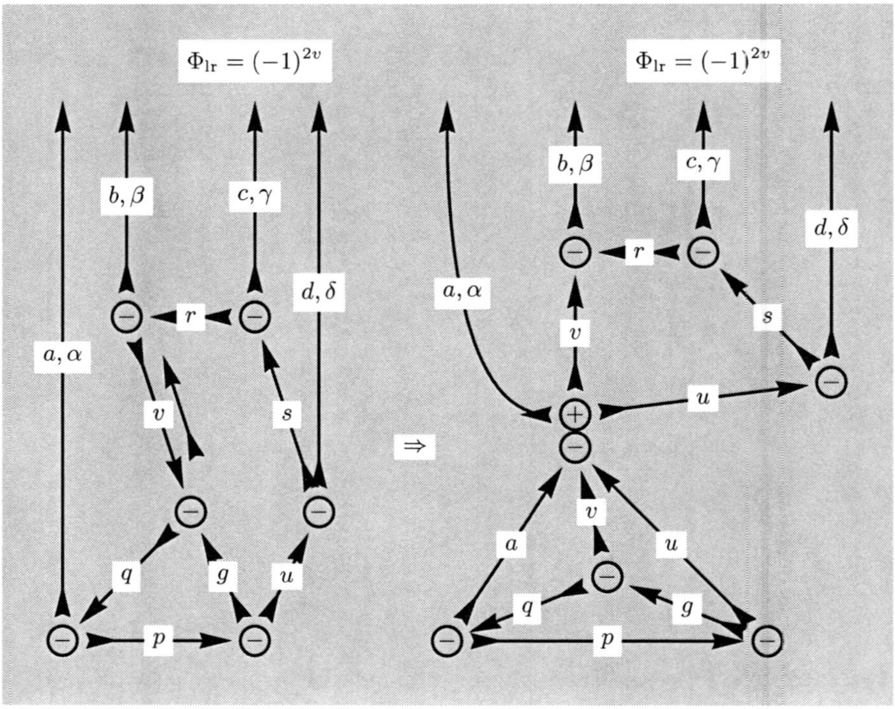

Fig. 9.25. After reversing the v-line the left-hand diagram has three lines in the same direction allowing a three-line separation as indicated in the right-hand part

The left-hand diagram needs further line reversals to approach the form of a $6j$-symbol. The remaining diagram is similar to the diagrams treated in Sect. 9.4 and leads to similar results.

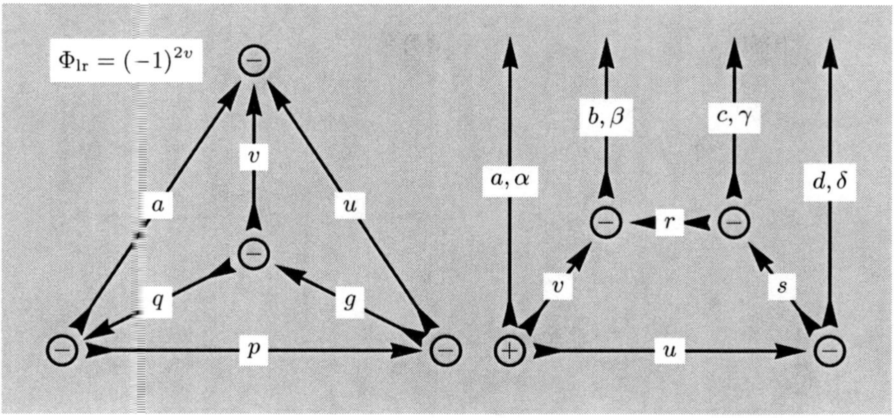

Fig. 9.26. The separated four node diagram is in the left part of the figure and the part with all the external lines is shown at the right-hand side

A sum over x, ξ is used to reduce the number of lines in preparation for a further three-line separation.

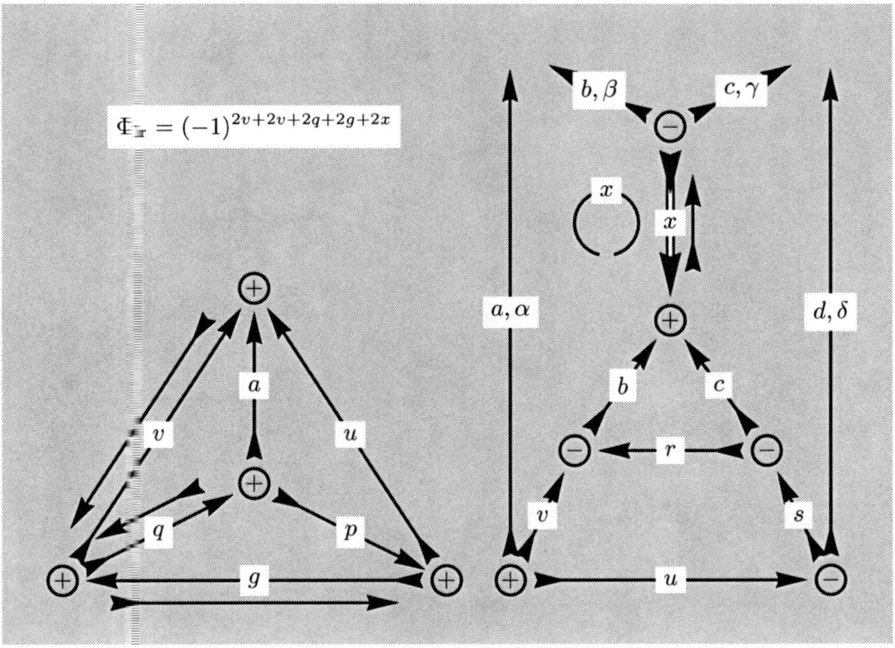

Fig. 9.27. Three line reversals are performed to achieve the appropriate definition for the separated $6j$-symbol

Line reversal of the x-line aligns the a, d and x-line in the same direction.

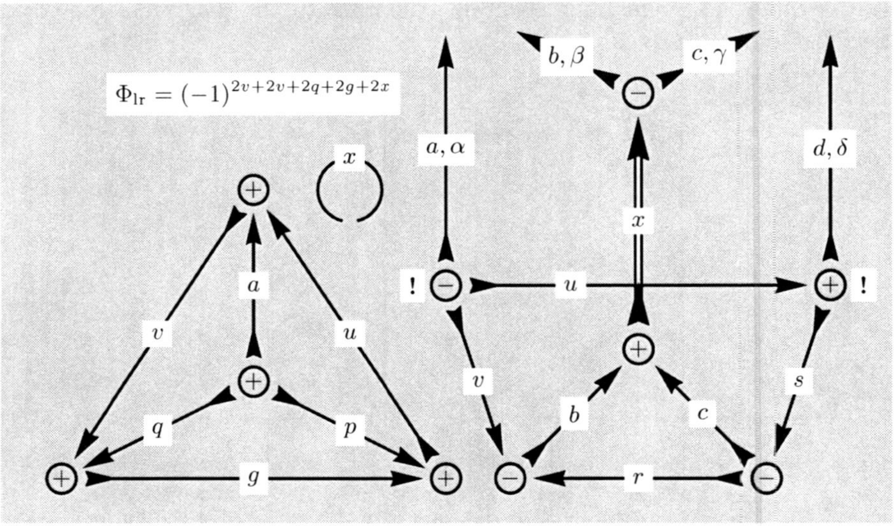

Fig. 9.28. Note the node sign changes brought about by moving the r-line across the u-line

The three-line separation creates a second closed diagram.

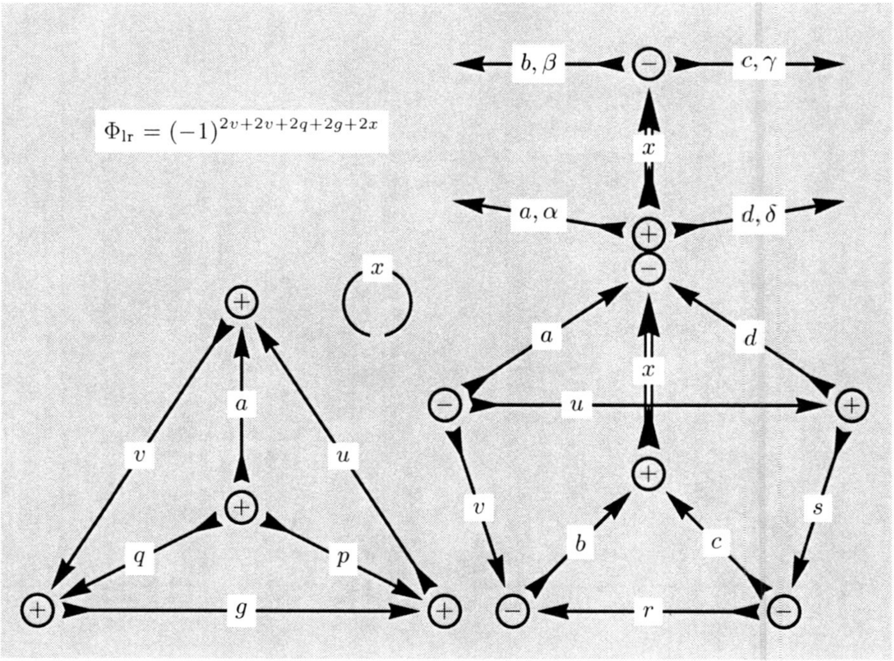

Fig. 9.29. The part with all external lines has two nodes and corresponds to two $3jm$-symbols

The second closed diagram allows a further separation.

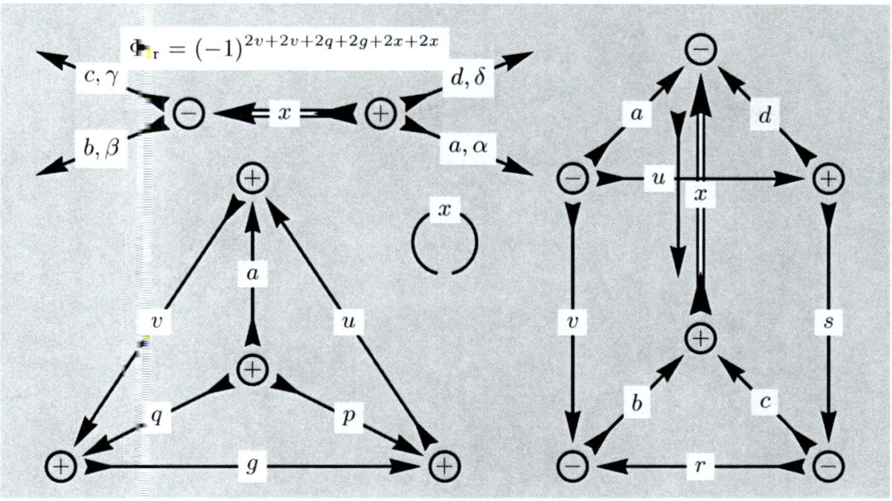

Fig. 9.30. The reversal of the x-line creates three lines in the same direction suitable for separation. The concomitant phase is added to the existing phase

The right-hand side diagram is ready to be separated into two parts.

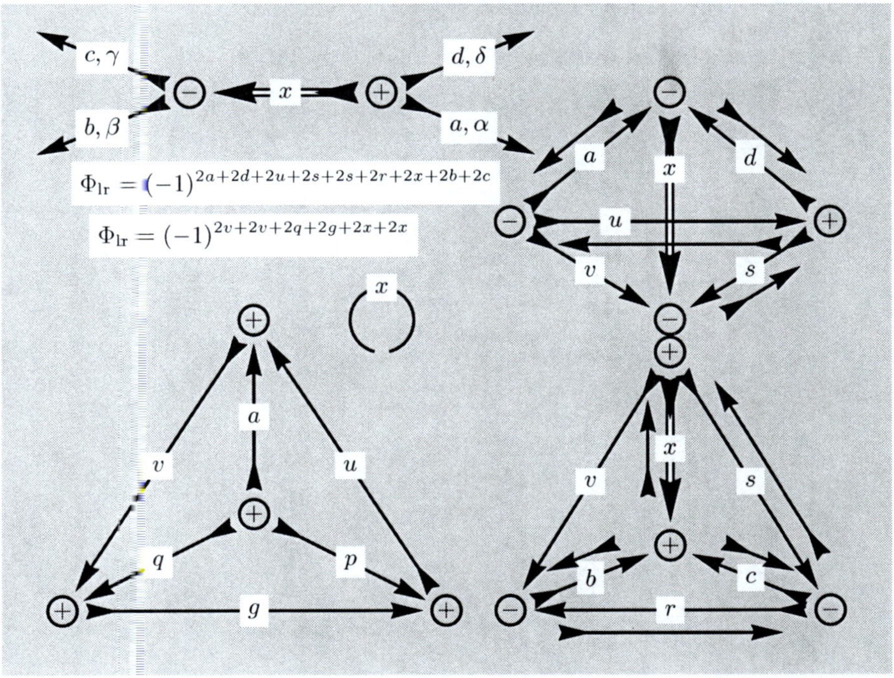

Fig. 9.31. Several line reversals together with the corresponding phases prepare transition to $6j$-symbols

The effect of the line reversals will be visible in the next figure.

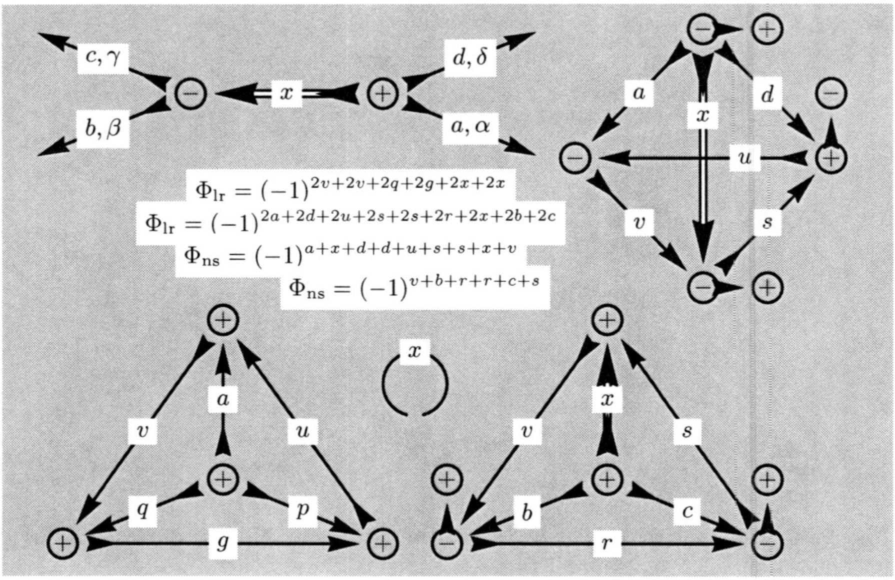

Fig. 9.32. Three direct node sign changes and the corresponding phases are introduced to reach the form of $6j$-symbols in the next figure

With just one more adjustment we reach our destination.

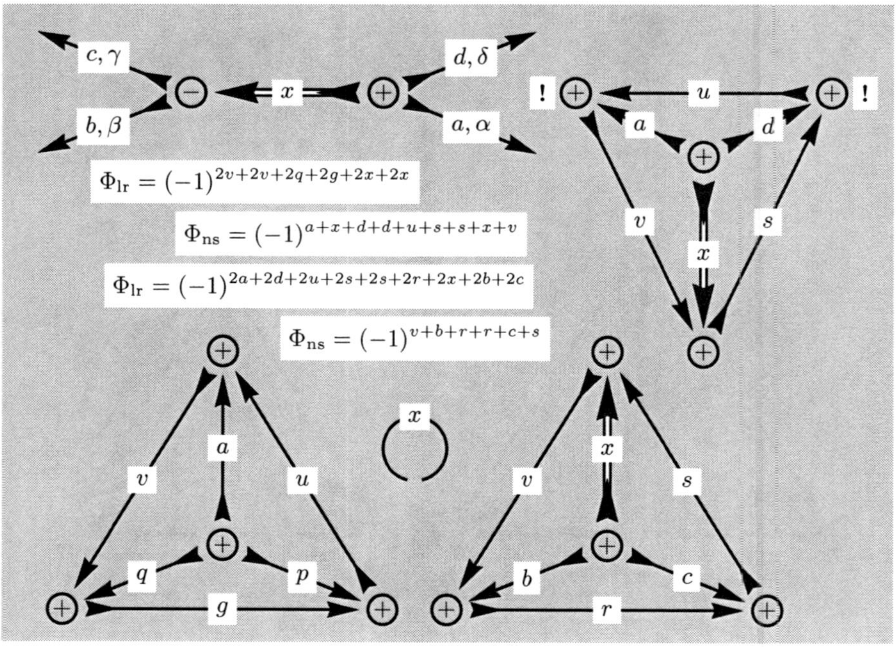

Fig. 9.33. Note the node sign changes due to the new position of the u-line

The line reversals and node sign changes performed in view of the desired final form of the diagrams have produced a complicated expression for the phase of the result.

$$\begin{aligned}\Phi &= (-1)^{2v+2v+2q+2g+2x+2x}(-1)^{2a+2d+2u+2s+2s+2r+2x+2b+2c}\\ &\quad\times(-1)^{a+x+d+d+u+s+s+x+v}(-1)^{v+b+r+r+c+s}\\ &= (-1)^{2q+2g+2a+2d+2u+2r+2x+2b+2c}(-1)^{a+2x+2d+u+2s+2v+b+2r+c+s}\\ &= (-1)^{2q+2g+2a+2u+(2b+2x+2c)+(2s+2x+2v)}(-1)^{a+b+c+s+u}\\ &= (-1)^{a+b+c+s+u}\ .\end{aligned}$$

We remove fourfold appearances of angular momentum variables and enclose in the exponents groups of angular momenta belonging to a single node in parentheses because $(-1)^{2b+2x+2c}=1$ and $(-1)^{2s+2x+2v}=1$. In addition to that, we find $(-1)^{2a+2u}=(-1)^{2v}$ and thus $(-1)^{2q+2g+2v}=1$.

The diagrams are now translated into their algebraic form and we obtain perfect agreement with the formula given by Varshalovich et al. [12].

$$\begin{aligned}&\sum_{\substack{\psi,\kappa,\mu,\rho\\ \sigma,\mu,\lambda}}(-1)^{p-\kappa}\begin{pmatrix}p & a & q\\ \psi & \alpha & -\kappa\end{pmatrix}(-1)^{v-\mu+g-\lambda}\begin{pmatrix}q & v & g\\ \kappa & -\mu & -\lambda\end{pmatrix}(-1)^{p-\psi}\begin{pmatrix}g & u & p\\ \lambda & \nu & -\psi\end{pmatrix}\\ &\qquad\times(-1)^{r-\rho}\begin{pmatrix}v & b & r\\ \mu & \beta & -\rho\end{pmatrix}(-1)^{s-\sigma}\begin{pmatrix}r & c & s\\ \rho & \gamma & -\sigma\end{pmatrix}(-1)^{u-\nu}\begin{pmatrix}s & d & u\\ \sigma & \delta & -\nu\end{pmatrix}\\ &= (-1)^{a+b+c+s+u}\begin{Bmatrix}a & p & q\\ g & v & u\end{Bmatrix}\sum_{x,\xi}(2x+1)(-1)^{x-\xi}\begin{pmatrix}a & d & x\\ \alpha & \delta & \xi\end{pmatrix}\begin{pmatrix}x & b & c\\ -\xi & \beta & \gamma\end{pmatrix}\\ &\qquad\qquad\times\begin{Bmatrix}a & d & x\\ s & v & u\end{Bmatrix}\begin{Bmatrix}x & b & c\\ r & s & v\end{Bmatrix}\ .\end{aligned}$$

A Variation of this Formula

As in Sect. 9.4, graphical analysis may proceed with a slight variation but we will not repeat this analysis and instead make use of results given in Sect. 9.4.

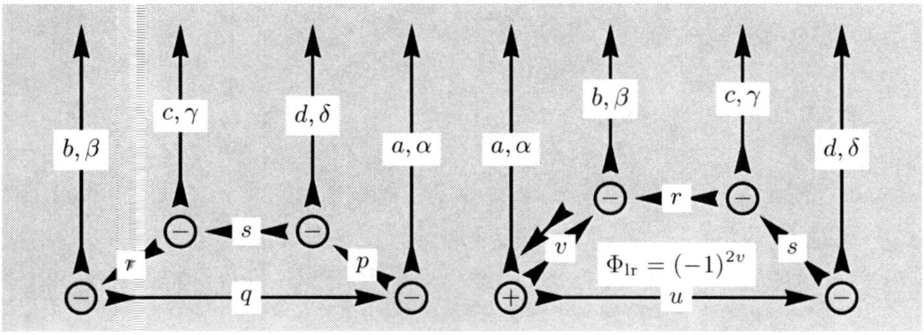

Fig. 9.34. The diagram on the left-hand side is taken from Fig. 9.19 and the right-hand side from 8.75. The figure indicates the variable changes required to translate one diagram into the other

The result given in Fig. 9.21 can be used here with the following variable changes: $b \to a$, $c \to b$, $d \to c$, $a \to d$, $r \to v$, $s \to r$, $p \to s$, and $q \to u$. Note that the required reversal of the v-line introduces an extra phase $(-1)^{2v}$.

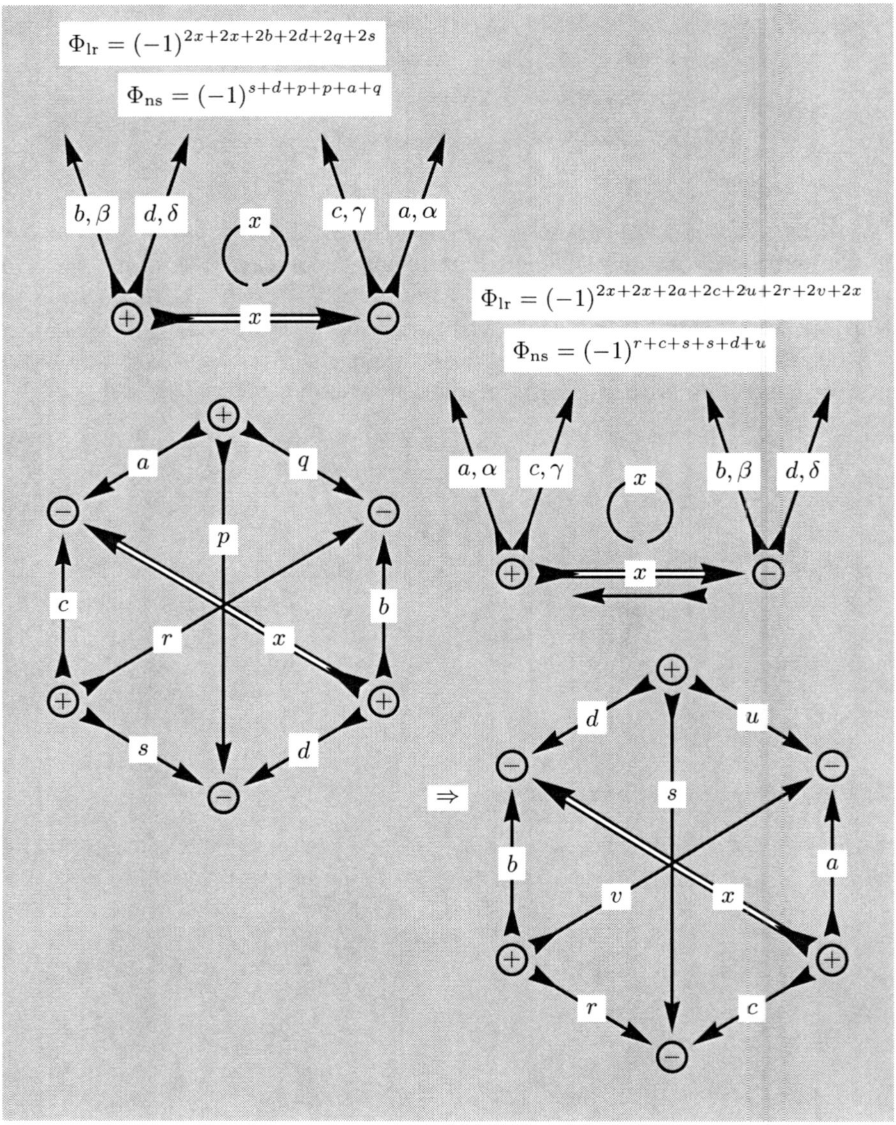

Fig. 9.35. The figure shows the result from Fig. 9.22 and in the lower part the renamed version relevant here. Note the phase stemming from the reversal of the x-line

The final diagram is a combination of the left part of Fig. 9.28 with the phase $(-1)^{2v+2v+2q+2g}$ and the right part of Fig. 9.35.

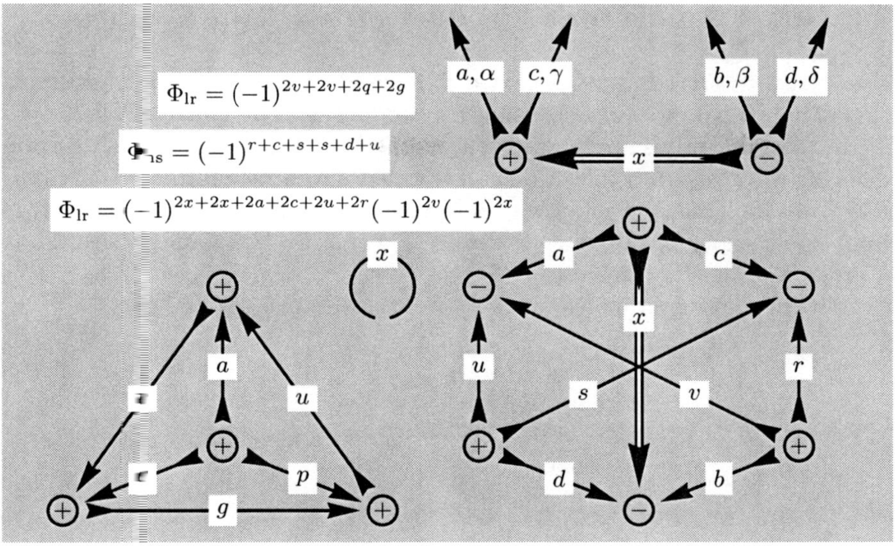

Fig. 9.36. Note that the diagram for the 9j-symbol has been rotated

The total phase Φ can be simplified if we keep in mind that $(-1)^{2v+2q+2g} = 1$ and also $(-1)^{2x+2a+2c} = 1$.

$$
\begin{aligned}
\Phi &= (-1)^{2v+2v+2q+2g}(-1)^{2x+2x+2a+2c+2u+2r}(-1)^{2v}(-1)^{2x} \\
&\quad \times(-1)^{r+c+s+s+d+u} \\
&= (-1)^{2v+2x+2u+2r+2v+2x+2s}(-1)^{r+c+d+u} \\
&= (-1)^{2u+2r+2s}(-1)^{c+d+r+u} \\
&= (-1)^{2d+2r}(-1)^{c+d+r+u} \\
&= (-1)^{c-d-r+u} \, .
\end{aligned}
$$

The final result requires $(-1)^{2u+2s} = (-1)^{-2d} = (-1)^{2d}$ and also $(-1)^{3d} = (-1)^{-d}$ and $(-1)^{3r} = (-1)^{-r}$. The transcription of the diagrams leads to an equation which, for instance, is not listed Varshalovich et al. [12].

$$
\begin{aligned}
&\sum_{\substack{\psi,\kappa,\mu,\rho \\ \sigma,\mu,\lambda}} (-1)^{q-\kappa+v-\mu+g-\lambda+p-\psi} \begin{pmatrix} p & a & q \\ \psi & \alpha & -\kappa \end{pmatrix} \begin{pmatrix} q & v & g \\ \kappa & -\mu & -\lambda \end{pmatrix} \begin{pmatrix} g & u & p \\ \lambda & \nu & -\psi \end{pmatrix} \\
&\quad \times(-1)^{r-\rho} \begin{pmatrix} v & b & r \\ \mu & \beta & -\rho \end{pmatrix} (-1)^{s-\sigma} \begin{pmatrix} r & c & s \\ \rho & \gamma & -\sigma \end{pmatrix} (-1)^{u-\nu} \begin{pmatrix} s & d & u \\ \sigma & \delta & -\nu \end{pmatrix} \\
&= (-1)^{c-d-r+u} \begin{Bmatrix} a & p & q \\ g & v & u \end{Bmatrix} \sum_{x,\xi} (2x+1) \\
&\quad \times(-1)^{x-\xi} \begin{pmatrix} a & x & c \\ \alpha & \xi & \gamma \end{pmatrix} \begin{pmatrix} d & x & b \\ \delta & \xi & \beta \end{pmatrix} \begin{Bmatrix} a & x & c \\ v & b & r \\ u & d & s \end{Bmatrix} .
\end{aligned}
$$

9.6 New Formula for a Product of two $3jm$-Symbols

In contrast to preceding sections, which focused on the graphical transformation of sums, here we consider simply a product of two $3jm$-symbols.

In this section we apply graphical methods to demonstrate that the product of two $3jm$-symbols can be transformed into a new formula containing four $3jm$- and one $9j$-symbol with several summations. Such a product arises from applications of the Wigner–Eckart Theorem to two tensor operators in a product of matrix elements.

In the following graphical analysis we consider two $3j$-symbols,

$$(-1)^{a-\alpha}\begin{pmatrix} a & b & c \\ -\alpha & \beta & \gamma \end{pmatrix}(-1)^{d-\delta}\begin{pmatrix} d & e & f \\ -\delta & \epsilon & \phi \end{pmatrix} .$$

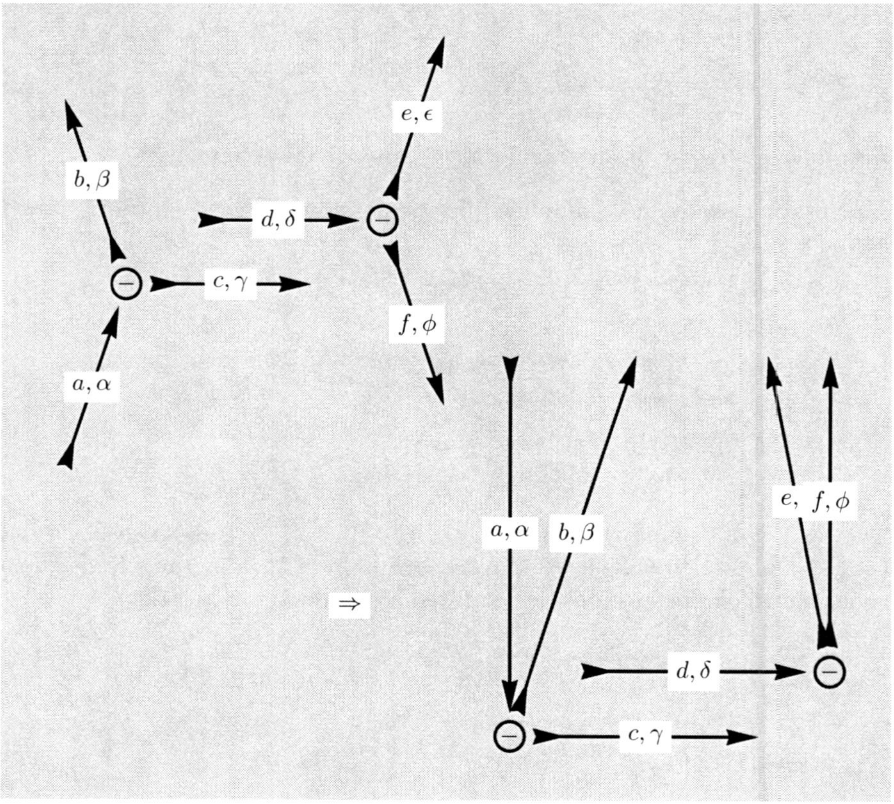

Fig. 9.37. Both nodes have one ingoing and two outgoing lines and both node signs are negative

The first step introduces a sum over z, ζ which reduces the b- and the e-line. For the next lines to be combined, the c- and the d-line, the latter has to be reversed temporarily.

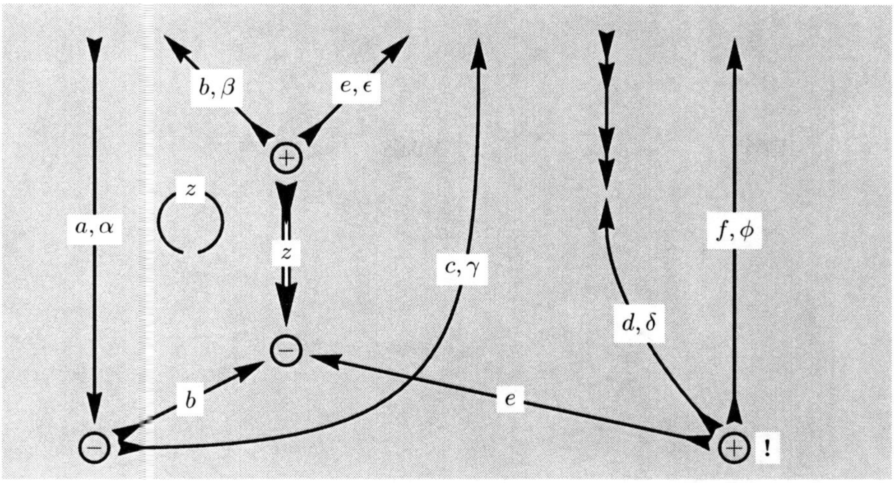

Fig. 9.38. Note the sign change in the bottom right node due to the movement of the d-line across the e-line

A further sum over y, η combines the c- and the d-line. The circular diagrams indicate the factors to be included in a sum over an angular momentum value.

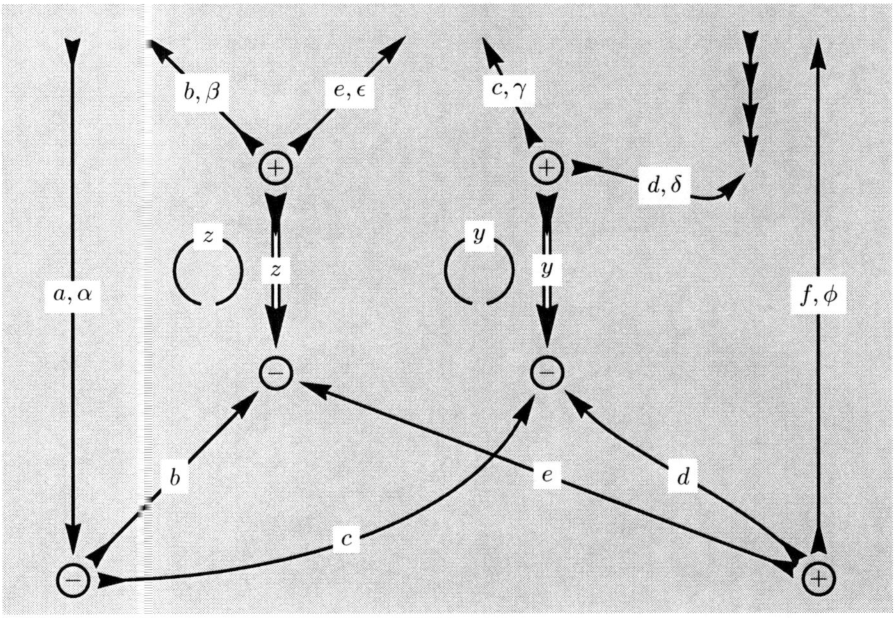

Fig. 9.39. Note a temporary change in the direction of the d-line to give it the same direction as the c-line

After the introduction of the y, η-sum the external d-line returns to its original direction.

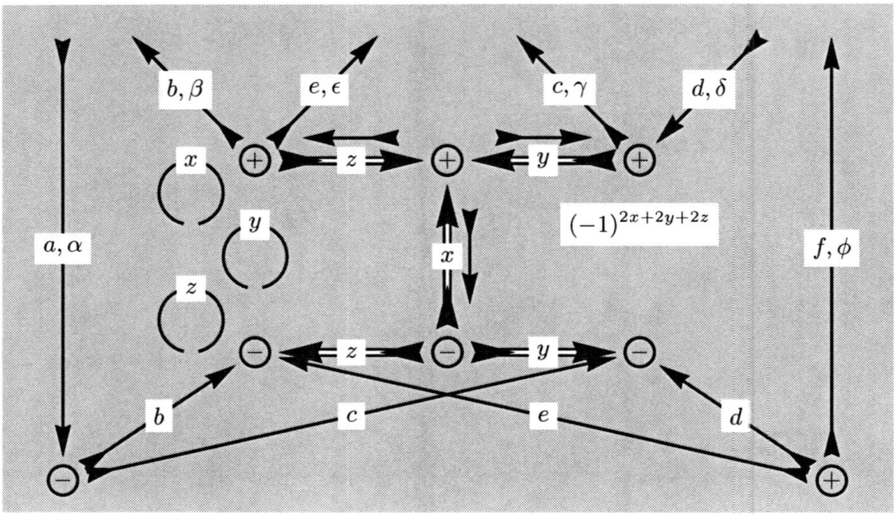

Fig. 9.40. A third sum over x, ξ is introduced here. Note the line reversals for x, y- and z-lines joined in one node. The phase factor is 1 and displayed for completeness

x- and a-line are now aligned in the same direction but the direction of the f-line is now reversed temporarily to prepare a three line separation.

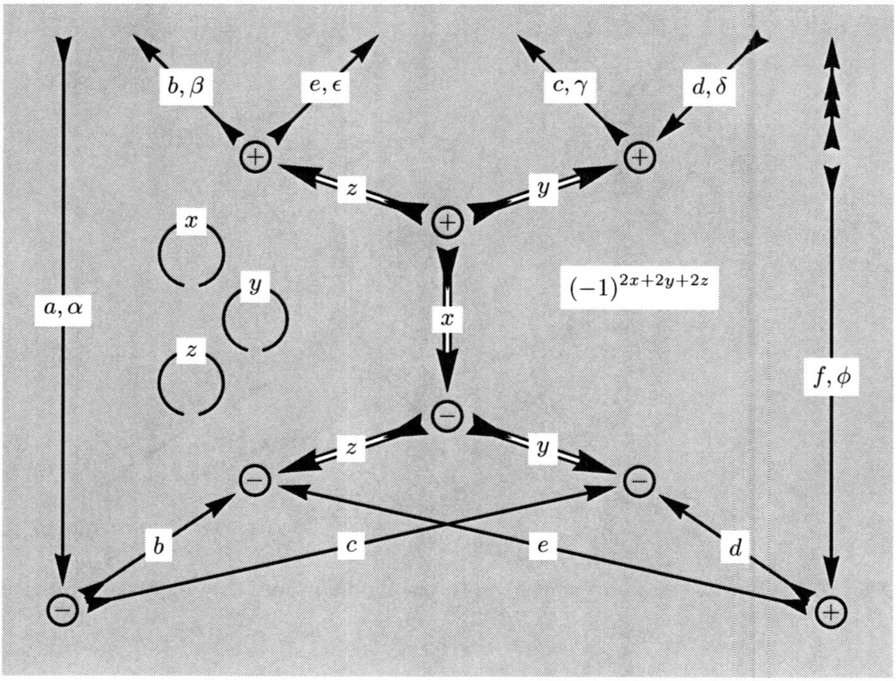

Fig. 9.41. Note a temporary change in direction for the f-line

The ground has been cleared for a three line separation which creates a closed diagram in the bottom of the figure.

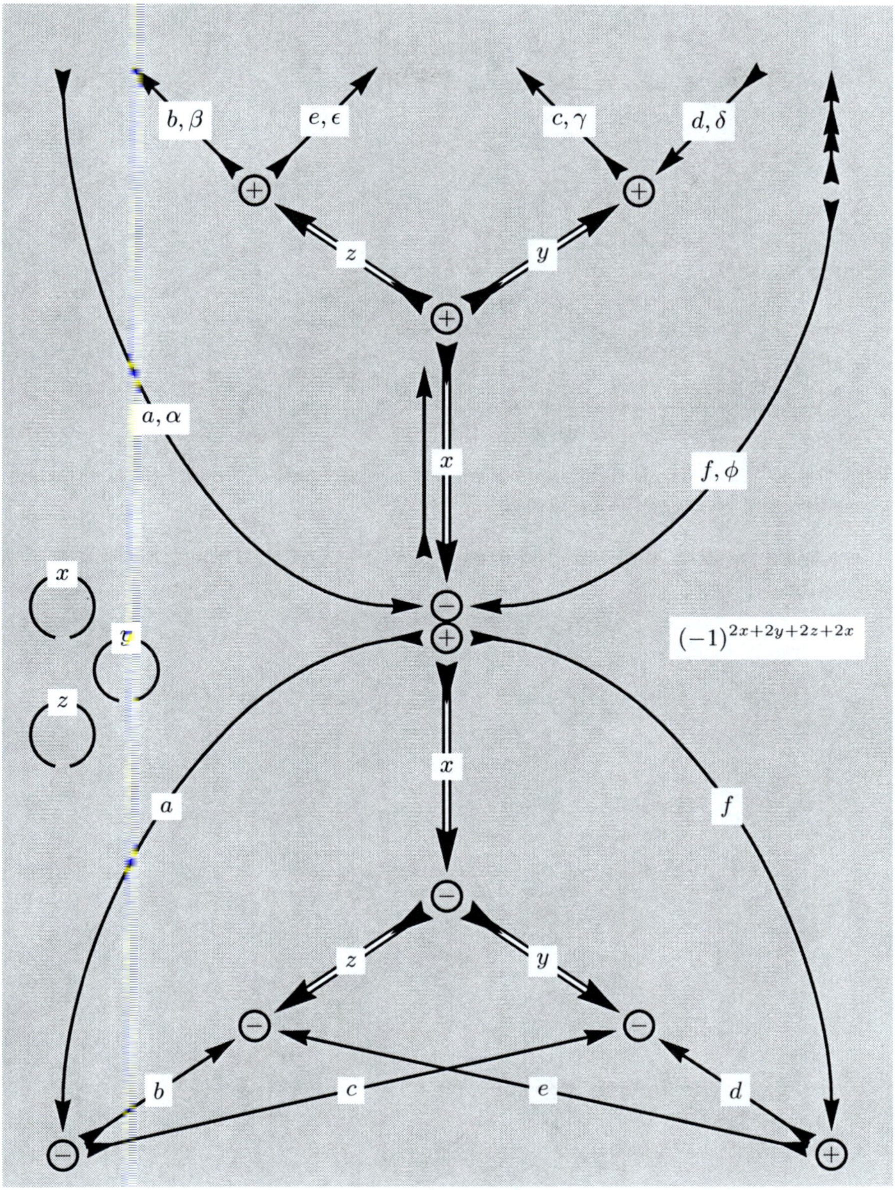

Fig. 9.42. Two nodes of different node sign are inserted for the separation of the diagram into two parts. Note a further change in the direction of the upper x-line

The external f-line returns to its original direction and the closed diagram is transformed.

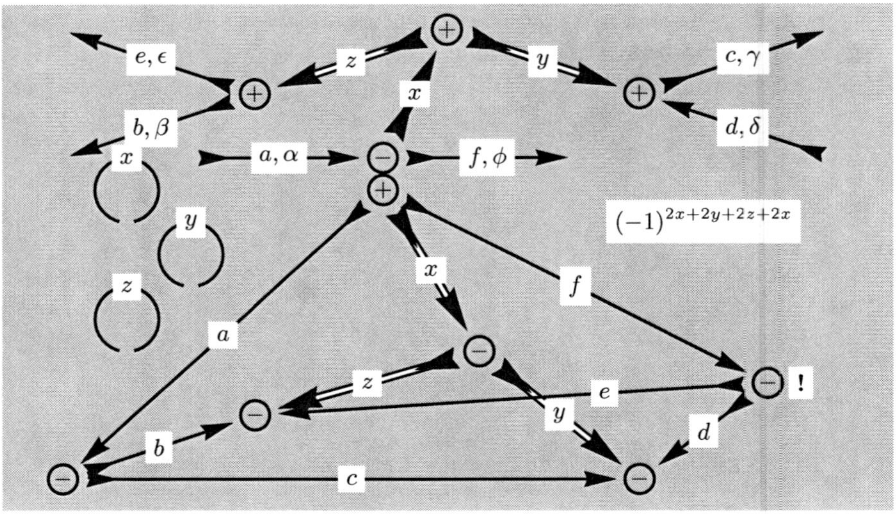

Fig. 9.43. The bottom right node is moved upward and the e-line passes through the d-line and changes the node sign

The closed bottom diagram requires a few changes to achieve the form of a $9j$-symbol.

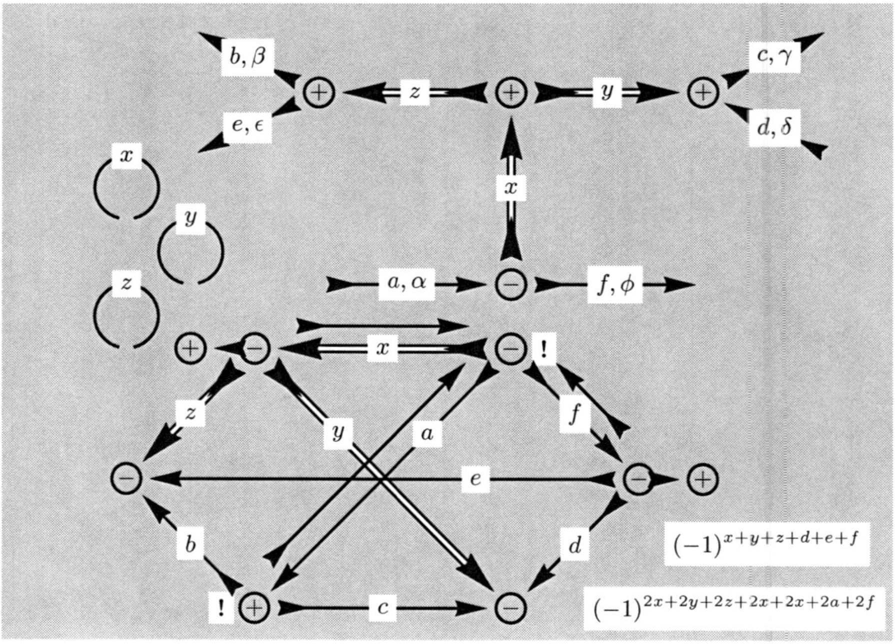

Fig. 9.44. The x-, the z- and the b-lines are moved across the a-line and the adjoining nodes change sign. In addition, two more node sign changes and three line reversals are required

The changes have transformed the closed diagram into a $9j$-symbol.

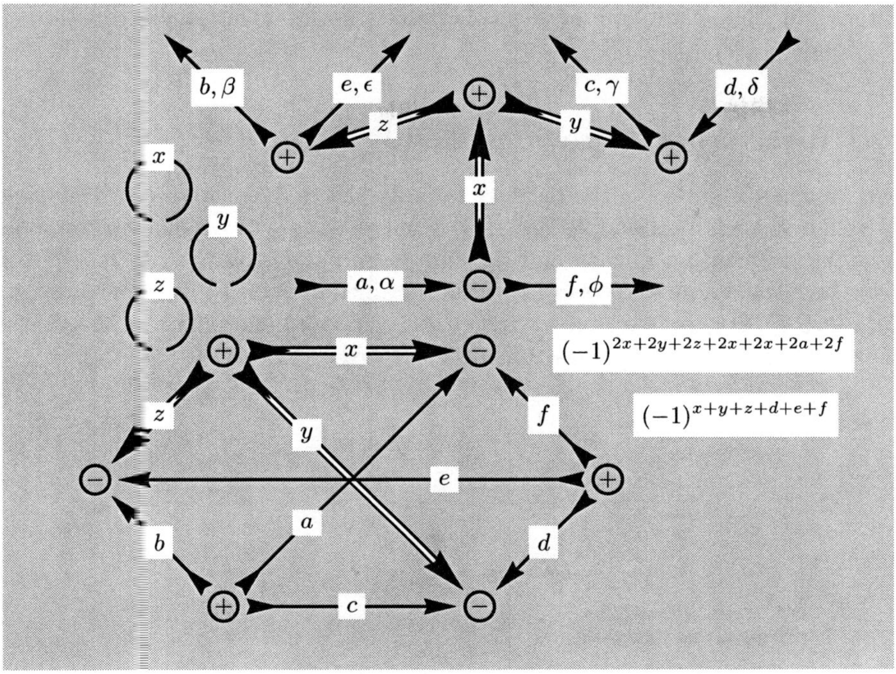

Fig. 9.45. The diagram consists of four $3jm$-symbol and a $9j$-symbol. The phase can be simplified

The accumulated phase is

$$\Phi = (-1)^{2z+2y+2x}(-1)^{2x}(-1)^{2x+2a+2f}(-1)^{x+y+z+e+d+f} ,$$

and this phase may be reduced to $(-1)^{-x+y+z+e+d+f}$.

The application of the methods of graphical analysis have produced a new formula for the product of two $3jm$-symbols.

$$(-1)^{a-\alpha}\begin{pmatrix} a & b & c \\ -\alpha & \beta & \gamma \end{pmatrix}(-1)^{d-\delta}\begin{pmatrix} d & e & f \\ -\delta & \epsilon & \phi \end{pmatrix} =$$

$$= \sum_{z,\zeta}(2z+1)\sum_{y,\eta}(2y+1)\sum_{x,\xi}(2x+1)$$

$$\times(-1)^{z-\zeta}\begin{pmatrix} b & z & e \\ \beta & -\zeta & \epsilon \end{pmatrix}(-1)^{x-\xi}\begin{pmatrix} z & x & y \\ \zeta & -\xi & \eta \end{pmatrix}$$

$$\times(-1)^{y-\eta}(-1)^{d-\delta}\begin{pmatrix} d & c & y \\ -\delta & \gamma & -\eta \end{pmatrix}(-1)^{a-\alpha}\begin{pmatrix} a & x & f \\ -\alpha & \xi & \phi \end{pmatrix}\begin{Bmatrix} z & y & x \\ e & d & f \\ b & c & a \end{Bmatrix}$$

$$\times(-1)^{-x+y+z+e+d+f} .$$

To the best of our knowledge, this formula has never before appeared in the literature. The result can, however, be linked to the orthogonality relation of $9j$-symbols [32].

9.7 Recovering a Previous Result

Seeking confirmation of the product formula just derived, we recall analysing a sum over such a product in Sect. 9.1. The two $3jm$-symbols share a common angular momentum q there and the sum is over the projection κ. It seems a worthwhile exercise to recover the result obtained in Sect. 9.1 from the diagram in Fig. 9.45. For this we have to change four variable names in order to achieve correspondence between Figs. 9.1 and 9.37.

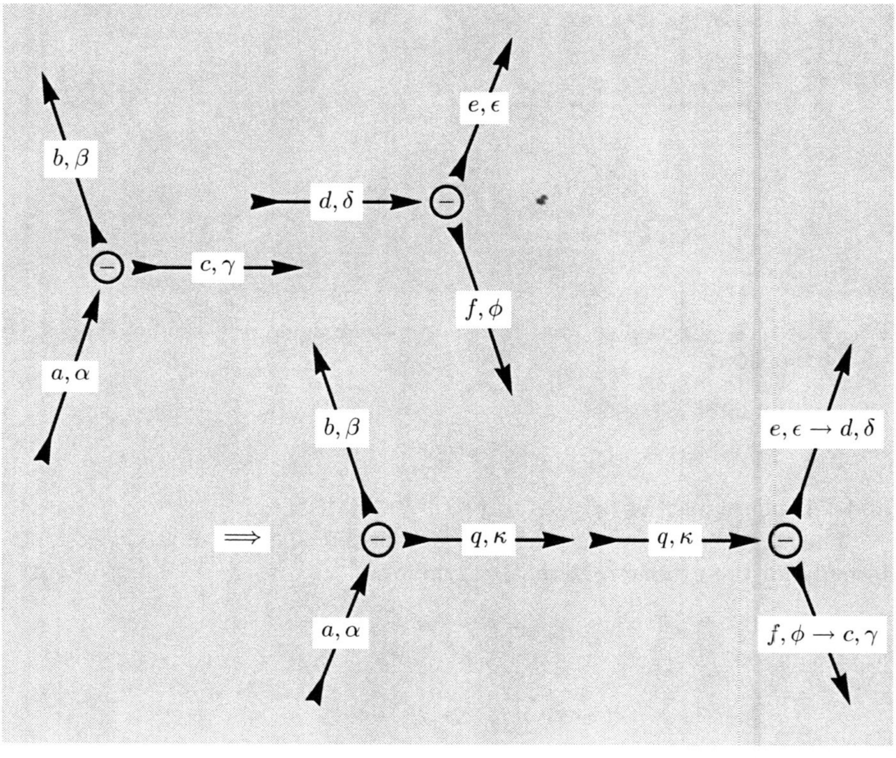

Fig. 9.46. In the transition from top to bottom we note that, both c and d have been replaced by q. For correspondence with Fig. 9.1 we also replace $e \to d$ and $f \to c$

The positive node sign in Fig. 9.1 can be achieved by changing the cyclic order at both nodes. The only remaining differences are the different orientation of the q-line when we sum over κ and that the a-line is ingoing instead of outgoing. These differences are taken into account by an extra phase $(-1)^{2q}$ and a different direction of the a-line in Fig. 9.45.

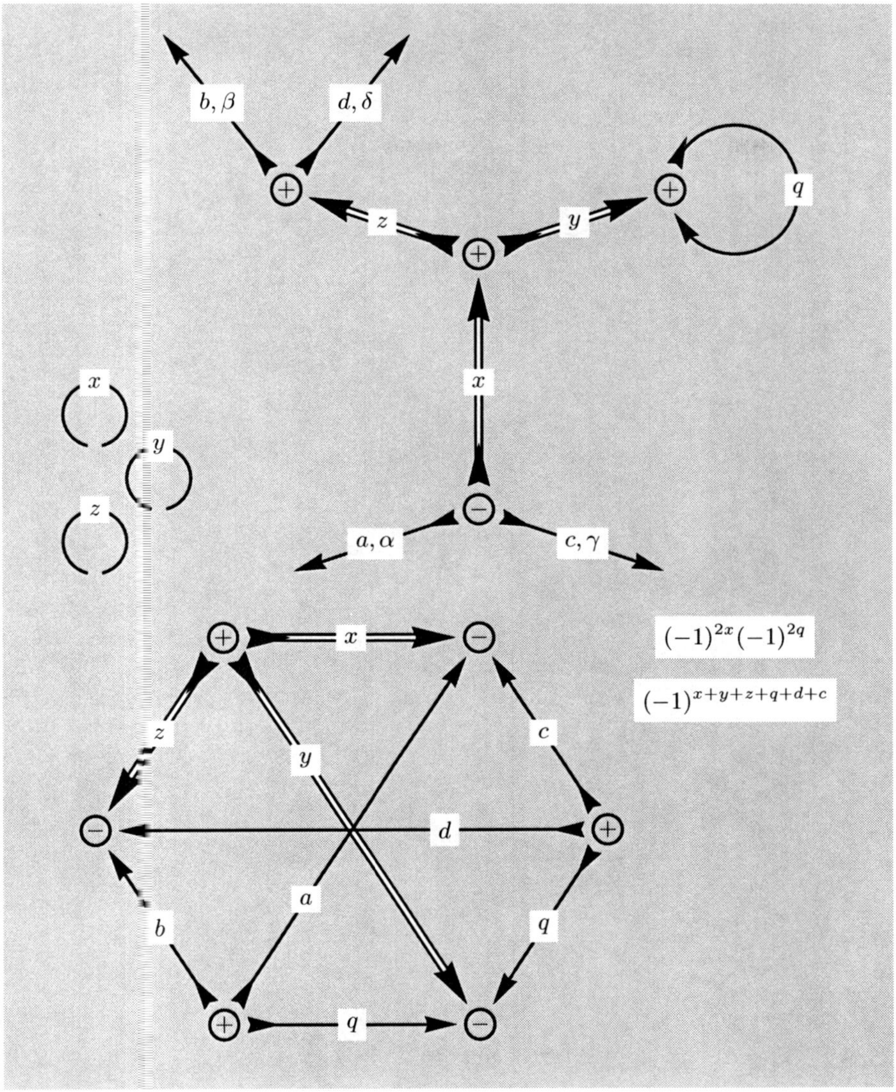

Fig. 9.47. The figure is the adapted version of Fig. 9.45. Note the phase has been simplified and a phase $(-1)^{2q}$ inserted. The phase has been adjusted to the new angular momentum variable names. The sum over κ connects two external lines as indicated by the circular q-line

The circular q-line represents a closed diagram connected to the rest by only one line and this means, as we have seen, that the single connecting line can only have one value, namely zero.

Setting $q = 0$ has the consequence $x = z$ and we are left with four removable nodes.

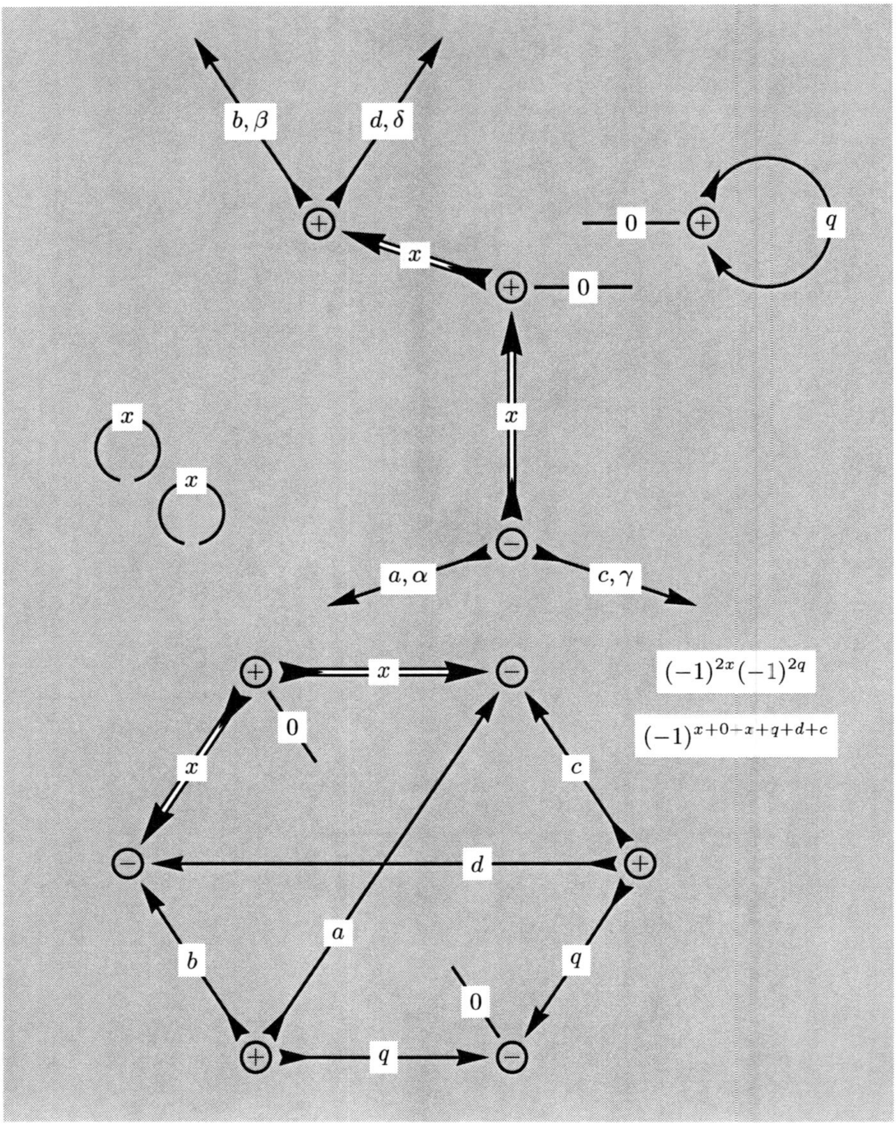

Fig. 9.48. Note that with $y = 0$ the factor $(2y + 1) = 1$ can be dropped. Some nodes need appropriate line reversals or, equivalently, node sign changes to become removable

The nodes attached to the 0-lines will now be removed following the procedure described in Sect. 3.5. The requirements for removing the nodes in the $9j$-symbol call for line reversals of an x- and a q-line which decide, of course, the final direction of these lines.

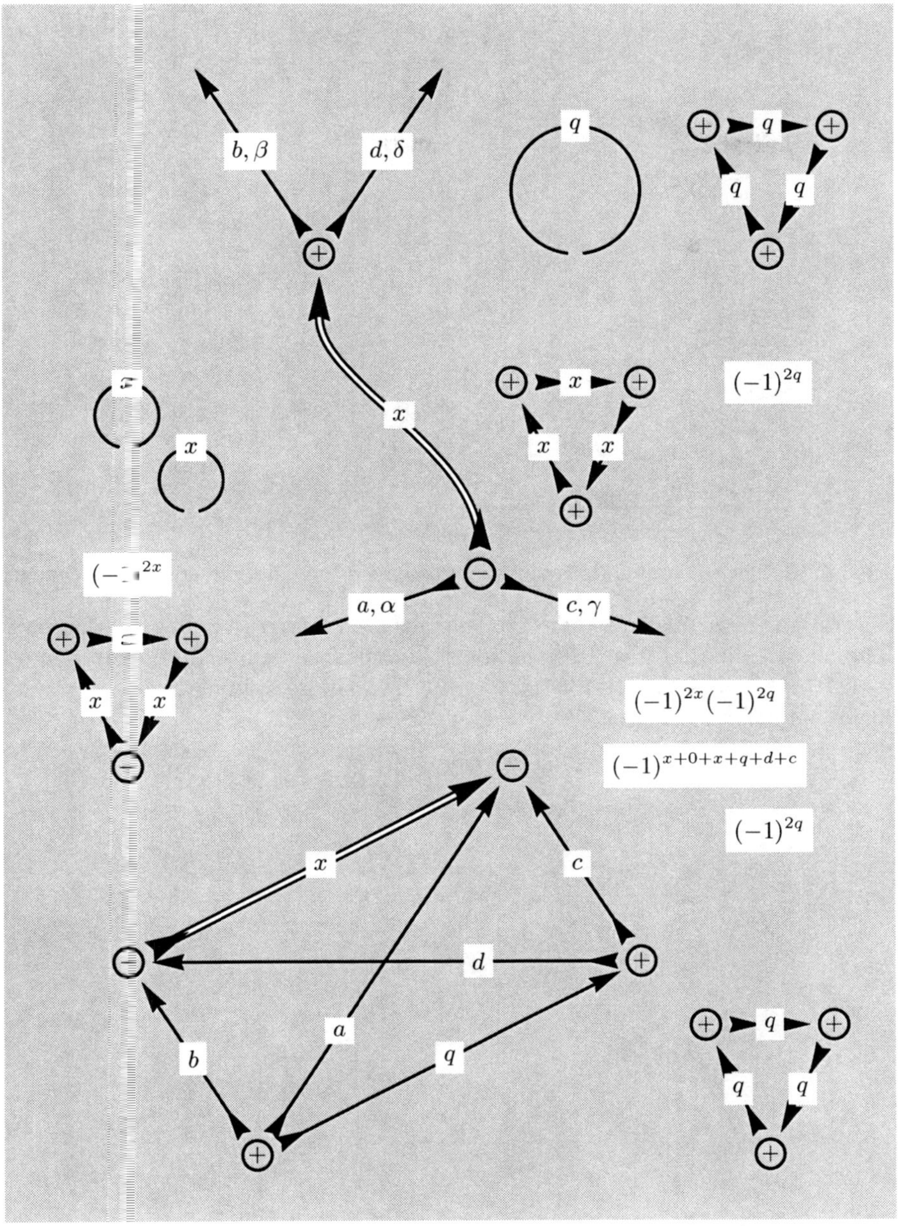

Fig. 9.49. Removal of the nodes attached to a 0-line creates the triangular diagrams. Two of these cancel the circular q-line. In the same way one circular x-line is cancelled by two triangle diagrams

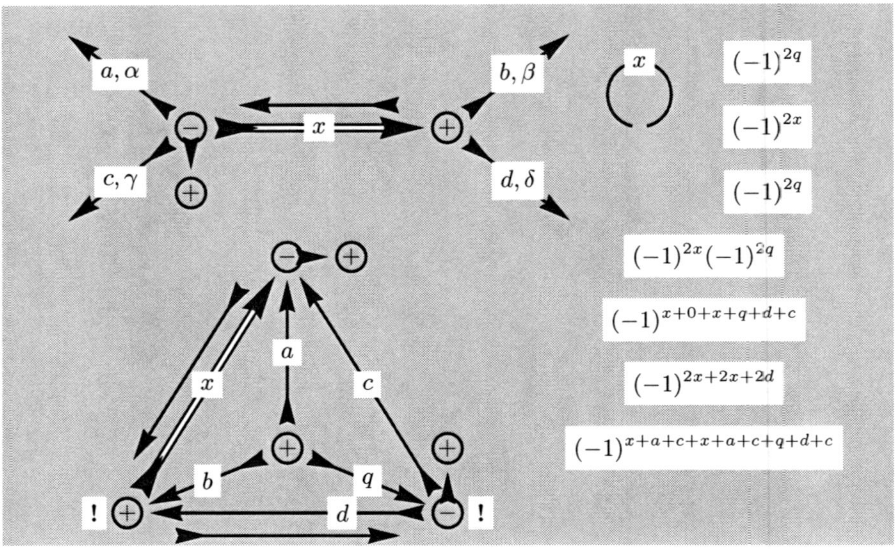

Fig. 9.50. The bottom node is moved across the d-line and two node signs change

Several line reversals and node sign changes are required for the final diagram. The phase simplifies if we drop all fourfold appearances of x and q and remove $(-1)^{2x+2a+2c} = 1$ and $(-1)^{2q+2d+2c} = 1$. For the remaining phase factor we have $(-1)^{2x+2q+2d} = (-1)^{2x+2c} = (-1)^{-2a} = (-1)^{2a}$.

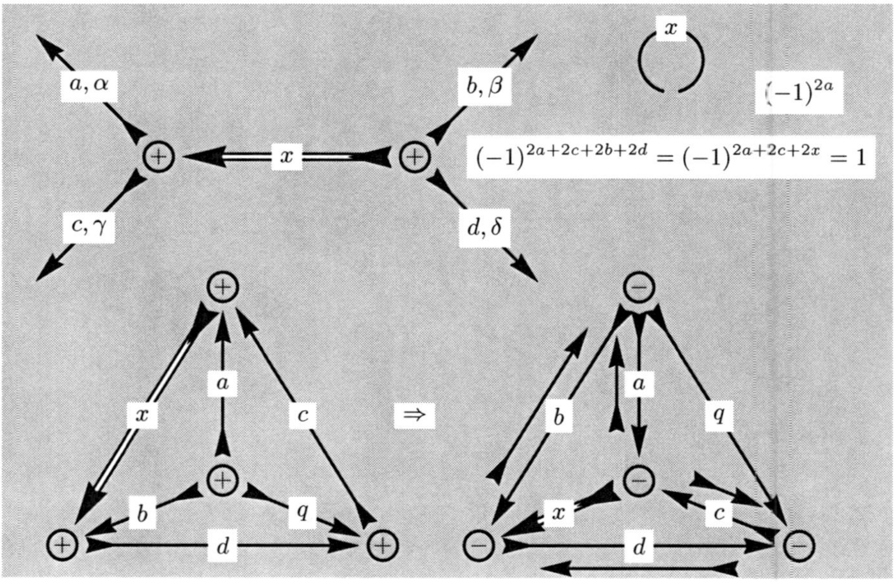

Fig. 9.51. In the bottom part of the figure the $6j$-symbol is subjected to a transformation which exchanges the two nodes connected by the a-line. This changes all node signs and requires more line reversals

To achieve the required positive node signs for the $6j$-symbol, the diagram needs to be flipped around the c-line.

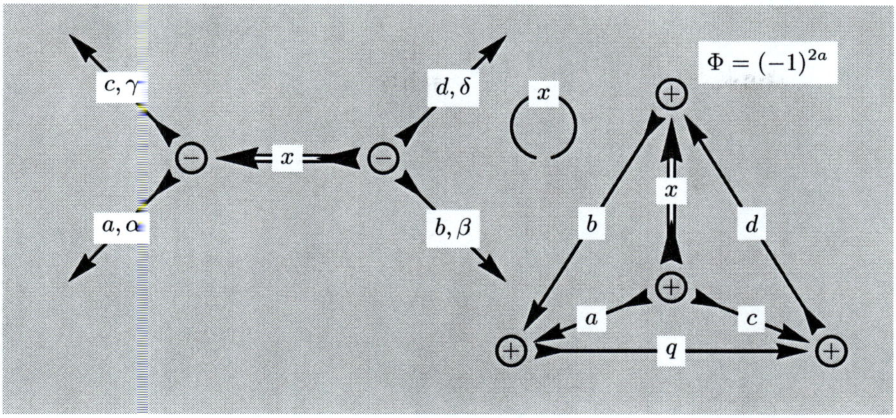

Fig. 9.52. Note that the external lines have changed position thus changing the cyclic order and the node sign

The last figure shows the same diagram as given in Fig. 9.4. That we have recovered the result of Sect. 9.2 is an indication for the correctness of the new product formula for two $3jm$-symbols. derived and analysed in Sect. 9.6.

10

Application in Atomic Physics

In preceding sections rules and methods required for the graphical theory of angular momentum have been introduced and exercised in various examples dealing with closed or open diagrams in Sects. 8 and 9, respectively. It is now time to consider an example in which open and closed diagrams are combined.

Most of the results found are well known but it is also quite feasible to search for new relations as, for example, in Sect. 9.6 and Sect. 8.3.3. It is thus worthwhile to study the application of the graphical method on hand of an example plucked from atomic physics. There is, however, a drawback in that we have to insert a short introduction to concepts related to spherical tensor operators and their matrix elements for many electron atoms before we return to the graphical method. In this context we wish to draw attention to the contents of App. A.

10.1 Spherical Tensor Operators

Spherical, or irreducible, tensors are usually defined by their commutation relations or through their properties under proper rotations, which should be like those for spherical harmonics, as defined in [4; 9]. A tensor operator T^K_Q has rank K and projections specified by Q with $-K \leq Q \leq K$. In the following we will always consider symmetric tensor operators built from sums of tensor operators $(t^K_Q)_i$ for single electrons denoted by subscript i.

$$T^K_Q = \sum_i (t^K_Q)_i \ . \tag{10.1}$$

In general any tensor operator may be coupled from two parts, u^a_α and v^b_β, which operate on spin and orbital components, respectively. We will use the notation $\{u^a_\alpha \otimes v^b_\beta\}^K_Q = t^K_Q$ for the following definition

$$t^K_Q = \sum_{\alpha,\beta} u^a_\alpha v^b_\beta (a\alpha b\beta|KQ) \ . \tag{10.2}$$

E. Balcar, S.W. Lovesey, *Introduction to the Graphical Theory of Angular Momentum*, STMP 234, 159–200, DOI 10.1007/978-3-642-03118-2_10,

Although the coupling formula uses Clebsch–Gordan coefficients, we have seen that there is a close connection between these coefficients and the $3jm$-symbols, see Sect. 2.2.

In a quantum mechanical calculation the goal is usually the calculation of matrix elements of T^K_Q.

$$\langle JM|T^K_Q|J'M'\rangle = \langle \alpha SLJM|T^K_Q|\alpha' S'L'J'M'\rangle = \quad ? \tag{10.3}$$

The coupling used for the initial and final state function is the Russell–Saunders or SL-coupling. The additional quantum numbers α and α' cover those cases in which the values S and L are not sufficient to uniquely label the term. These α-labels should not be confused with the magnetic quantum numbers α, β in the coupling equation.

The properties of the spherical tensor operators have a very important consequence for their matrix elements and this is the content of the Wigner–Eckart Theorem: the dependence of the matrix element above on M, Q and M' is described by an $3jm$-symbol, namely

$$\langle JM|T^K_Q|J'M'\rangle = (-1)^{J-M} \begin{pmatrix} J & K & J' \\ -M & Q & M' \end{pmatrix} \times (\text{red. mat. el.})\,. \tag{10.4}$$

The factor on the right-hand side is usually, but not quite fittingly, called the reduced matrix element and comprises the remaining part of the matrix element, once the $3jm$-symbol with the phase is split off. The formidable power of the Wigner–Eckart Theorem is that the projections are solely in the $3jm$-symbol representing the geometric part of the matrix element.

10.2 Reduced Matrix Elements

A reduced matrix element is independent of the magnetic quantum numbers M, Q and M' and this fact is also displayed in the notation.

$$\langle JM|T^K_Q|J'M'\rangle = (-1)^{J-M} \begin{pmatrix} J & K & J' \\ -M & Q & M' \end{pmatrix} \times (\alpha SLJ\|T^K\|\alpha' S'L'J')\,. \tag{10.5}$$

Racah has shown that a reduced matrix element may be separated into a general and a specific part. Assuming that the states labelled by JM and $J'M'$ belong to the same electron configuration (angular momentum l and spin $s = 1/2$) the reduced matrix element can be separated into a unit tensor $(\alpha SLJ\|W^{(a,b)K}\|\alpha' S'L'J')$ and those factors that depend only on the specific characteristics of the one electron operators or rather on their individual reduced matrix elements. Here, a denotes the rank of the spin part, u^a_α, and b the rank of the orbital part, v^b_β, of a coupled spherical tensor operator, t^K_Q, defined in 10.2.

$$\begin{aligned}(\alpha SLJ\|T^K\|\alpha' S'L'J') &= (\alpha SLJ\|W^{(a,b)K}\|\alpha' S'L'J') \\ &\quad \times (s\|u^{(a)}\|s)(l\|v^{(b)}\|l)\,. \end{aligned} \tag{10.6}$$

It is possible to factorize the reduced matrix element of the unit tensor $W^{(a,b)K}$ into a $9j$-symbol and the reduced matrix element of a second, simpler unit tensor $W^{(a\ b)}$.

$$(\alpha SLJ\|W^{(a,b)K}\|\alpha' S'L'J') = \left[\frac{(2J+1)(2K+1)(2J'+1)}{(2a+1)(2b+1)}\right]^{1/2} \times \left\{\begin{matrix} S & L & J \\ S' & L' & J' \\ a & b & K \end{matrix}\right\} (\alpha SL\|W^{(a,b)}\|\alpha' S'L') \,. \tag{10.7}$$

The form of the reduced matrix element of $W^{(a,b)}$ has been given by Racah and is based on the use of Racah's coefficients of fractional parentage. These coefficients allow the construction of an antisymmetrised state for n electrons from a linear combination of products of parent states with $n-1$ electrons and a single electron state, provided all electrons belong to one shell of angular momentum l.

$$(\alpha SL\|W^{(a,b)}\|\alpha' S'L') = \sum_{\alpha^* S^* L^*} \left\{\begin{matrix} S & a & S' \\ s & S^* & s \end{matrix}\right\} \left\{\begin{matrix} L & b & L' \\ l & L^* & l \end{matrix}\right\} \times (-1)^{S^*+s+S+a+L^*+l+L+b} \times (l^n\alpha SL\{\leftarrow |l^{n-1}\alpha^* S^* L^*)(l^{n-1}\alpha^* S^* L^*| \rightarrow\} l^n\alpha' S'L') \,. \tag{10.8}$$

Here the last line contains the coefficients of fractional parentage, or cfp, for brevity, that are common to the initial and the final state.

With the last equation we have assembled all the general ingredients for the matrix element of a, possibly coupled, tensor operator.

$$\langle SLJM|T^K_Q|S'L'J'M'\rangle = (-1)^{J-M}\begin{pmatrix} J & K & J' \\ -M & Q & M' \end{pmatrix} \tag{10.9}$$
$$\times \frac{[J,K,J']^{1/2}}{[a,b]^{1/2}} \left\{\begin{matrix} S & L & J \\ S' & L' & J' \\ a & b & K \end{matrix}\right\}$$
$$\times \sum_{\alpha^* L^* S^*} \left\{\begin{matrix} S & a & S' \\ s & S^* & s \end{matrix}\right\} \left\{\begin{matrix} L & b & L' \\ l & L^* & l \end{matrix}\right\} (-1)^{S^*+s+S+a+L^*+l+L+b}$$
$$\times (l^n\alpha SL\{|l^{n-1}\alpha^* S^* L^*)(l^{n-1}\alpha^* S^* L^*|\} l^n\alpha' S'L')$$
$$\times (s\|u^{(a)}\|s)(l\|v^{(b)}\|l) \,.$$

From the point of graphical analysis we note four characteristic features of a matrix element of a tensor operator: The $3jm$-symbol from the Wigner–Eckart Theorem, a $9j$-symbol containing the ranks a, b and K, and, finally two $6j$-symbols related to the passage from parent to daughter state. One should, however, not expect the same characteristics for a matrix element between differing configurations, as we shall see.

10.3 X-ray Absorption: an $E1$ Process

The example considered here is based on the description of an x-ray absorption process in which an electron from the d-shell is excited into the f-shell, say.

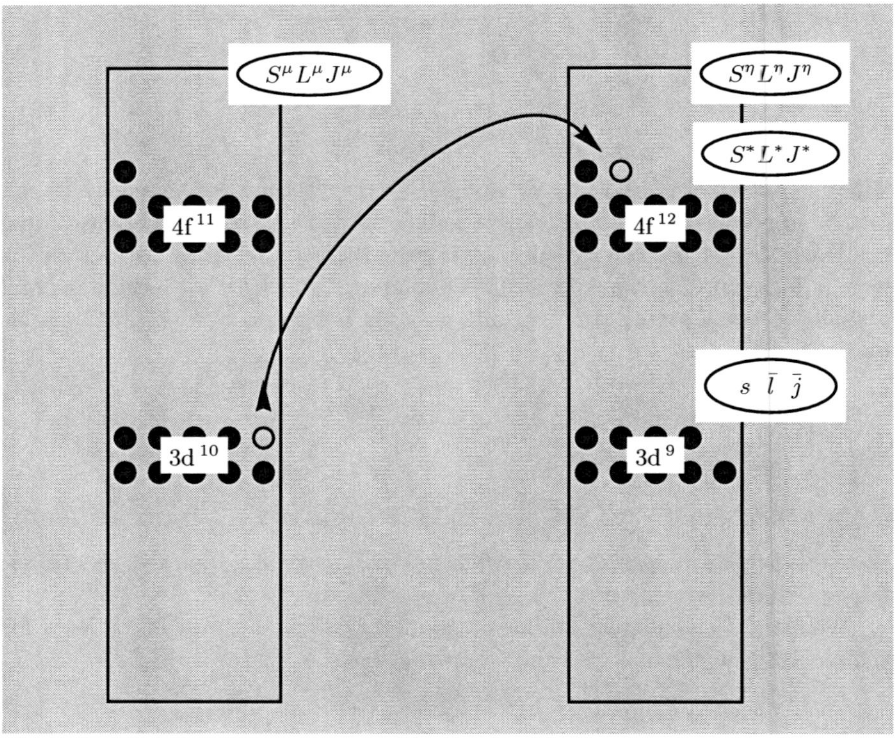

Fig. 10.1. Symbolic sketch of the electron shells in the initial state and the excited state with the corresponding quantum numbers

The calculation of such matrix elements is given in detail in [15], formula (14.88), for a transition involving two active electron shells. This formula simplifies, to some extent, because for this case considered here the electron is excited from a full shell with orbital and spin quantum number zero.

The drawing in Fig. 10.1 displays the variables used to describe the initial state, $S^\mu L^\mu J^\mu$, where the d-shell is closed and has no angular momentum. In the photon absorption process an electron is excited from the d-shell and leaves a hole, $s\bar{l}$, which, subject to spin-orbit coupling, is characterized by $\bar{j}$. The photo-excited electron becomes part of the f-shell with new angular momenta $S^* L^* J^*$. The quantum numbers for the complete, f^{12}, excited state, $S^\eta L^\eta J^\eta$, result from coupling the hole and the f-shell.

The spherical tensor operator T_Q^K contains in this case the electric dipole (E1) operator $\boldsymbol{r}^{(1)} = r^{(1)} C^{(1)}$. The reduced matrix element has the form

$$
\begin{aligned}
&((00, l^{k-1}\alpha^\mu S^\mu L^\mu)S^\mu L^\mu, J^\mu \| \sum_i r^{(i)} C^{(i)} \| (s, \bar{l}, l^k \alpha^* S^* L^*) S^\eta L^\eta, J^\eta) \\
&= \delta_{S^\mu S^\eta} (-1)^{k+L^*+\bar{l}+S^\mu+J^\mu} \frac{(nk[L^*, S^*, L^\eta, J^\mu, J^\eta])^{1/2}}{[\bar{l}, s, S^\mu]^{1/2}} \qquad (10.10) \\
&\times \begin{Bmatrix} L^\mu & S^\mu & J^\mu \\ J^\eta & 1 & L^\eta \end{Bmatrix} \begin{Bmatrix} 1 & l & \bar{l} \\ L^* & L^\eta & L^\mu \end{Bmatrix} \\
&\times (l^{k-1}\alpha^\mu S^\mu L^\mu |\} l^k \alpha^* S^* L^*)(\bar{l} \| r^{(1)} C^{(1)} \| l) \,.
\end{aligned}
$$

The bottom line contains $(l^{k-1}\alpha^\mu S^\mu L^\mu |\} l^k \alpha^* S^* L^*)$, a coefficient of fractional parentage connecting the parent state of l^{k-1} electrons to the state of l^k with the extra excited electron.

The reduced matrix element in (10.10) is derived in the SL-coupling scheme and it would make sense to recouple the angular momenta to express the dependence on $\bar{\jmath}$ of the electron hole. The recoupling formula entails a $9j$-symbol and a sum over S^η, L^η.

$$
\begin{aligned}
&|[(s, \bar{l})\bar{\jmath}, (S^*, L^*)J^*]J^\eta M^\eta\rangle \qquad (10.11) \\
&= \sum_{S^\eta L^\eta} [((s, S^*)S^\eta, \bar{l}, L^*)L^\eta]J^\eta M^\eta\rangle [\bar{\jmath}, J^*, S^\eta, L^\eta]^{1/2} \begin{Bmatrix} s & \bar{l} & \bar{\jmath} \\ S^* & L^* & J^* \\ S^\eta & L^\eta & J^\eta \end{Bmatrix} .
\end{aligned}
$$

Because the full d-shell has neither spin nor orbital moment, the recoupling has no effect for the initial state. Due to the Kronecker-δ, $\delta_{S^\mu S^\eta}$, the sum over S^η is trivial and only the sum over L^η remains in the reduced matrix element.

$$
\begin{aligned}
&(S^\mu L^\mu J^\mu \| \sum_i r^{(i)} C^{(i)} \| \bar{\jmath} J^* J^\eta) \\
&= (-1)^{k+L^*+\bar{l}+S^\mu+J^\mu} \frac{(k[L^*, S^*, J^\mu, J^\eta])^{1/2}}{[S^\mu]^{1/2}} [\bar{\jmath}, J^*, S^\mu]^{1/2} \qquad (10.12) \\
&\times \sum_{L^\eta} (2L^\eta + 1) \begin{Bmatrix} s & \bar{l} & \bar{\jmath} \\ S^* & L^* & J^* \\ S^\mu & L^\eta & J^\eta \end{Bmatrix} \begin{Bmatrix} \bar{l} & L^* & L^\eta \\ L^\mu & 1 & l \end{Bmatrix} \begin{Bmatrix} J^\eta & S^\mu & L^\eta \\ L^\mu & 1 & J^\mu \end{Bmatrix} \\
&\times (l^{k-1}\alpha^\mu S^\mu L^\mu |\} l^k \alpha^* S^* L^*)(\bar{l} \| r^{(1)} C^{(1)} \| l) \,.
\end{aligned}
$$

In (10.12) we recognize a sum over a $9j$- and two $6j$-symbols with the summation variable L^η. With some rearrangements in the $9j$- and $6j$-symbols we find correspondence with the sum from Sect. 8.3.2:

$$
\begin{aligned}
&\sum_x (2x+1) \begin{Bmatrix} e & f & x \\ a & b & p \\ c & d & q \end{Bmatrix} \begin{Bmatrix} f & e & x \\ l & k & h \end{Bmatrix} \begin{Bmatrix} p & q & x \\ k & l & g \end{Bmatrix} \\
&\Rightarrow \sum_{L^\eta} (2L^\eta + 1) \begin{Bmatrix} J^\eta & S^\mu & L^\eta \\ \bar{\jmath} & s & \bar{l} \\ J^* & S^* & L^* \end{Bmatrix} \begin{Bmatrix} S^\mu & J^\eta & L^\eta \\ 1 & L^\mu & J^\mu \end{Bmatrix} \begin{Bmatrix} \bar{l} & L^* & L^\eta \\ L^\mu & 1 & l \end{Bmatrix} . \qquad (10.13)
\end{aligned}
$$

This correspondence also defines the variable substitutions that have to take place if we want to retrieve the graphical part of the matrix element from the corresponding figure in Sect. 8.3.2.

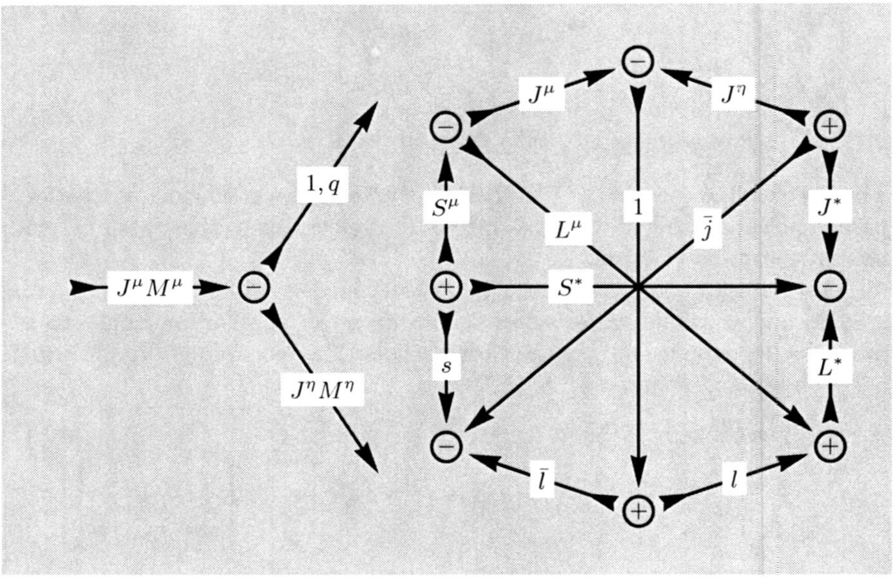

Fig. 10.2. In order to obtain this diagram from Fig. 8.62 several line reversals and node sign changes are required. In addition to that, the following variable names are replaced: $a \to \bar{j}$, $b \to s$, $c \to J^*$, $d \to S^*$, $e \to J^\eta$, $f \to S^\mu$, $g \to l$, $h \to J^\mu$, $k \to L^\mu$, $l \to 1$, $p \to \bar{l}$, and $q \to L^*$. The variable q will in the following describe the projection of the dipole operator. The phase is not taken into account in this chapter

The $3jm$-symbol completes the matrix element of a vector operator of rank 1 (dipole operator) as required by the Wigner–Eckart Theorem. Thus, based on prior results we have recovered the graphical parts of the matrix element describing the absorption of an x-ray photon. An equivalent $12j$-symbol has been used by Thole and van der Laan in their analysis of x-ray absorption in [16].

10.4 Resonance-Enhanced X-Ray Scattering: *E*1-*E*1-Process

Proceeding from absorption to scattering we note a weak contribution to the x-ray scattering amplitude f that is described by a term from second-order perturbation theory. Due to the energy denominator present in this term, the weak contribution may be strongly enhanced if the x-ray photon is in resonance with the absorbing and emitting states of the atom [37].

$$f(\mu,\mu') \propto \sum_{\eta} \frac{\langle\mu|\sum_i \boldsymbol{\epsilon}\cdot\boldsymbol{r_i}|\eta\rangle\langle\eta|\sum_i \boldsymbol{\epsilon}\cdot\boldsymbol{r_i}|\mu'\rangle}{\hbar\omega-(E^{\eta}-E^{\mu})+\mathrm{i}\Gamma_{\eta}} . \tag{10.14}$$

If the photon energy $\hbar\omega \sim (E^{\eta}-E^{\mu})$ is close to the energy difference of the levels described by μ and η the contribution to the atomic scattering amplitude will show a resonance with a width determined by the life-time $\hbar/\Gamma_{\eta}$ of the virtual, intermediate state. The product of matrix elements in the numerator refer to the absorption and emission process depicted in the following figure. d- and f-states are used as a concrete example, our calculation is for general equilibrium and virtual states.

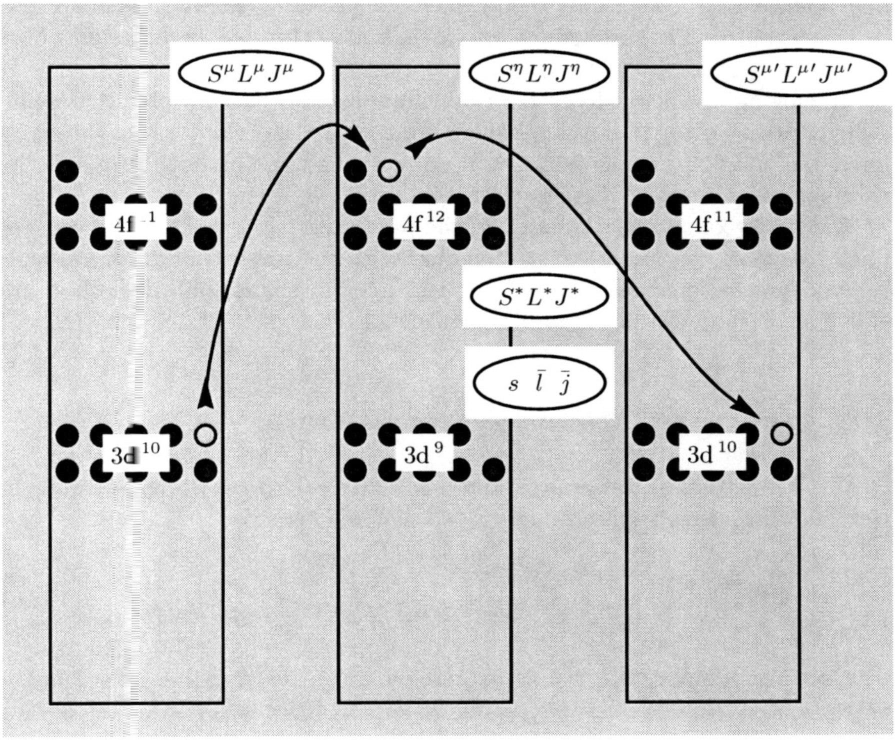

Fig. 10.3. Symbolic sketch for a second-order process. One electron from the d-shell absorbs a photon and is excited into an intermediate state. The photon is re-emitted by an electron returning to the d-shell of the final state

Depending on the number of electrons in the f-shell there may be a great number of states $S^{*}L^{*}J^{*}M^{*}$ available for coupling with the hole $s\bar{l}\bar{j}\overline{m}$ in the d-shell into an even greater number of states $S^{\eta}L^{\eta}J^{\eta}M^{\eta}$.

$$f(\mu,\mu') \propto \sum_{\eta?} \frac{\left\{\begin{array}{l}\langle\mu|\sum_i \boldsymbol{\epsilon}\cdot\boldsymbol{r_i}|(s\bar{l})\bar{j}(S^{*}L^{*})J^{*}J^{\eta}M^{\eta}\rangle \\ \quad\times\langle(s\bar{l})\bar{j}(S^{*}L^{*})J^{*}J^{\eta}M^{\eta}|\sum_i \boldsymbol{\epsilon}\cdot\boldsymbol{r_i}|\mu'\rangle\end{array}\right\}}{\hbar\omega-(E^{\eta}-E^{\mu})+\mathrm{i}\Gamma_{\eta}} . \tag{10.15}$$

In principle, all virtual, intermediate states have separate energy levels, in reality, however, many energies will be close to each other and may be grouped together and represented by a few values. Thus it seems a sensible approximation to separate the analysis of the numerator and denominator in such a way as to perform a *selective* sum, indicated by the question mark in (10.15), over the intermediate states which retains dependence on just a few quantum numbers of physical relevance.

In (10.15) we have specified the angular momentum variables present in the intermediate state. From these $\bar{j}$, the total angular momentum of the hole, has two values $\bar{l} \pm 1/2$ for non-zero orbital angular momentum $\bar{l}$. If the dependence on $\bar{j}$ is kept this is equivalent to gathering the energy levels into just two groups. The restricted summation will in this case include sums over M^η, J^η and J^*.

Removing the dependence on $\bar{j}$ by summing over this variable represents all energies by a single mean value for the energy denominator. On the other hand, we will find also an option to keep a dependence on both $\bar{j}, \overline{m}$ and the corresponding number of energy denominators.

The approximations discussed in the preceding paragraphs have been applied frequently in the literature together with the method of graphical analysis. A few selected citations which also refer to the graphical method are Carra et al. [24], van der Laan [20], and Thole et al. [21].

10.5 Equivalent Atomic Tensor Operator

In the framework of the approximations discussed above we concentrate in the following just on the product of two matrix elements

$$Z(\mu,\mu') \propto \sum_{\eta?} \left\{ \begin{array}{l} \langle\mu|\sum_i \boldsymbol{\epsilon}\cdot\boldsymbol{r_i}|(s\bar{l})\bar{j}(S^*L^*)J^*J^\eta M^\eta\rangle \\ \qquad \times\langle(s\bar{l})\bar{j}(S^*L^*)J^*J^\eta M^\eta|\sum_i \boldsymbol{\epsilon}\cdot\boldsymbol{r_i}|\mu'\rangle \end{array} \right\} . \tag{10.16}$$

The, rather complicated, scalar expression in (10.16) can be recast into a *single* matrix element between states μ and μ' for a scalar product of two tensor operators. The first tensor $X^K_Q(\boldsymbol{\epsilon}',\boldsymbol{\epsilon}) = (\boldsymbol{\epsilon}' \otimes \boldsymbol{\epsilon})^K_Q$ depends on the photon properties and the second, usually called atomic tensor, on the atomic electron state. Specific values of X are found in [37].

$$\begin{aligned} &\sum_{\eta?} \left\{ \begin{array}{l} \langle\mu|\sum_i \boldsymbol{\epsilon}\cdot\boldsymbol{r_i}|(s\bar{l})\bar{j}(S^*L^*)J^*J^\eta M^\eta\rangle \\ \qquad \times\langle(s\bar{l})\bar{j}(S^*L^*)J^*J^\eta M^\eta|\sum_i \boldsymbol{\epsilon}\cdot\boldsymbol{r_i}|\mu'\rangle \end{array} \right\} \\ &\equiv \sum_{KQ} (-1)^Q X^K_{-Q}(\boldsymbol{\epsilon}',\boldsymbol{\epsilon})\langle\mu|T^K_Q|\mu'\rangle \,. \end{aligned} \tag{10.17}$$

The equivalence in (10.17) can be viewed as a statement that the left-hand side is a scalar quantity of the form $\{X^K \otimes T^K\}^0_0$. This equivalence has first been explored by Luo et al. [33]. In the following we will undertake the task of confirming the properties of the equivalent atomic tensor with methods of the graphical theory. The tensor properties, as specified in (10.9), require the characteristic $3jm$-symbol of the Wigner–Eckart Theorem, a $9j$-symbol and two $6j$-symbols for the unit tensor $W^{(a,b)}$. For this purpose it will be sufficient to follow the transformations of the diagrams *without* keeping track of the phases introduced by node sign changes or line reversals. As an additional simplification, the circular diagrams representing factors like $(2x + 1)$ etc., see e.g. Fig. 7.4, will be left out from the diagrams. Of course, in any translation into algebraic expressions these factors have to be removed or inserted, whether the summation is carried out or remains in the final result, respectively.

For the graphical representation of the product of two matrix elements we use the result for a single matrix element as given in Fig. 10.2.

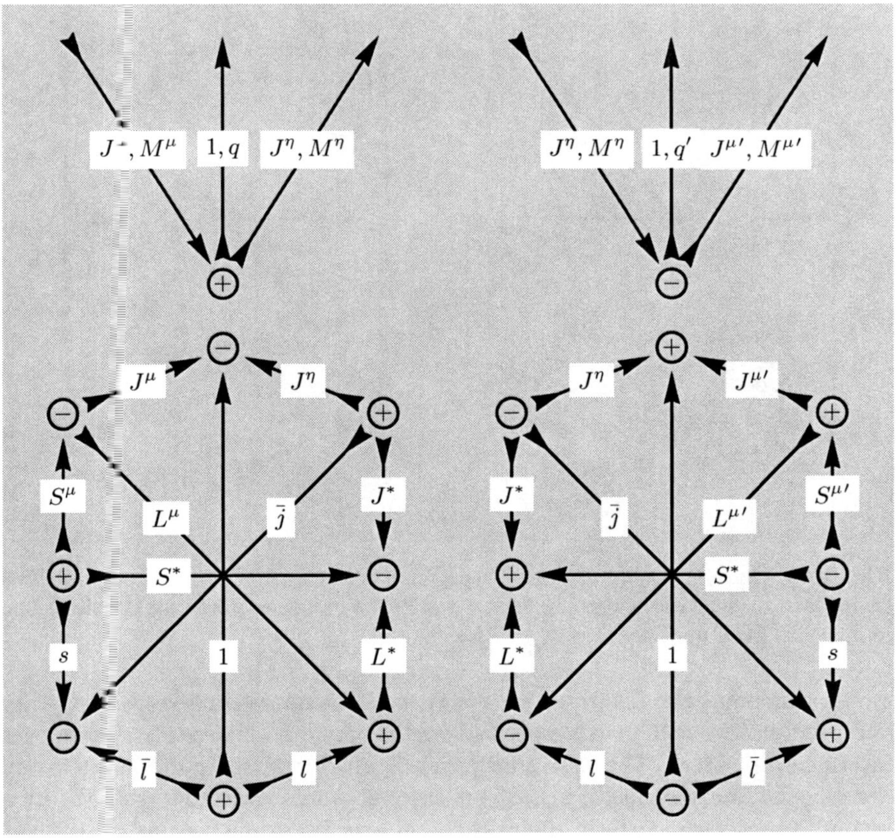

Fig. 10.4. Note the node sign change in the left $3jm$-symbol. The $12j$(I)-symbols have been rotated. The right $12j$-symbol is obtained as a mirror image from the left symbol. The direction of the lines labelled 1 has been reversed in the closed diagrams

The diagrams in Fig. 10.4 follow from Fig. 10.2 and represent the product of the two matrix elements displayed in (10.17). In the right part of the diagram L^μ, S^μ and J^μ are replaced by the primed angular momentum variables of the final state μ'.

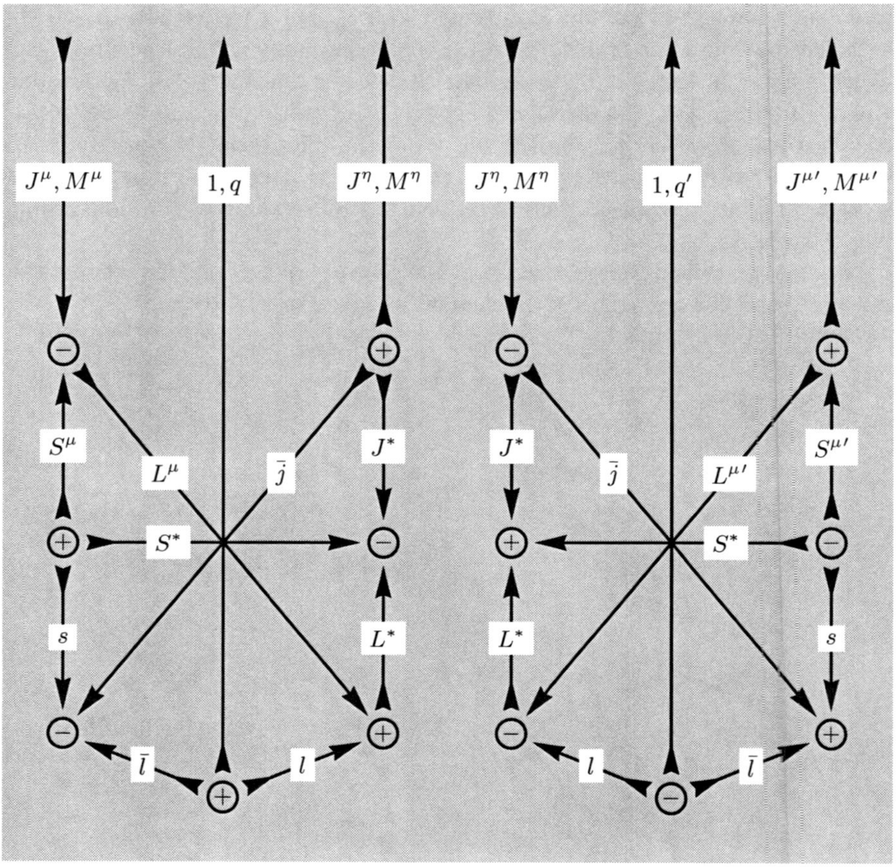

Fig. 10.5. Both $3jm$-symbols have been joined to their $12j$(I)-symbol partner. The temporary reversal of the ingoing lines of the $3jm$-symbols required for the three-line joining has been undone

At this point the diagrams are ready for the summation over η and other variables and we will have to specify how to carry out the 'selective' sum over intermediate states. The first step removes all dependence on the quantum numbers of the intermediate state η through sums over M^η and J^η. In a second step we deal with the quantum numbers of the shell holding the excited electron. We have, however, to retain the dependence on S^* and Ls because the graphical method does not include the coefficients of fractional parentage seen, e.g., in (10.8) or (10.9). A sum over M^* and J^* leaves just the dependence on $\bar{j}$ from the properties of the intermediate state.

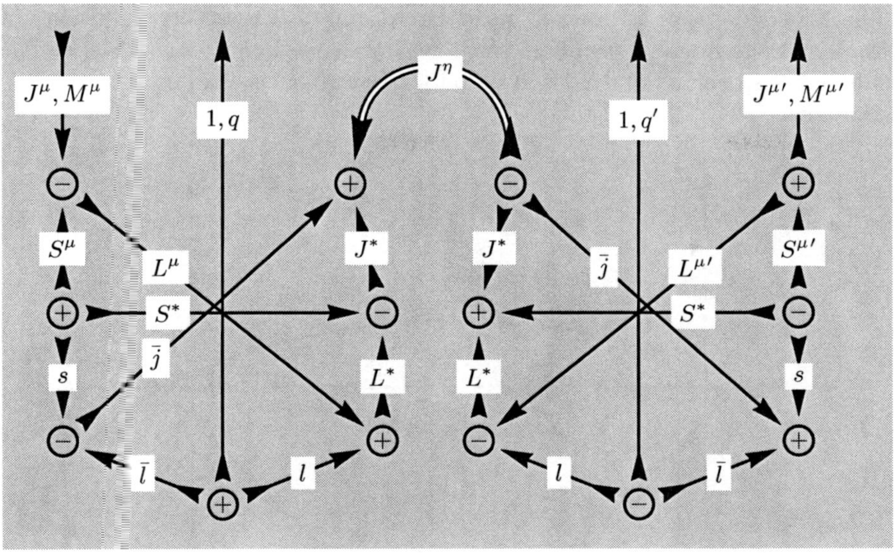

Fig. 10.6. In the left diagram the directions of the $\bar{j}$- and the J^*-line are reversed in preparation for the sum over J^η. The double line indicates the double sum over M^η and J^* and is also reversed

The circular diagrams representing, say, the factor $(2J^\eta+1)$ required for a sum over J^η, as well as similar factors in the following figures, are not displayed.

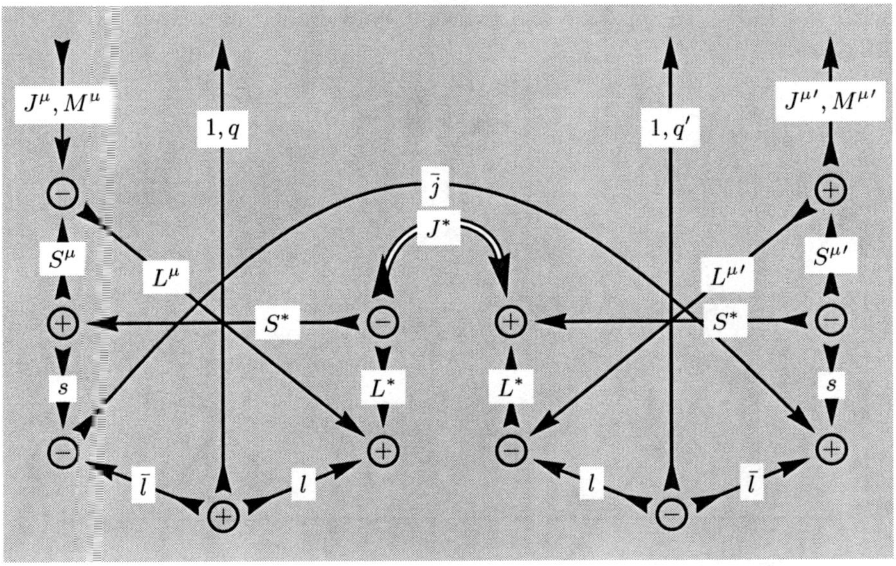

Fig. 10.7. The previously separate diagrams are joined to a single diagram. There is no dependence on J^η, M^η. The double line indicates the next sum over J^*. The directions of the L^*- and S^*-line are reversed in the left part of the diagram

The dependence on L^* and S^* must be retained because the coefficients of fractional parentage, invisible in the diagram, depend on these variables. The summations over S^* and L^* will be performed after the graphical analysis is completed.

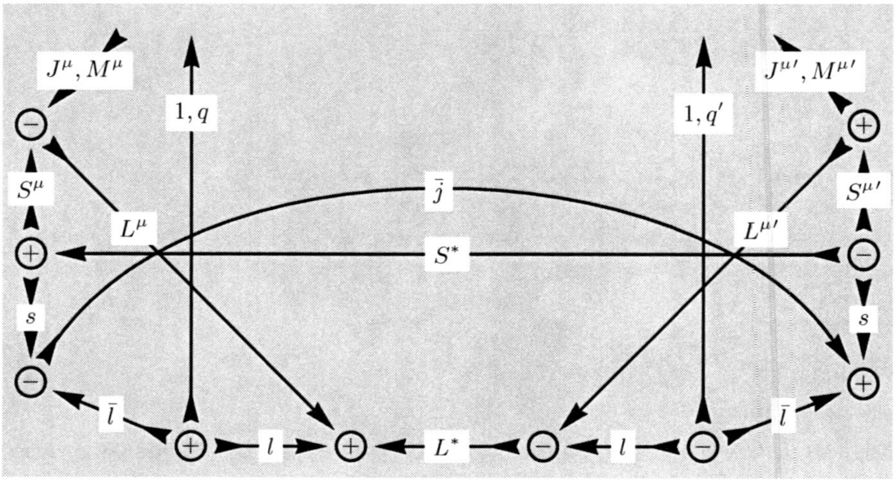

Fig. 10.8. The summations have removed several nodes and simplified the diagram considerably

Further progress may be made through the introduction of new summations with the intent to separate the external lines from the main body of the diagram.

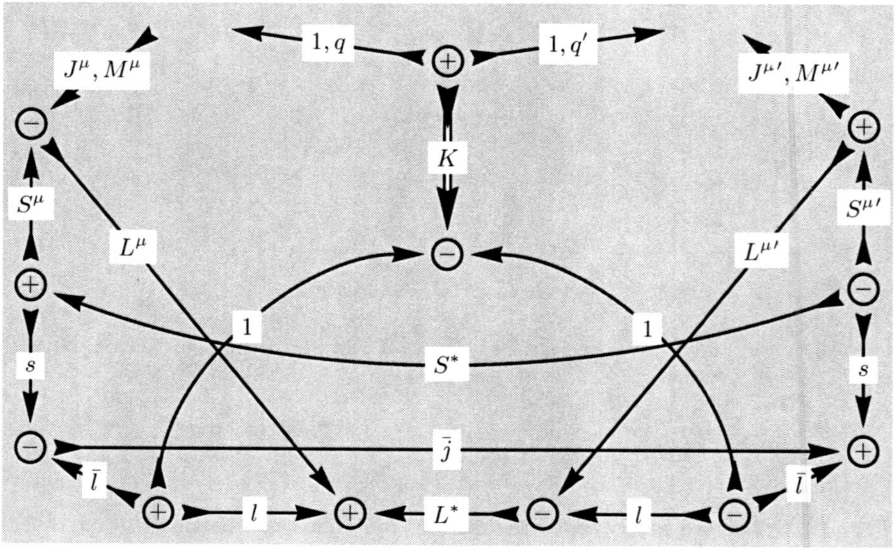

Fig. 10.9. Two new nodes, joined by a double line K, are introduced and contract the two external lines representing the rank of the vector operator $r^{(1)}C^{(1)}$

At this stage several options are available. It is possible to perform a three line separation into a diagram with the external lines and a closed diagram. One of the resulting $3jm$-symbols has the form required by the Wigner–Eckart Theorem for a tensor operator of rank K. In this case the diagram will retain a dependence on the total angular momentum $\bar{j}$ of the core hole. On the other hand, it seems worthwhile, from a physical point of view, to pursue an attempt to keep not only the dependence on $\bar{j}$ but to display the additional dependence on the magnetic quantum number $\overline{m}$. This will be the most general result that contains all the simpler options as we will find in subsequent sections.

10.6 Retaining Dependence on $\bar{j}$ and $\overline{m}$

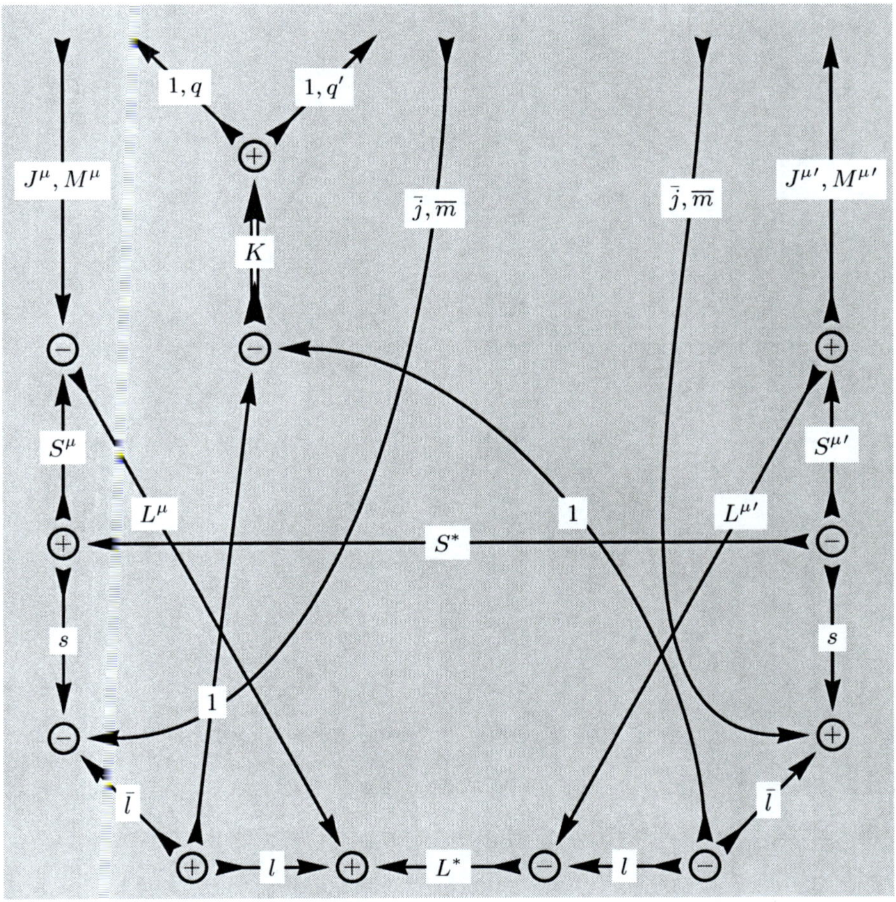

Fig. 10.10. The direction of the K-line and one external $\bar{j}$-line is reversed. By breaking the internal $\bar{j}$-line into two parts the diagram recovers the dependence on $\overline{m}$

Introducing the dependence on $\overline{m}$ has led to an increase of the number of external lines. The diagram is now more general at the price of higher complexity. The separation of a closed diagram will require more auxiliary summations, as we shall see.

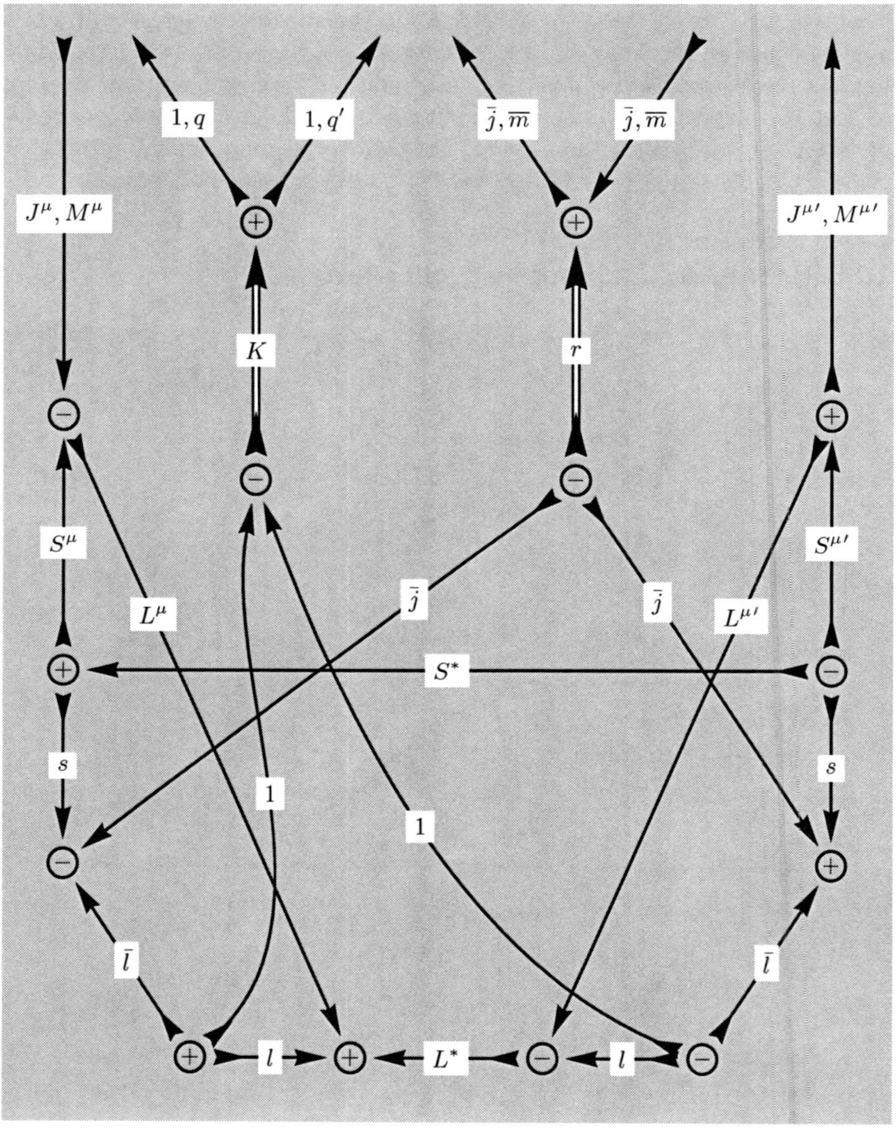

Fig. 10.11. The external $\bar{j}$-line has been returned to its previous direction

The introduction of the r-line and the corresponding sum over r is a first step. Another auxiliary sum is required to reduce the number of lines and to prepare the diagram for a three-line separation.

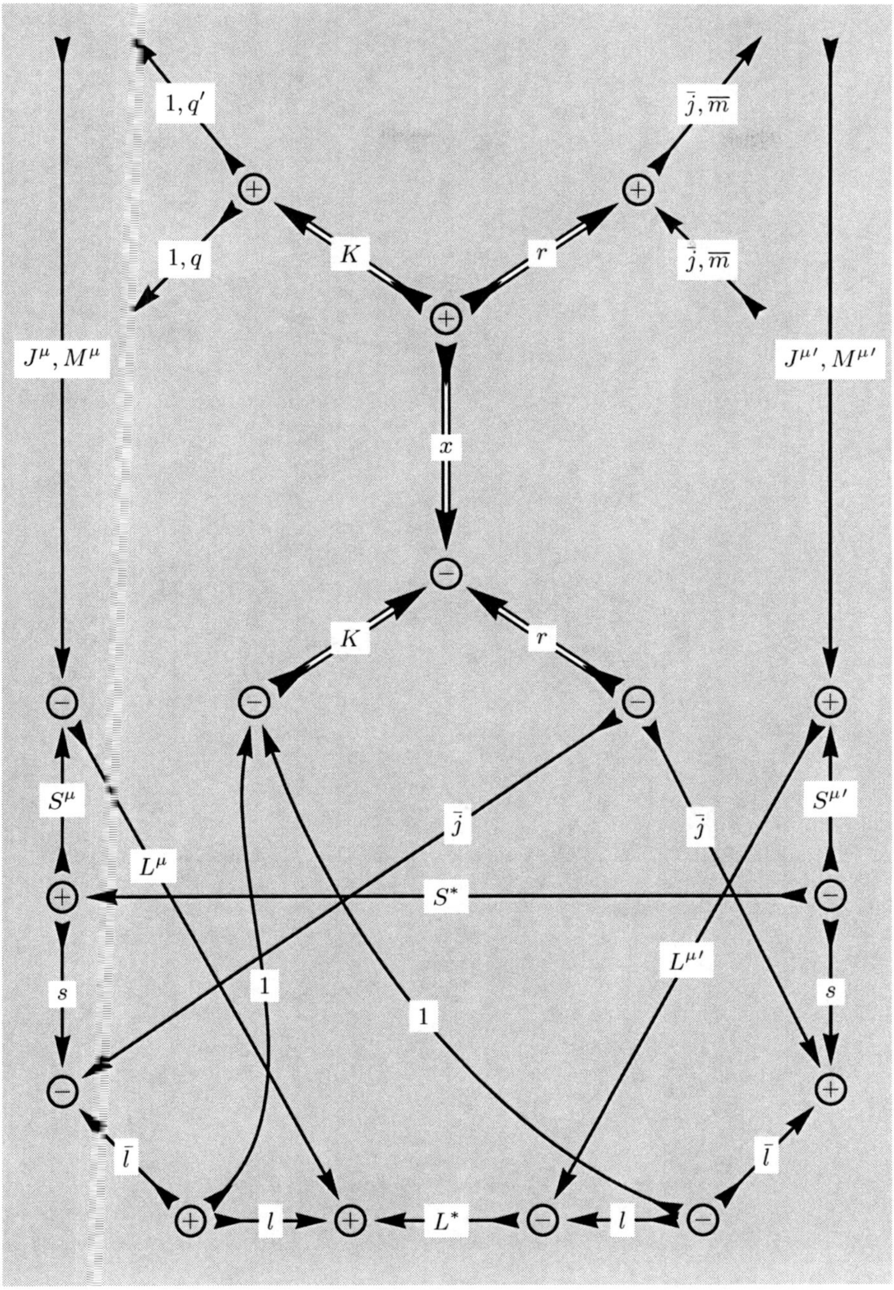

Fig. 10.12. The external $J^{\mu\prime}$-line is temporarily reversed as required to allow a three-line separation

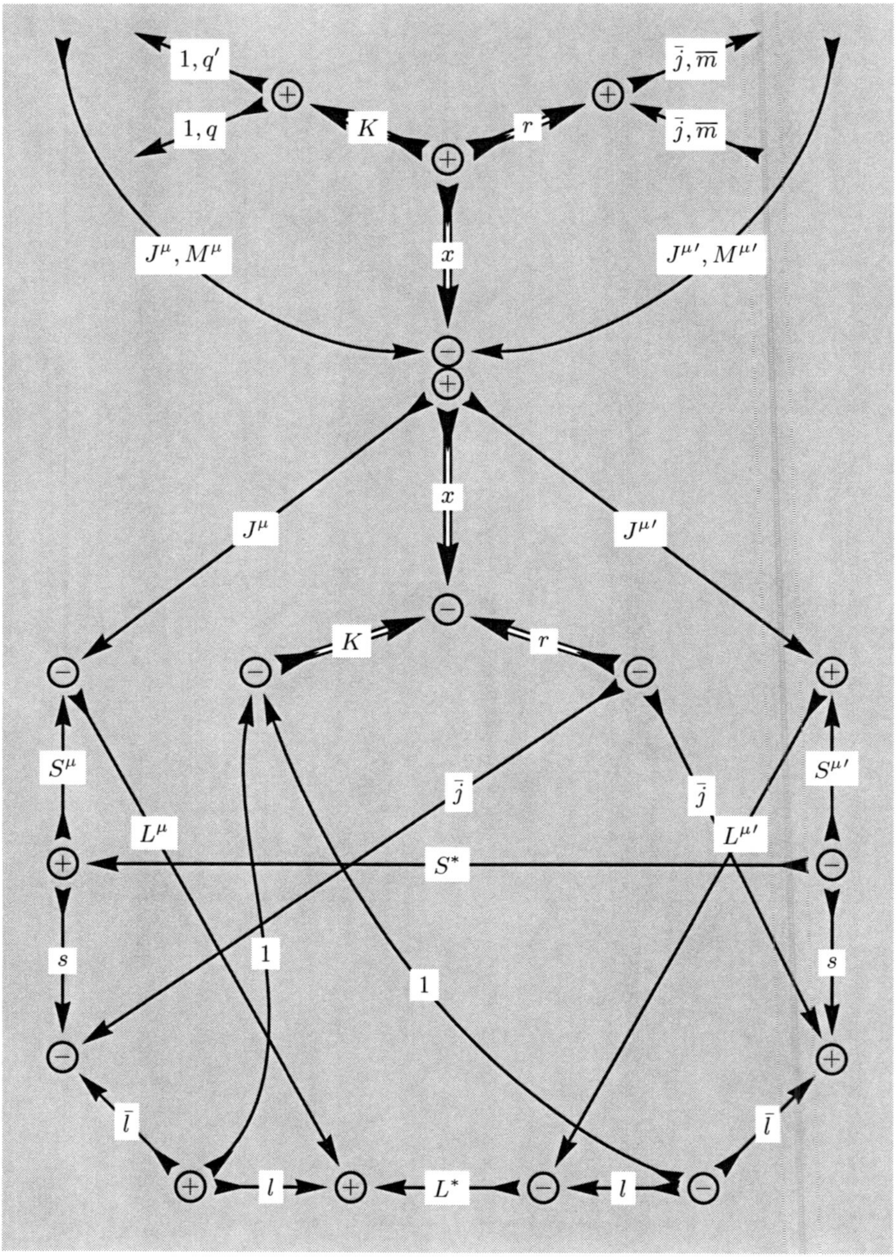

Fig. 10.13. The three-line separation collects all external lines in the upper diagram. Both diagrams are subject to the sums over K, r and x

Among emerging $3jm$-symbols one finds the signature of a Wigner–Eckart $3jm$-symbol for a tensor operator of rank x. The bottom diagram, which corresponds to a unspecific $21j$-symbol, not previously encountered, is rearranged to prepare it for subsequent transformations.

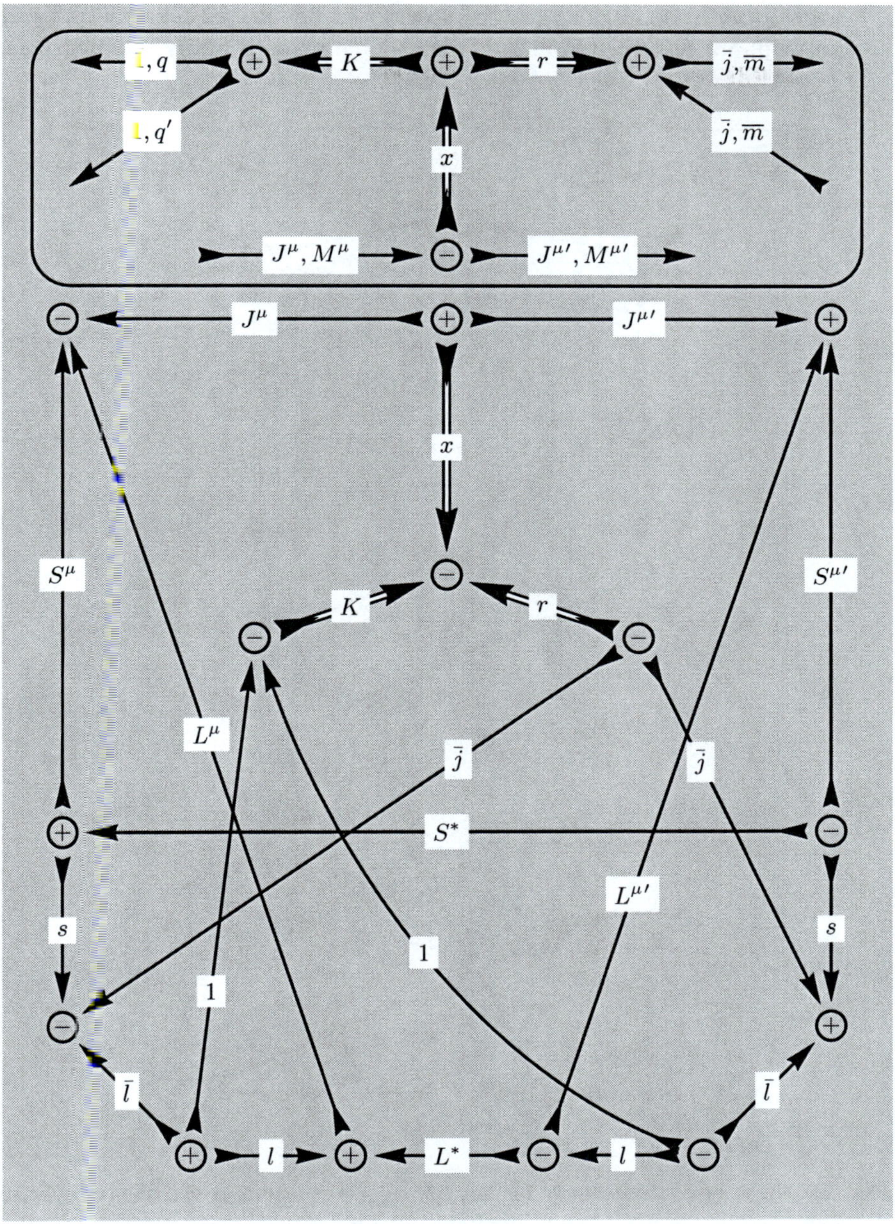

Fig. 10.14. Note that the external $J^{\mu\prime}$-line has returned to its previous direction. The external lines are collected in a group of four $3jm$-symbols. The directions of the L^{μ}, $L^{\mu\prime}$ and one x-line have been changed

Two more auxiliary sums will allow the separation of a $9j$-symbol present in the unit tensor $W^{(a,b)x}$ for a tensor operator of rank x.

Starting with Fig. 10.15 the top part of the figure containing four $3jm$-symbols will not be repeated in the following figures to simplify the diagrams.

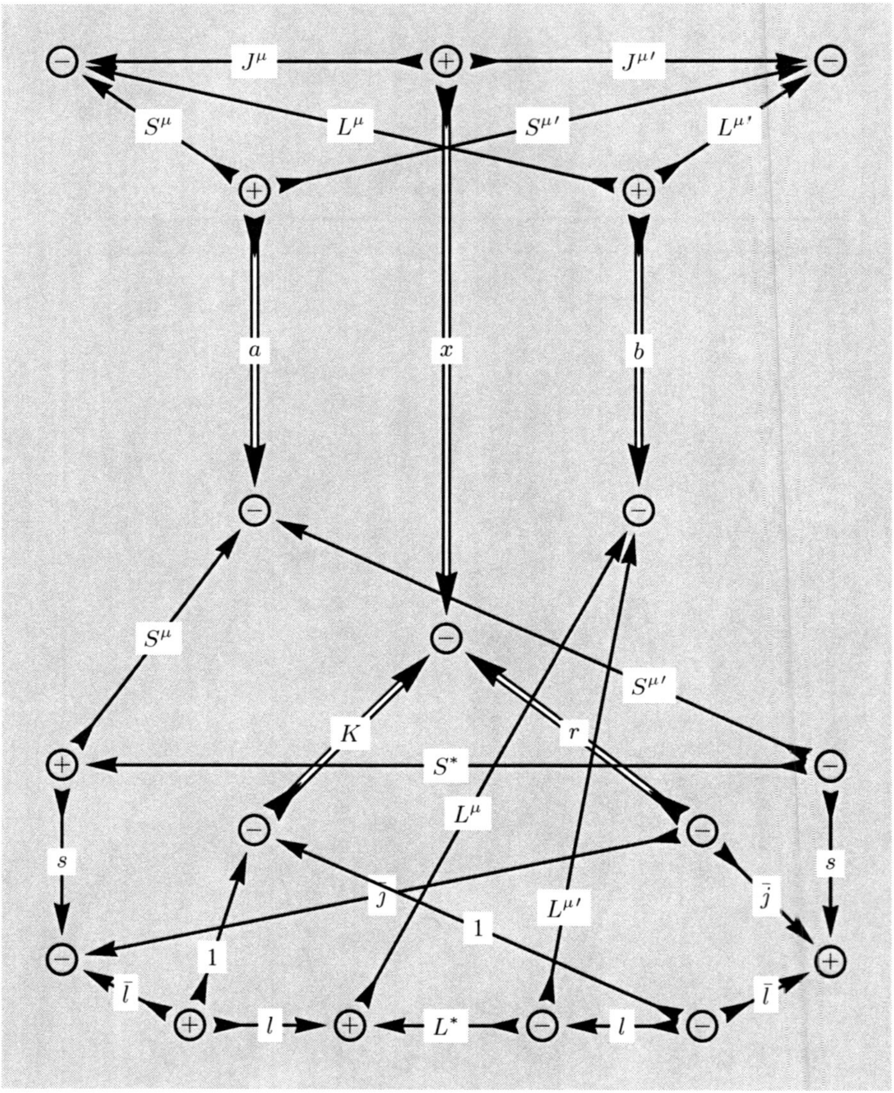

Fig. 10.15. A node sign change for the top right-hand node is not displayed here. As mentioned before, the circular diagrams for the factors $(2a + 1)$ or $(2b + 1)$ etc. are not explicitly included in the diagrams

The emerging $9j$-symbol is characteristic for the unit tensor $W^{(a,b)x}$ appropriate for a coupled tensor operator with spinor rank a and orbital rank b.

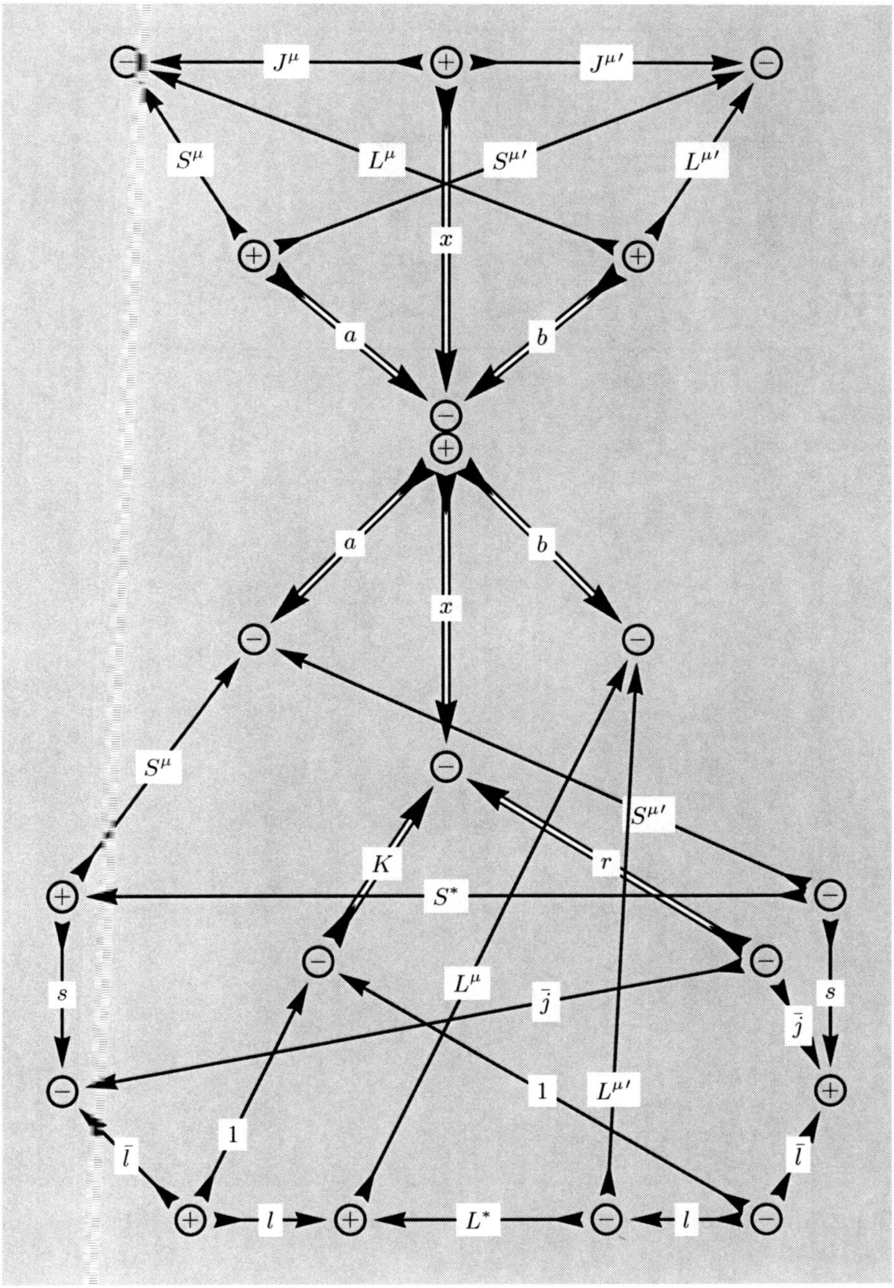

Fig. 10.16. A three-line separation leads to a $9j$-symbol and a general $21j$-symbol

Although not easily discernible, two triangular structures are contained in the bottom diagram and these will be moved to the left and to the right.

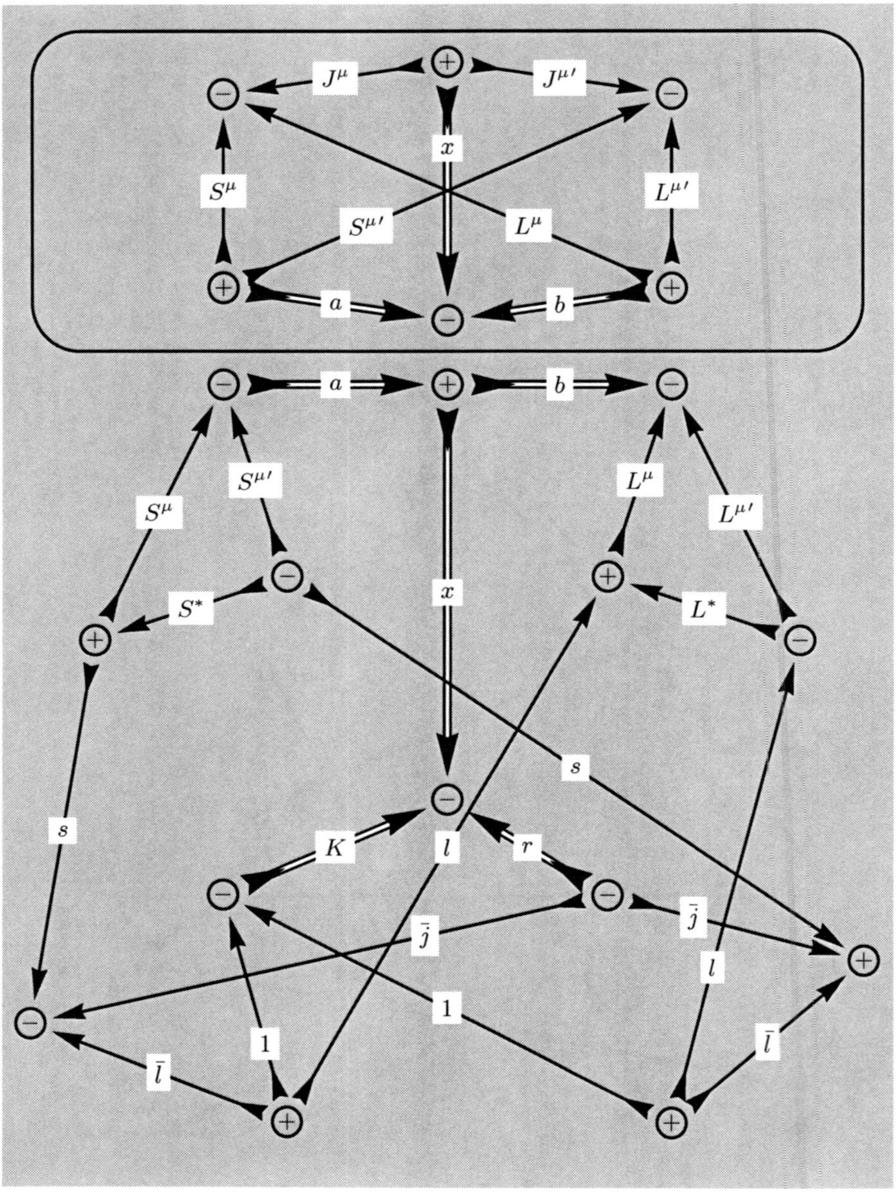

Fig. 10.17. the direction of one a-line is reversed here. The bottom right node sign has changed

Both triangles can be separated from the main diagram and will contain the dependence on L^* and S^*, respectively.

From now on the separated $9j$-symbol at the top of the figure will not be repeated in the following figures to simplify the diagrams.

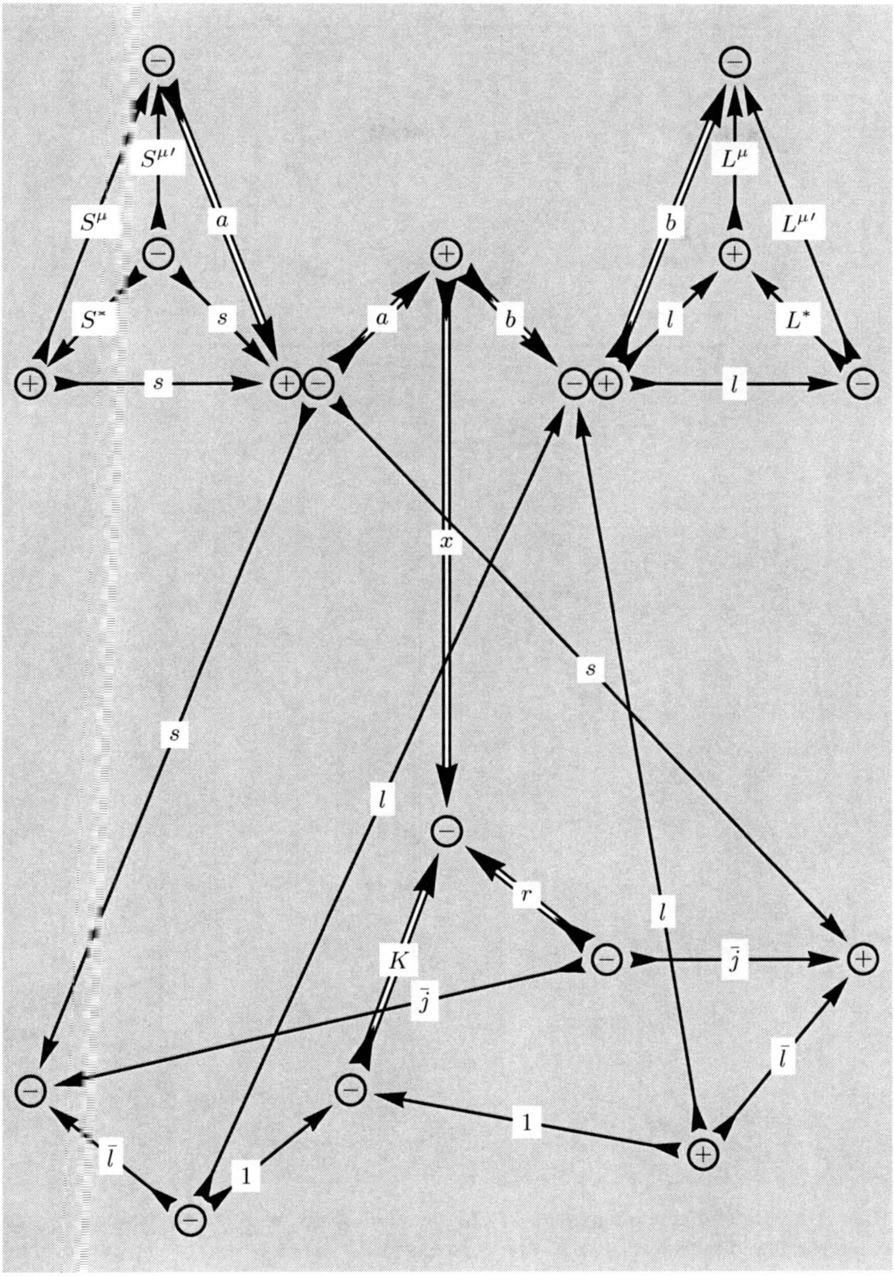

Fig. 10.18. Two $6j$-symbols are created by the two separations

The bottom diagram is rearranged in preparation for a line reduction that allows the separation of the variable x from the remaining diagram.

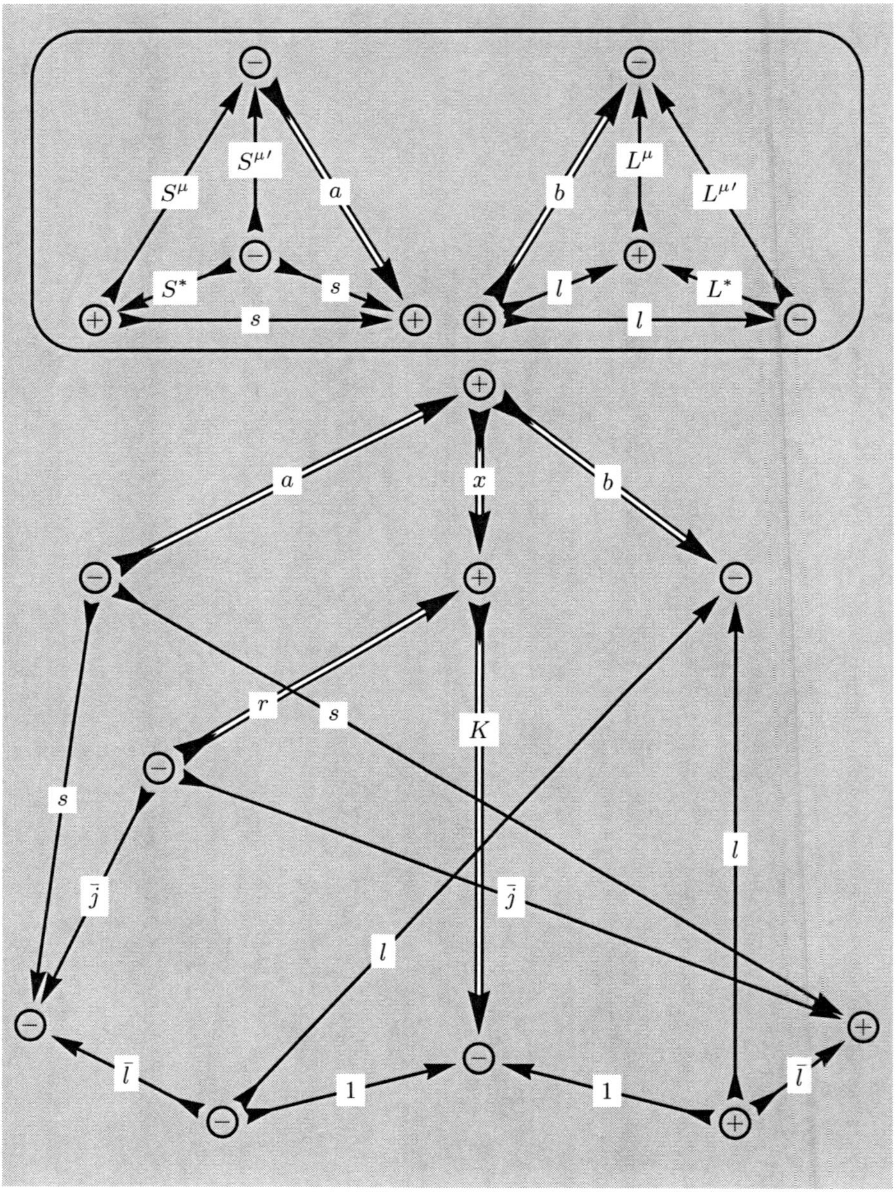

Fig. 10.19. The rearrangement of the bottom diagram effected two node sign changes. The direction of the K-line is reversed

In order to remove the dependence on x from the bottom diagram we contract the a- and r-line with a sum over the additional variable y.

In the following figures, the two separated $6j$-symbols at the top of Fig. 10.19 will no longer be repeated.

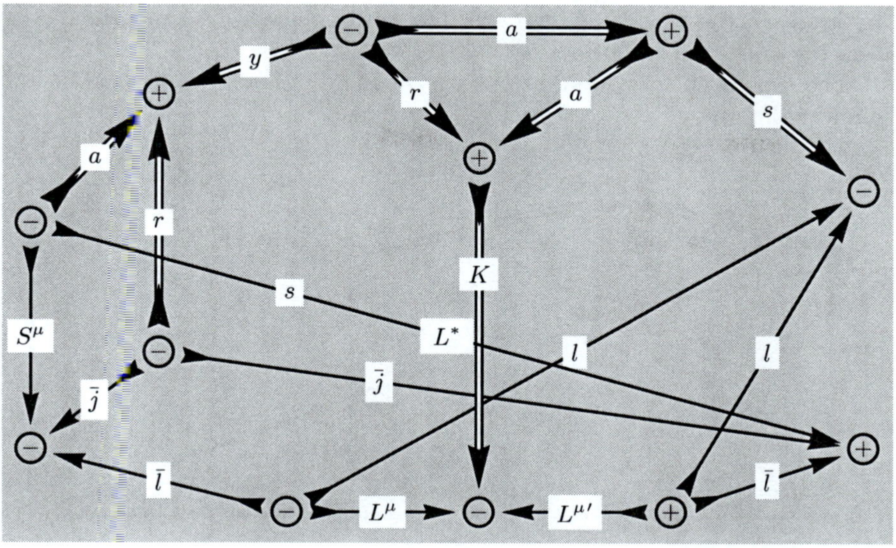

Fig. 10.20. The triangle with the r, x and a-line is now separable and leads to a new $6j$-symbol

The bottom diagram is ready for a three line separation creating an additional $6j$-symbol.

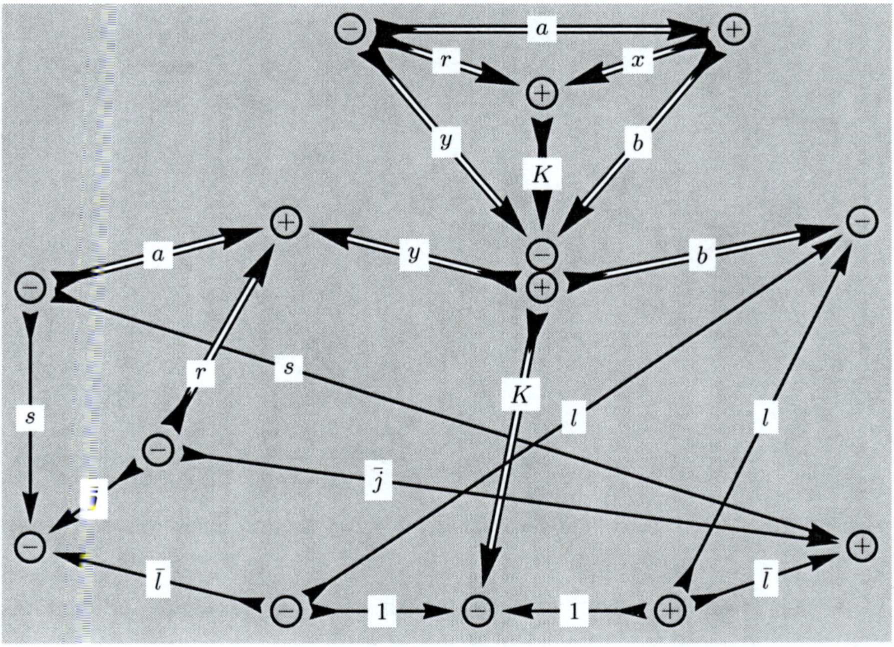

Fig. 10.21. The newly separated $6j$-symbol contains all auxiliary variables introduced so far

As intended, the new $6j$-symbol will remove the remaining dependence on x from the bottom diagram.

The separated $6j$-symbol at the top of Fig. 10.21 will be left out in the following figures.

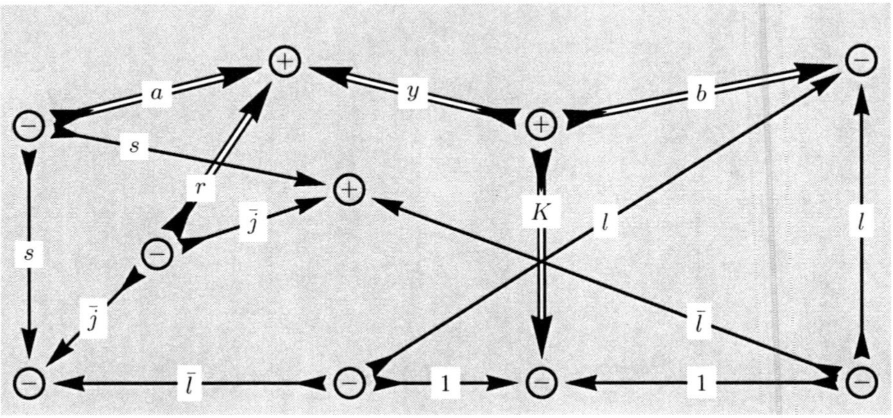

Fig. 10.22. By rearranging the remaining diagram we discover that it is separable. Note the node sign change in the bottom right node

This diagram is separable at the y- and two $\bar{l}$-lines without any further manipulation.

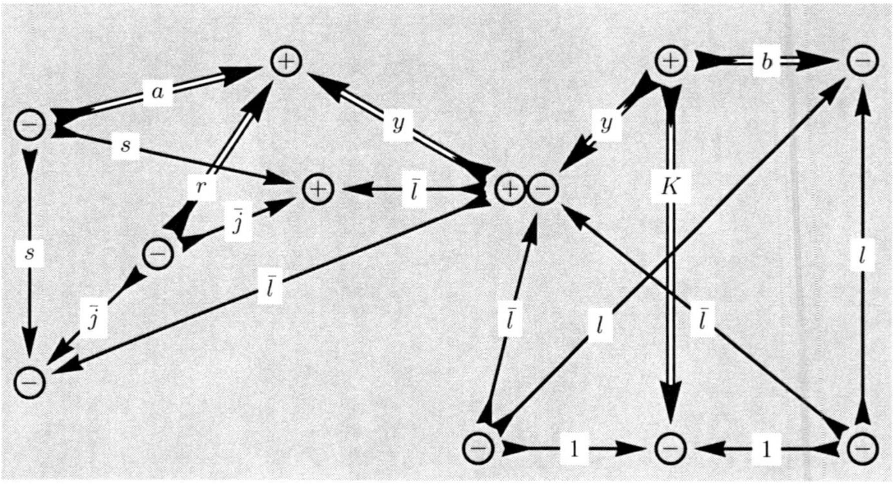

Fig. 10.23. A final three-line separation splits the bottom diagram into two $9j$-symbols

These are, essentially, the final diagrams which only need a few modifications to give the $6j$- and $9j$-symbols their proper form.

It is now time to collect all parts in Fig. 10.24 which have gradually been left out of the figures to simplify the diagrams during the separations.

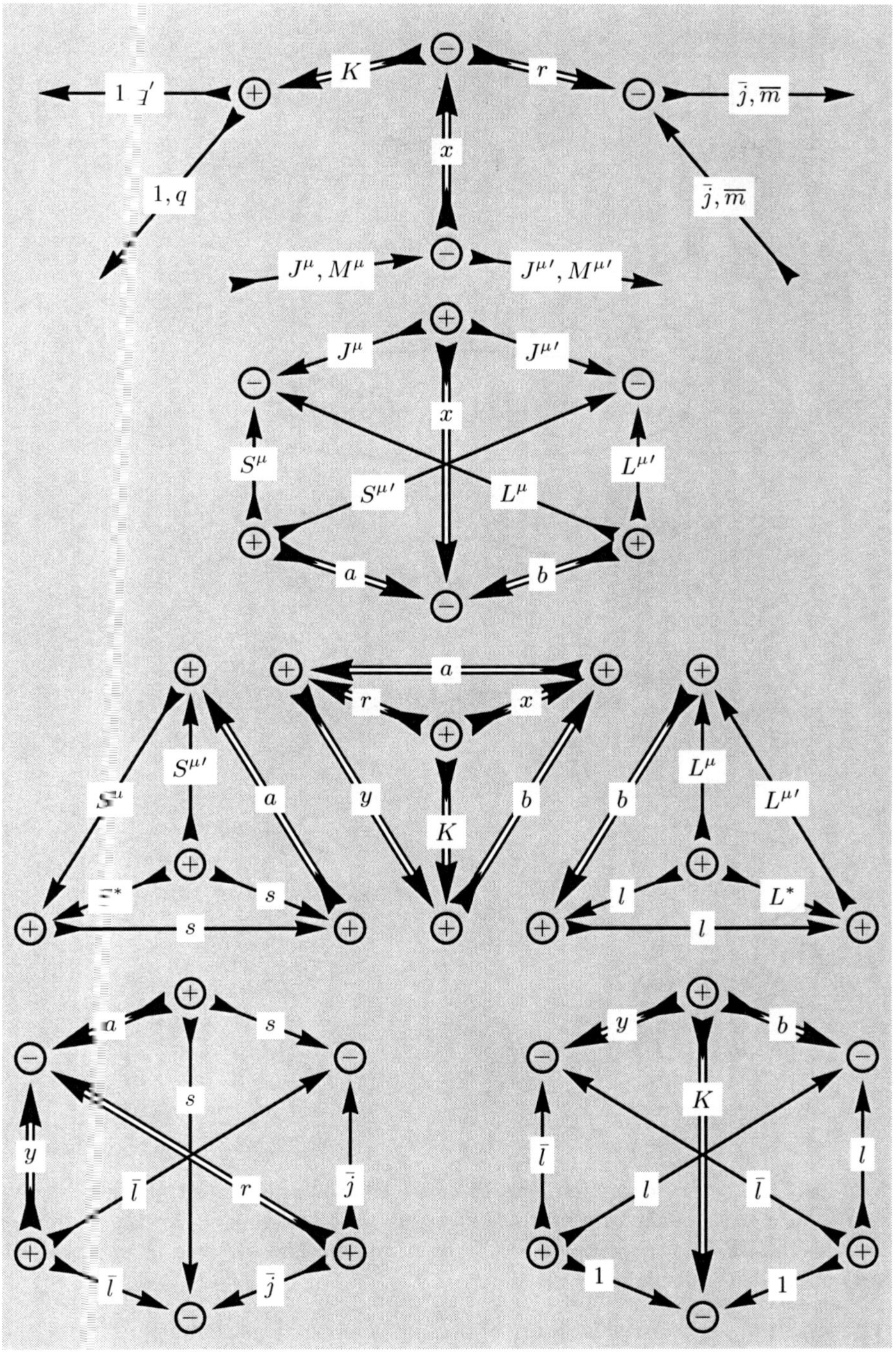

Fig. 10.24. Several node sign changes and line reversals are required to achieve the appropriate form for all $6j$- and $9j$-symbols

$$(-1)^{K-Q}\begin{pmatrix} 1 & K & 1 \\ q & -Q & q' \end{pmatrix} \qquad (-1)^{\bar{j}-\overline{m}+r-0}\begin{pmatrix} \bar{j} & r & \bar{j} \\ -\overline{m} & 0 & \overline{m} \end{pmatrix}$$

$$(-1)^{x-\xi}\begin{pmatrix} K & r & x \\ Q & 0 & -Q \end{pmatrix}$$

$$(-1)^{J^{\mu}-M^{\mu}}\begin{pmatrix} J^{\mu} & x & J^{\mu'} \\ -M^{\mu} & Q & M^{\mu'} \end{pmatrix}$$

$$\begin{Bmatrix} S^{\mu} & L^{\mu} & J^{\mu} \\ S^{\mu'} & L^{\mu'} & J^{\mu'} \\ a & b & x \end{Bmatrix}$$

$$\begin{Bmatrix} K & r & x \\ a & b & y \end{Bmatrix}$$

$$\begin{Bmatrix} s & a & s \\ S^{\mu} & S^{*} & S^{\mu'} \end{Bmatrix} \qquad \begin{Bmatrix} l & b & l \\ L^{\mu} & L^{*} & L^{\mu'} \end{Bmatrix}$$

$$\begin{Bmatrix} s & \bar{l} & \bar{j} \\ s & \bar{l} & \bar{j} \\ a & y & r \end{Bmatrix} \qquad \begin{Bmatrix} 1 & l & \bar{l} \\ 1 & l & \bar{l} \\ K & b & y \end{Bmatrix}$$

Fig. 10.25. The diagrams from Fig. 10.24 are translated into their algebraic equivalents. Properties of the $3jm$-symbols allow only projection $\rho = 0$ for the variable r and this implies also projection $\xi = -Q$ for x. Summations over r, x as well as a, b, y are not explicitly displayed

The marked terms in this figure represent the main contributions from the graphical analysis of the product of two matrix elements. They demonstrate that these terms can be interpreted as signature of a matrix element of an *equivalent* tensor operator between the states described by J^{μ}, M^{μ} and

$J^{\mu'}, M^{\mu'}$. This interpretation rests on the $3jm$-symbol prescribed by the Wigner–Eckart Theorem for a spherical tensor operator T^x_Q. The $9j$-symbol and the two $6j$-symbols are a signature of a corresponding unit tensor $W^{(a,b)x}$, provided one introduces sums over L^* and S^* into the expression and takes the coefficients of fractional parentage, cfp, into account.

The bottom part of Fig. 10.25 represents, in the framework of our interpretation, the one electron reduced matrix element of the *equivalent* tensor operator. We note that the two $9j$-symbols contain the single electron quantum numbers of the hole $\bar{l}, s, \bar{j}$ and the excited shell l. It should, however, be noted that this *equivalent* operator is a bit elusive in that it has no algebraic form but all characteristics properties manifest themselves in the formula.

The complete result for the quantity defined by the right-hand side of (10.16) is

$$\begin{aligned} Z(\mu,\mu') = {} & \sum_{KQ} X^K_{-Q}(2K+1)^{1/2}(-1)^K(2\bar{j}+1)\sum_r(2r+1) \\ & \times \sum_x (-1)^x(2x+1)^{1/2}(-1)^{\bar{j}-\overline{m}} \begin{pmatrix} \bar{j} & r & \bar{j} \\ -\overline{m} & 0 & \overline{m} \end{pmatrix} \\ & \times \begin{pmatrix} K & r & x \\ -Q & 0 & Q \end{pmatrix} (-1)^{J^\mu - M^\mu} \begin{pmatrix} J^\mu & x & J^{\mu'} \\ -M^\mu & Q & M^{\mu'} \end{pmatrix} \\ & \times \sum_{ab}(2a+1)(2b+1)(\alpha^\mu S^\mu L^\mu J^\mu \| W^{(a,b)x} \| \alpha^{\mu'} S^{\mu'} L^{\mu'} J^{\mu'}) \\ & \times \sum_y (2y+1) \begin{Bmatrix} K & r & x \\ a & b & y \end{Bmatrix} \begin{Bmatrix} s & \bar{l} & \bar{j} \\ s & \bar{l} & \bar{j} \\ a & y & r \end{Bmatrix} \begin{Bmatrix} t & l & \bar{l} \\ t & l & \bar{l} \\ K & b & y \end{Bmatrix} \\ & \times (\bar{l}\|r^{(1)}C^{(1)}\|l)(l\|r^{(1)}C^{(1)}\|\bar{l}) \,, \end{aligned} \tag{10.18}$$

and this formula has been given in [34]. With $s = 1/2$, $a = 0$ and 1, while $b = 0, 1 \ldots 2l$. The variable $y = 0, 1, \ldots 2\bar{l}$ obeys triangle conditions with a, r and also K, b. Subsequent simplifications in Sect. 10.7 give $r = 0$. The rank K in (10.18) is determined by the rank $t = 1$ of the dipole operators and the triangle condition gives $K = 0$ (scalar), 1 (dipole) and 2 (quadrupole). Properties of $9j$-symbols require $a+y+r$ and $K+b+y$ to be even for non-zero values.

In equality (10.18) we harvest the results of applying graphical methods and return to algebraic expressions, last used in (10.15) and (10.17). The full algebraic form of the diagrams in Fig. 10.4 is not given explicitly, but main ingredients are two $3jm$-symbols and a product of two sums as in (10.12) on p. 163 To estimate the value of the graphical method one should contemplate on which route an algebraic derivation could arrive at the same result. In this context, it is also not easy to see how to introduce the dependence on $\overline{m}$ in the algebra. The next sections are devoted to a gradual removal of dependence first on $\overline{m}$ in Sect. 10.7 and then on $\bar{j}$ in Sect. 10.8 leading to the simpler expressions (10.19) and (10.20), respectively.

10.7 The Sum Over $\overline{m}$

From the general result (10.18), or its graphical part Fig. 10.24, obtained in the previous section we will remove dependence on $\overline{m}$ by a sum over the same variable. In principle, we could return to Fig. 10.9 and proceed from there, deriving the new result by a series of similar steps. It is, however, much simpler and more instructive to use the general result as in Fig. 10.24 and study effects introduced by the sum over $\overline{m}$.

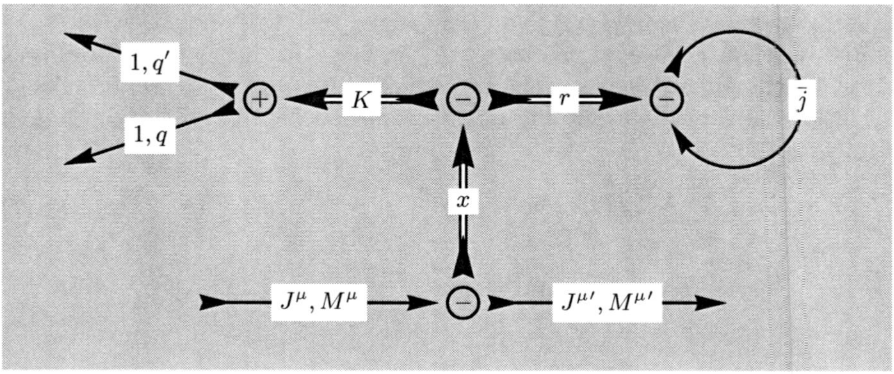

Fig. 10.26. The figure displays the top diagram of Fig. 10.24. The variable $\overline{m}$ appears only in one $3jm$-symbol

The diagram in Fig. 10.26 is a case of a closed diagram, the loop $\bar{j}$, connected by just one line to the rest, encountered in Sect. 7.4. The only value possible is $r = 0$ with a removable node that forces $x = K$ in other diagrams.

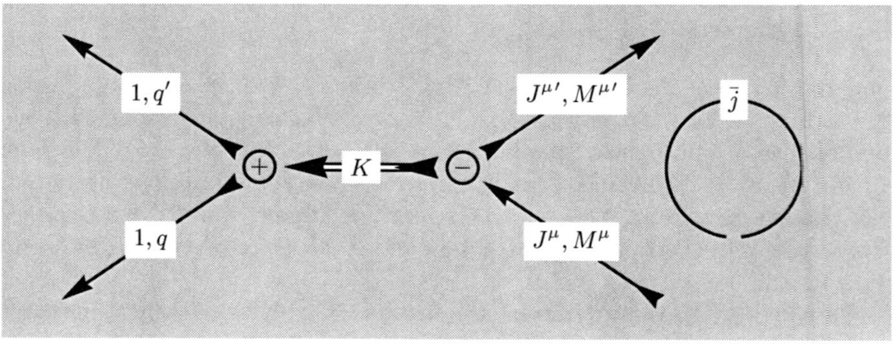

Fig. 10.27. Two nodes have been removed and possible phase changes are not recorded. The factor represented by the circular diagram will not be displayed in subsequent figures

Two diagrams in Fig. 10.24 are affected by $r = 0$, a $6j$-symbol and a $9j$-symbol. From Sect. 5.3 we gather that the $6j$-symbol collapses into a simple $3j$-symbol. Section 6.3 demonstrates that a $6j$-symbol will remain after $r = 0$ is taken into account for the $9j$-symbol.

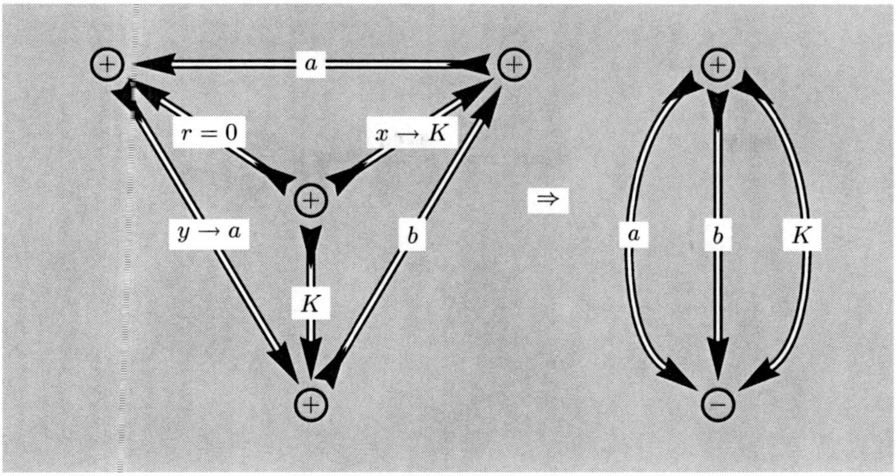

Fig. 10.28. The $r = 0$ and the corresponding two removable nodes are equivalent to $\delta_{x,K}$ and $\delta_{y,a}$. The sums over x and y lead to the replacements $x \to K$ and $y \to a$

The right-hand side diagram is a $3j$-symbol with the value 1, provided the three angular momenta a, b and K obey the triangle condition.

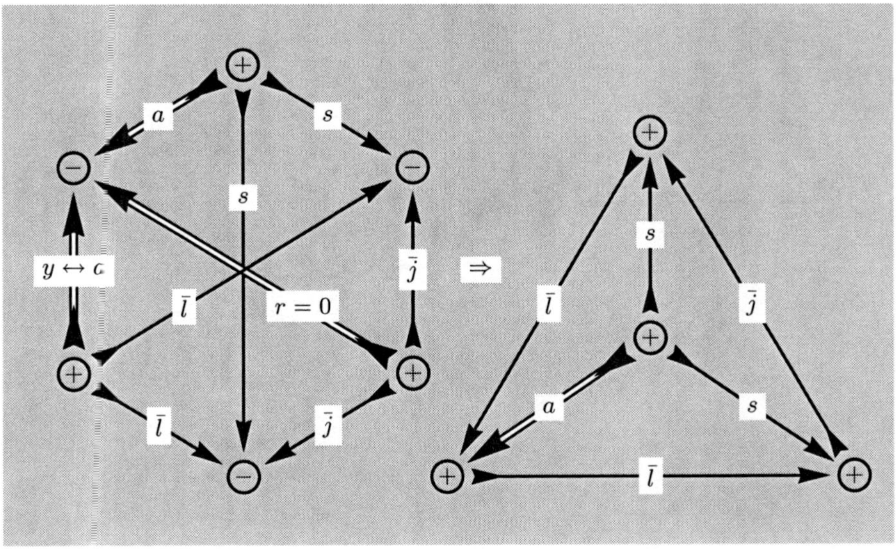

Fig. 10.29. Here we consider the changes for the $9j$-symbol. One removable node is equivalent to $\delta_{y,a}$ and requires the replacement $y \to a$. The $9j$-symbol collapses into a $6j$-symbol

The sums over the variables r, x and y have been performed and we have reached a simpler result, compared to the previous section, but the main features are quite similar.

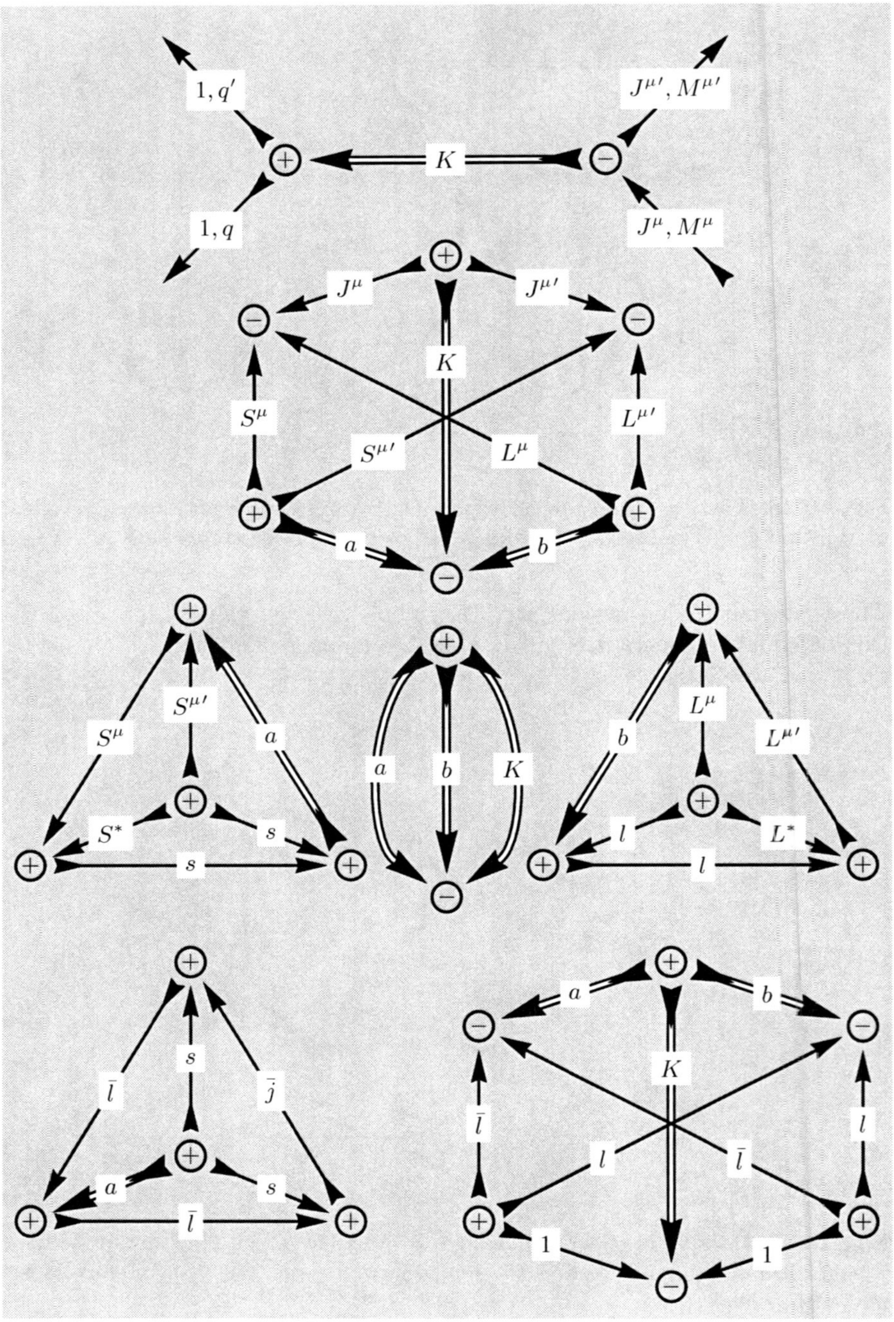

Fig. 10.30. Replacements $x \to K$ and $y \to a$ are performed in the diagrams. From the four $3jm$-symbols only two remain after the sum over $\overline{m}$. Note that we do not keep track of the phase and the various factors $(2a+1)$ etc.

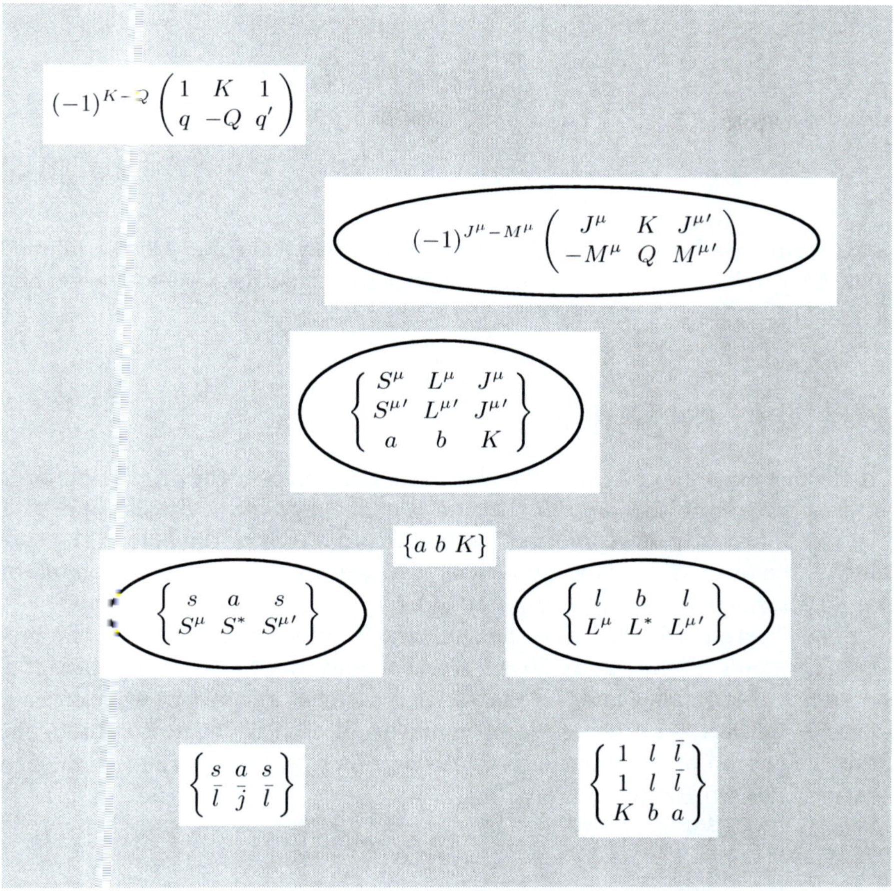

Fig. 10.31. The diagrams from Fig. 10.30 are translated into their algebraic form. Several $3nj$-symbols are rearranged using their symmetry properties. Summations over a and b are not displayed

From the marked terms in this figure we deduce that these terms can be interpreted as signature of a matrix element of an *equivalent* tensor operator between the states described by J^μ, M^μ and $J^{\mu'}, M^{\mu'}$. This interpretation rests on the $3jm$-symbol prescribed by the Wigner–Eckart Theorem for a tensor operator T^K_Q. The $9j$-symbol and the two $6j$-symbols are a signature of the corresponding unit tensor $W^{(a,b)K}$, provided one introduces sums over L^* and S^* into the expression.

The bottom part of the diagram in Fig. 10.31 represents, in the framework of our interpretation, the one electron reduced matrix element of the *equivalent* tensor operator. We note that the $6j$-symbol and the $9j$-symbol contain the single electron quantum numbers of the hole $\bar{l}, s, \bar{j}$ and the excited shell l.

The complete result for the quantity defined in (10.16) is

$$Z(\mu,\mu') = \sum_{KQ}(-1)^Q X^K_{-Q} \sum_{a,b}(2a+1)(2b+1)(-1)^b \tag{10.19}$$
$$\times (-1)^{J^\mu - M^\mu} \begin{pmatrix} J^\mu & K & J^{\mu'} \\ -M^\mu & Q & M^{\mu'} \end{pmatrix} (\alpha^\mu S^\mu L^\mu J^\mu \| W^{(a,b)K} \| \alpha^{\mu'} S^{\mu'} L^{\mu'} J^{\mu'})$$
$$\times (2\bar{j}+1)(-1)^{s+\bar{l}+\bar{j}} \begin{Bmatrix} \bar{l} & a & \bar{l} \\ s & \bar{j} & s \end{Bmatrix} \begin{Bmatrix} \bar{l} & l & 1 \\ \bar{l} & l & 1 \\ a & b & K \end{Bmatrix} (\bar{l}\|r^{(1)}C^{(1)}\|l)(l\|r^{(1)}C^{(1)}\|\bar{l}) \,.$$

This result has been used in [35] under the aspect that 0, 1 are the allowed values for a. The expression can be different from zero for $a+b+K$ even, and $b = 0, 1, \ldots, 2l$.

10.8 The Sum Over $\bar{j}$

As the last example of a choice for the restricted sum over the properties of the intermediate state we consider the sum over the variable $\bar{j}$. By this sum over the two values of $\bar{j}$ more information on the properties of the hole in the core shell is removed and this is equivalent to using a single energy denominator for all intermediate states, say in (10.15).

The shortest calculation of the sum over $\bar{j}$ employs the result of the preceding section. From Fig. 10.30 we see that only the bottom left $6j$-symbol contains the variable $\bar{j}$ and is, thus, the only element affected by a sum over $\bar{j}$. Material in Sect. 8.1.4 offers an independent, alternative route to obtain the result. The sum produces essentially a condition $\delta_{a,0}$, i.e., information on spin states is deleted by the sum on $\bar{j}$ and $\overline{m}$.

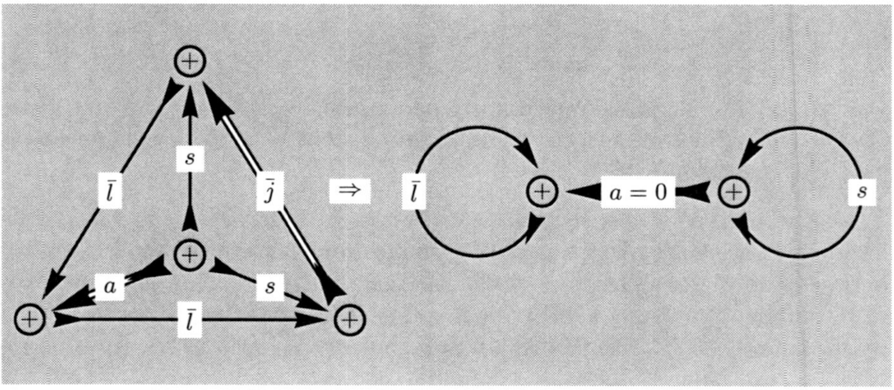

Fig. 10.32. The sum over $\bar{j}$ leads to the condition $a = 0$. The irrelevant details of factors and phase are not displayed

The $3j$-symbol in Fig. 10.30 for $a = 0$ signals another condition: as a consequence we have $b = K$. From one of the $6j$-symbols follows the condition $S^\mu = S^{\mu'}$.

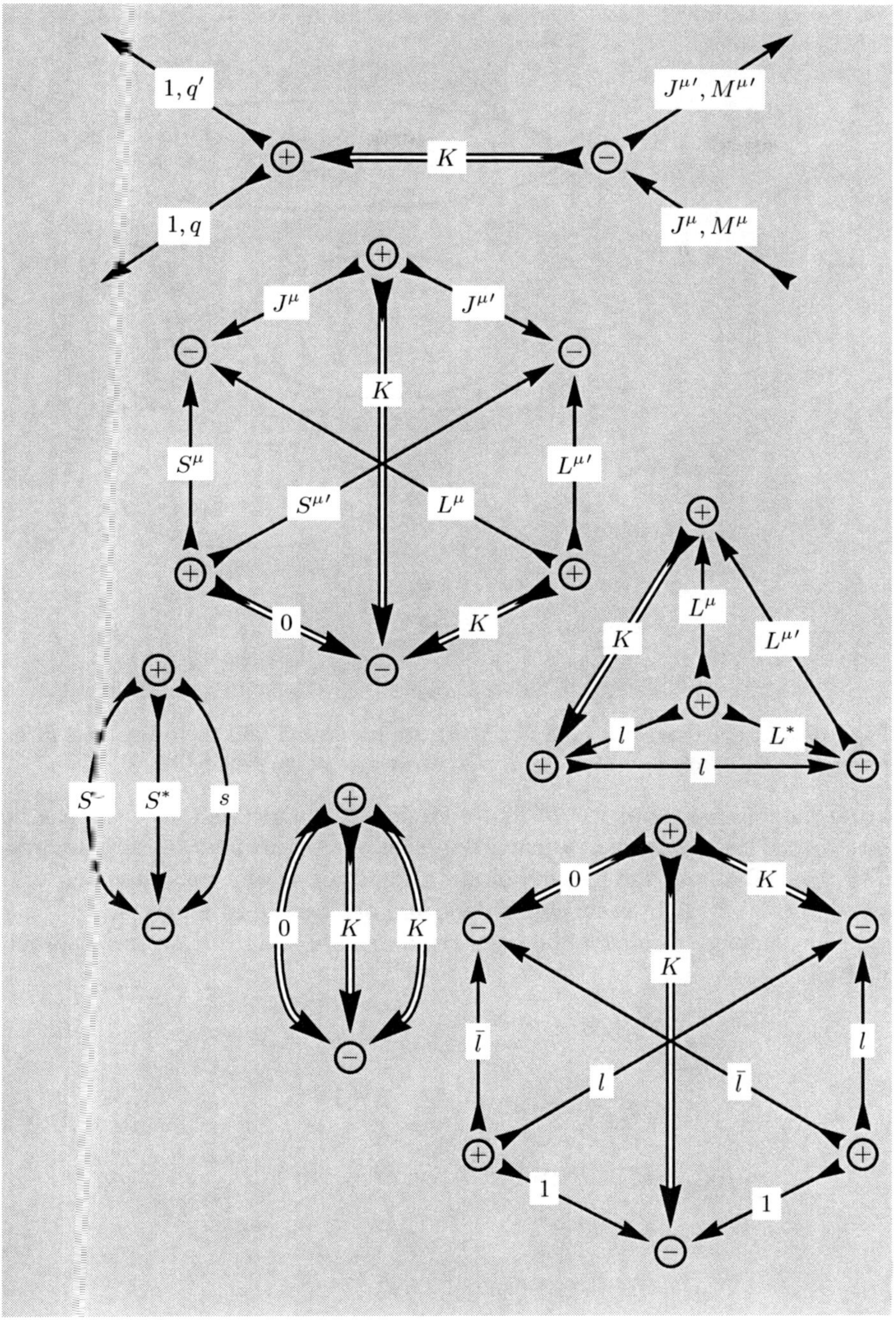

Fig. 10.33. This is the modified version of Fig. 10.30 and we have replaced a by 0 and b by K. In principle the two $9j$-symbols collapse into $6j$-symbols but we prefer the diagrams with one angular momentum zero here

This is the simplest result leading to an equivalent tensor operating only on orbital variables.

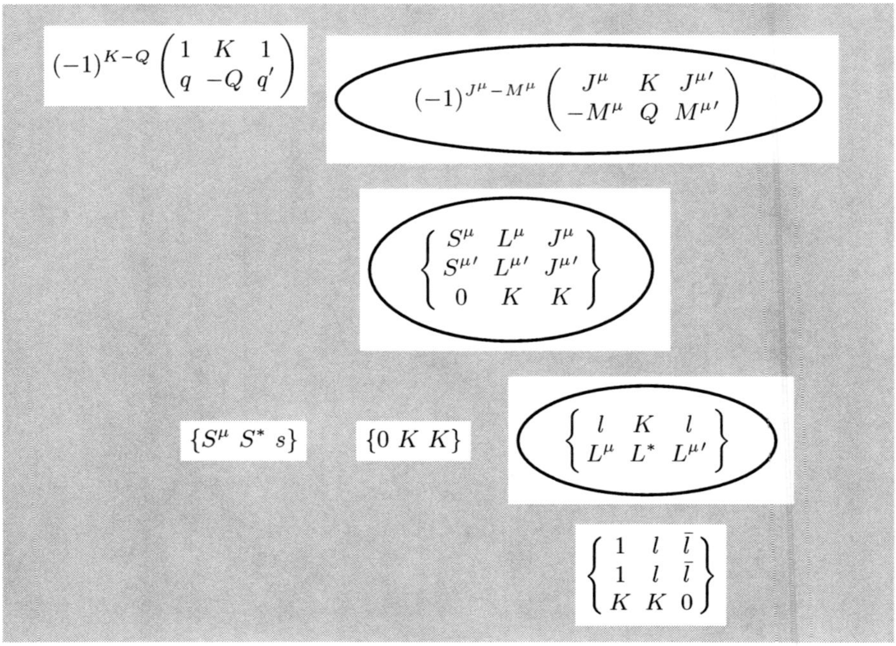

Fig. 10.34. The diagrams from Fig. 10.33 are translated into their algebraic form. Several $3nj$-symbols are rearranged using their symmetry properties

From the marked terms in this figure we deduce, as before, that these can be interpreted as signature of a matrix element of an *equivalent* tensor operator. The $9j$-symbol and the $6j$-symbol are a signature of the corresponding unit tensor $W^{(0,K)K}$. All properties of the core hole have been removed.

The complete result for the quantity defined in (10.16) is, in this approximation,

$$
\begin{aligned}
Z(\mu,\mu') = \sum_{KQ} & X^K_{-Q}(2K+1)^{1/2} \\
& \times(-1)^Q(-1)^{J^\mu - M^\mu}\begin{pmatrix} J^\mu & K & J^{\mu'} \\ -M^\mu & Q & M^{\mu'} \end{pmatrix} \\
& \times(\alpha^\mu S^\mu L^\mu J^\mu \| W^{(0,K)K} \| \alpha^{\mu'} S^{\mu'} L^{\mu'} J^{\mu'}) \\
& \times(2s+1)^{1/2}\begin{Bmatrix} 1 & K & 1 \\ l & \bar{l} & l \end{Bmatrix} \\
& \times(\bar{l}\|r^{(1)}C^{(1)}\|l)(l\|r^{(1)}C^{(1)}\|\bar{l}) \; .
\end{aligned}
\tag{10.20}
$$

This result is used in [36] but instead of the unit tensor $W^{(0,K)K}$ the derivation employs the closely related unit tensor $V(K)$ as given in [7].

Confirmation via a Different Route

In this section we return now to Fig. 10.9 and, as a final example of the graphical method, demonstrate the direct route by introducing the sum over $\bar{j}$ in this diagram.

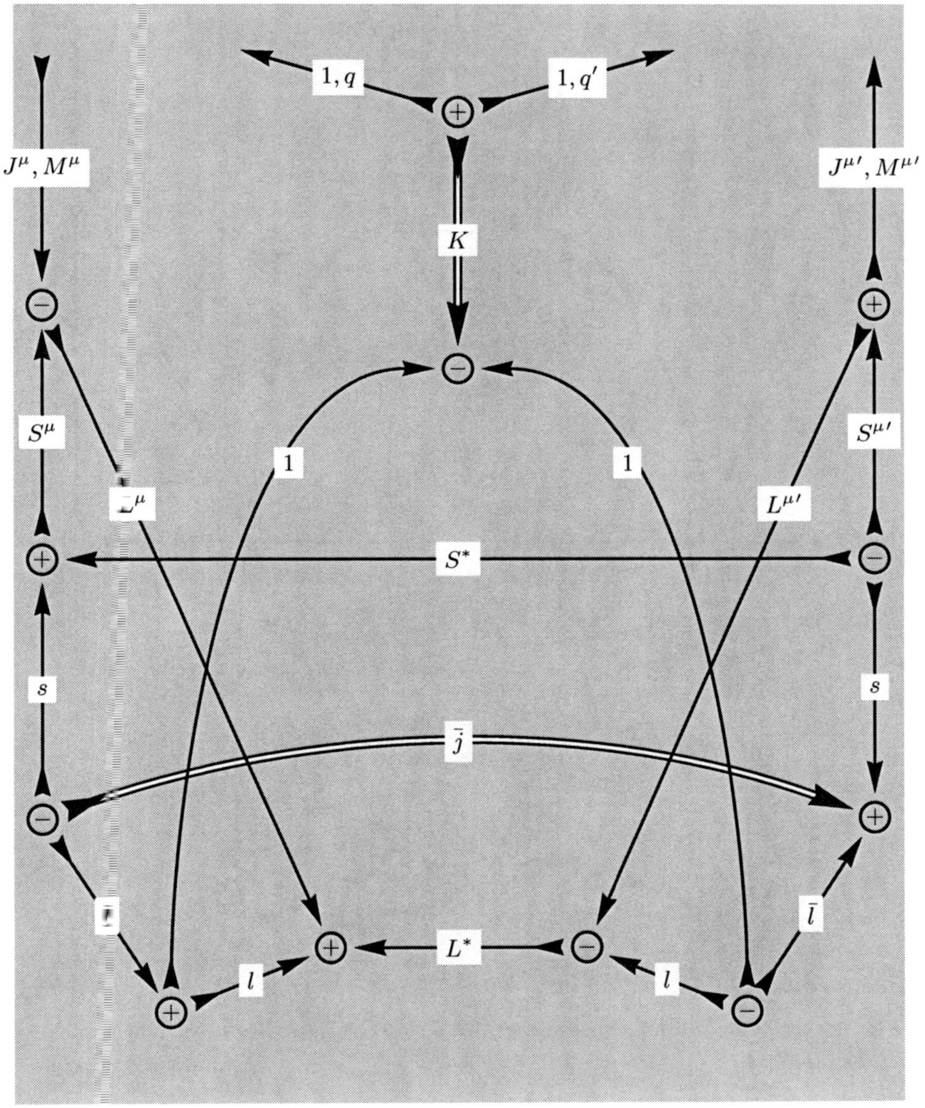

Fig. 10.35. In preparation for the sum over $\bar{j}$ the directions of the s- and the $\bar{l}$-line have been reversed on the left-hand side

To facilitate analysis of the changes in the diagram, all nodes remain at their positions in the next figure.

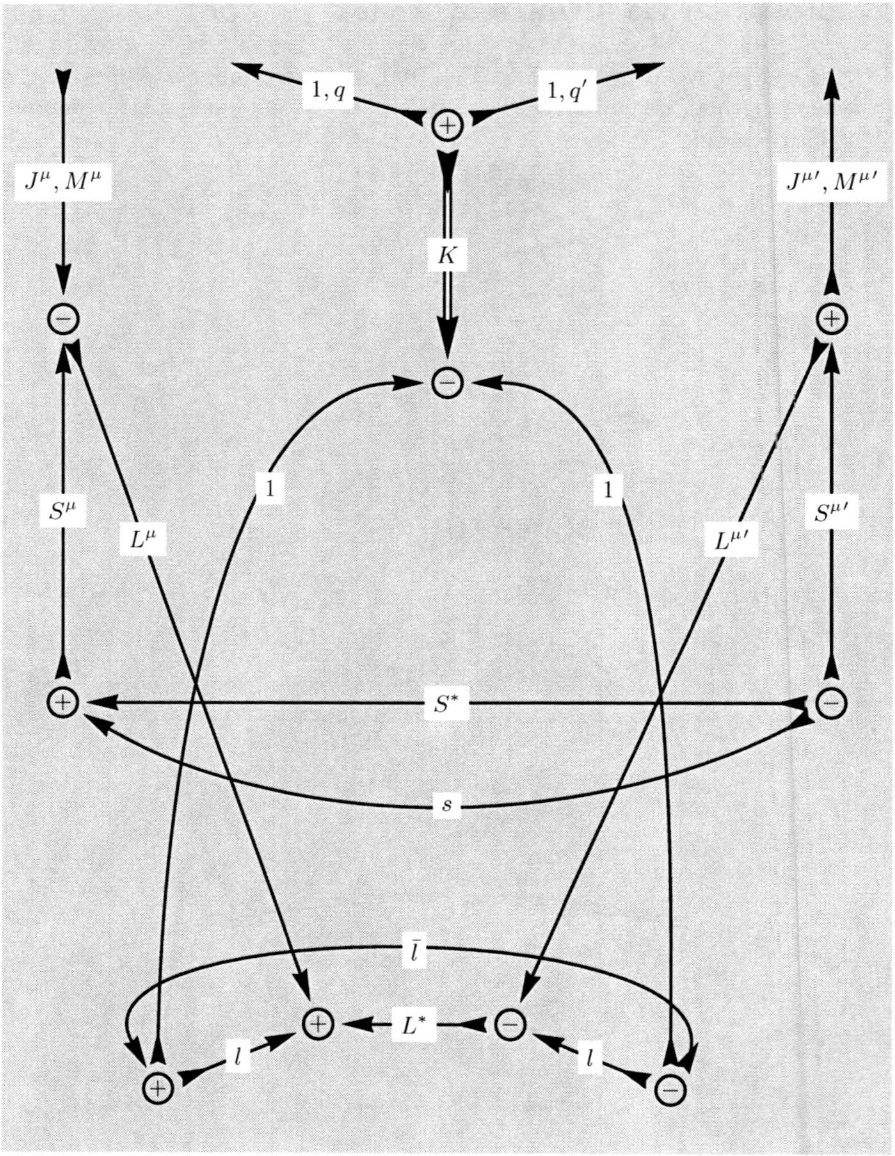

Fig. 10.36. The sum has reconnected the s- and the $\bar{l}$-line from the right to the left of the diagram

It is, perhaps, not easy to see at once, but the lines for the variables S^* and s form a sub-diagram connected to the main body by just two lines. The rule for such a separation has been given in Sect. 7.5, but the quadrangular is not kept here. The two-line separation will impose the condition $S^{\mu} = S^{\mu\prime}$ and will lead to a $3j$-symbol for the angular momentum lines of S^{μ}, S^* and s.

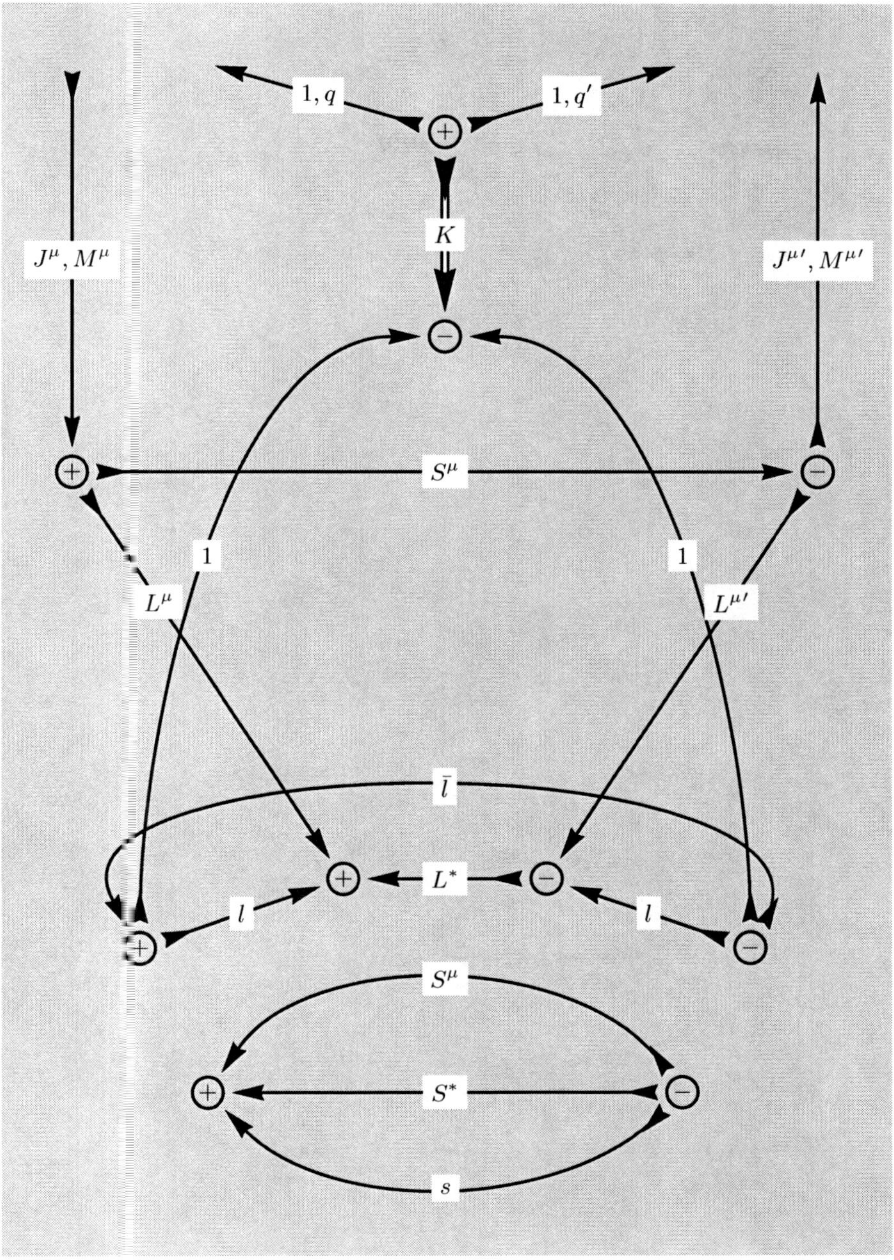

Fig. 10.37. The diagram resulting from the two-line separation is at the bottom right of the figure

A three-line separation for J^μ, K and $J^{\mu\prime}$ combines the external lines in two $3jm$-symbols. The resulting $12j$-diagram turns out to be separable.

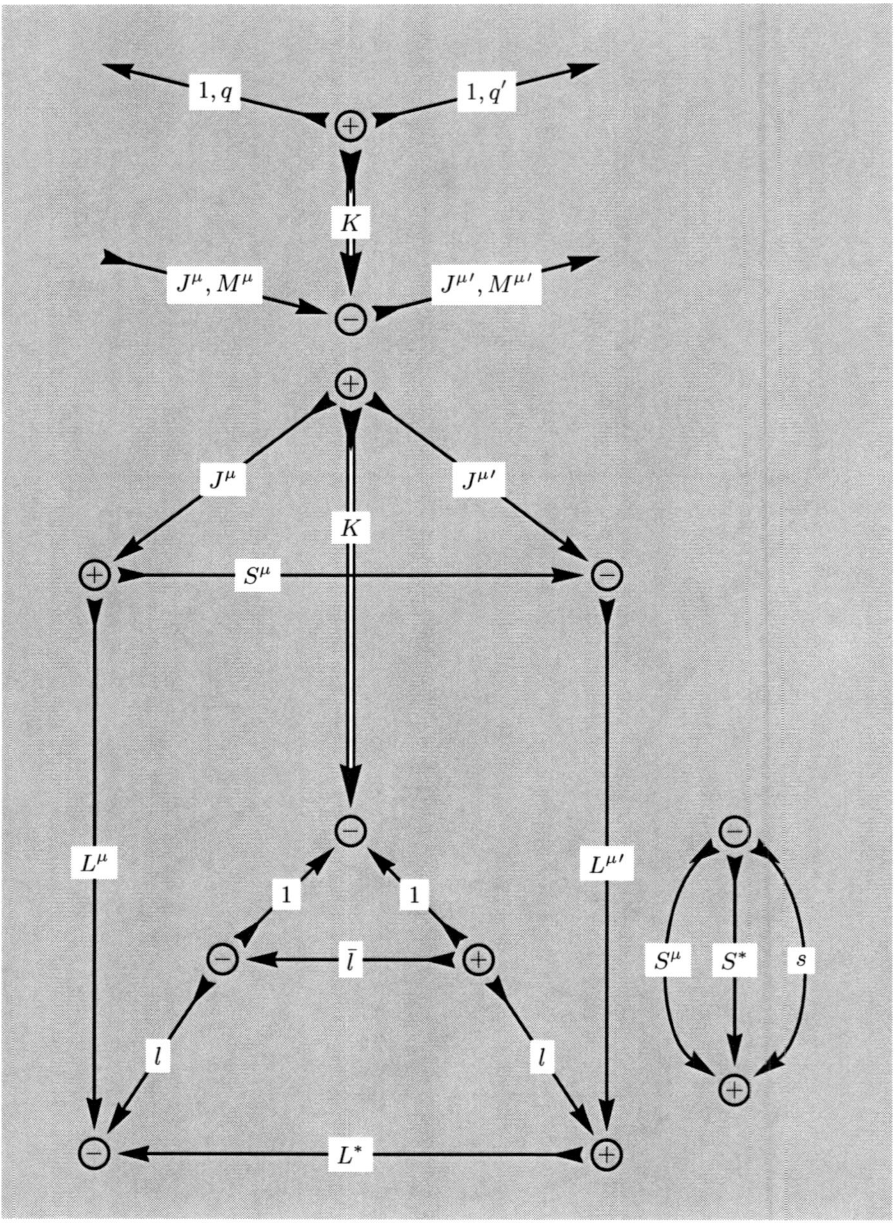

Fig. 10.38. After the three-line separation the lower part of the diagram in Fig. 10.37 is rearranged. Note the node sign changes in the lower part of the diagram

The three angular momentum lines L^{μ}, K and $L^{\mu'}$ are set for the next three-line separation.

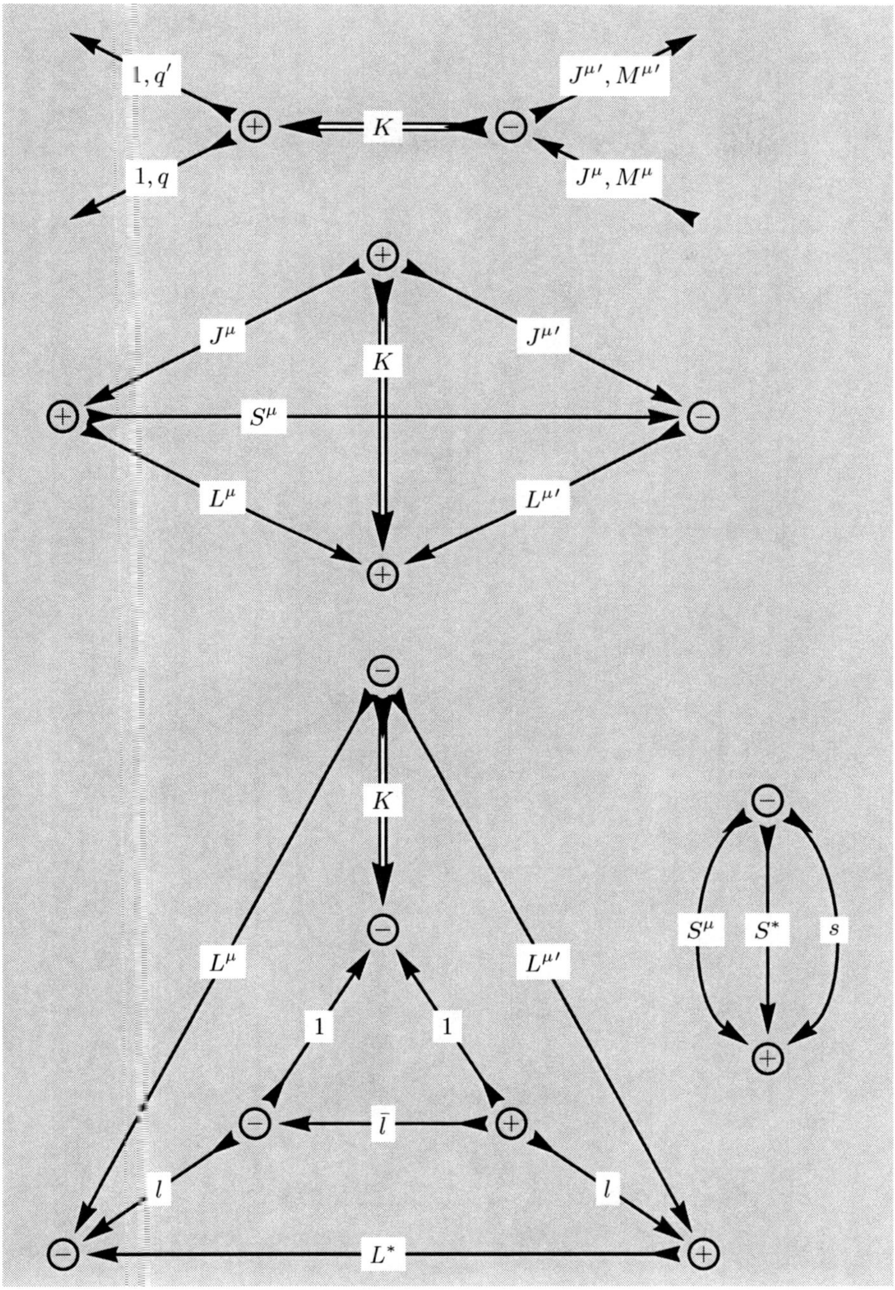

Fig. 10.39. A $6j$-symbol has been separated from the main diagram

The outer triangle is shifted upward in the bottom diagram consisting of two linked triangles.

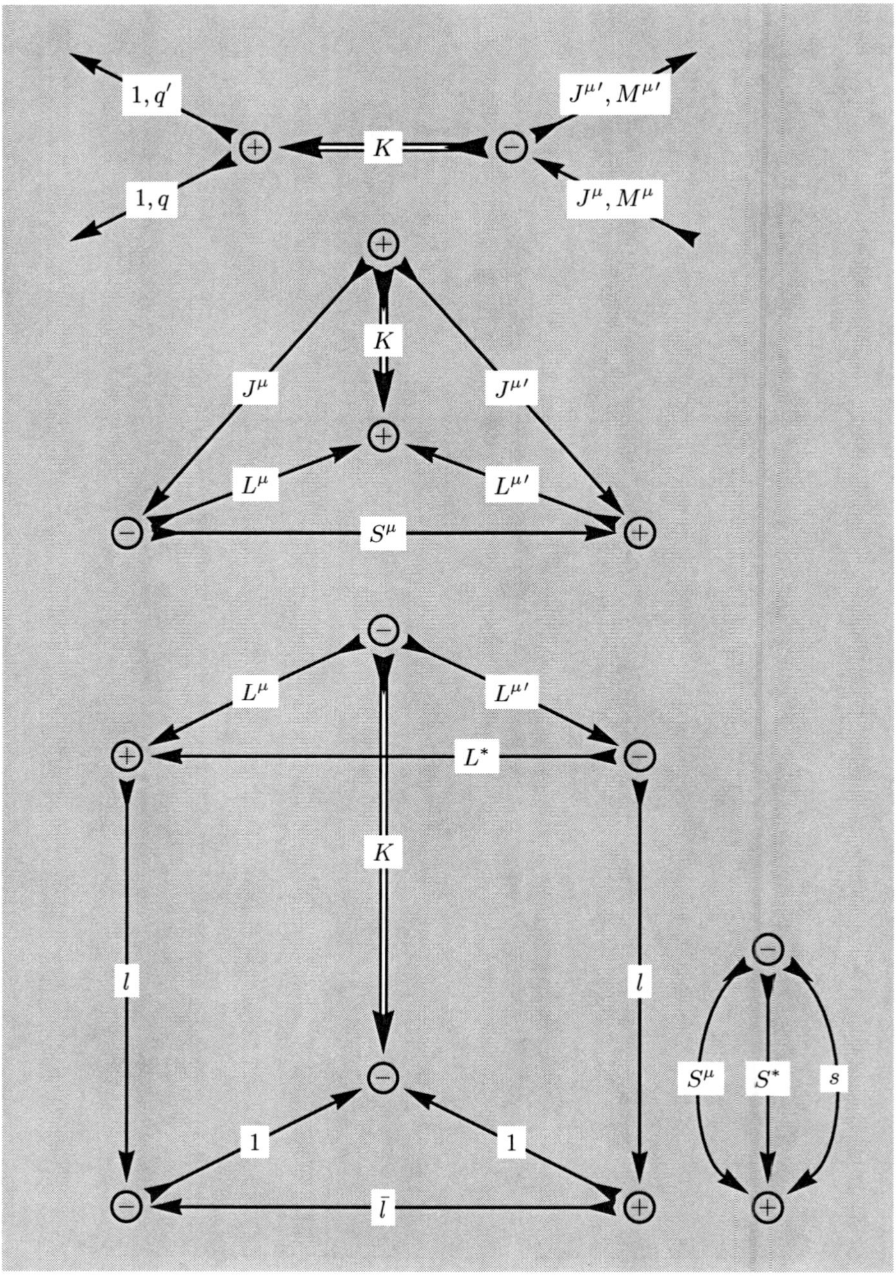

Fig. 10.40. The two triangles are joined by three lines and can be separated

At this point we can foresee that an additional three-line separation is the final simplification required for this diagram.

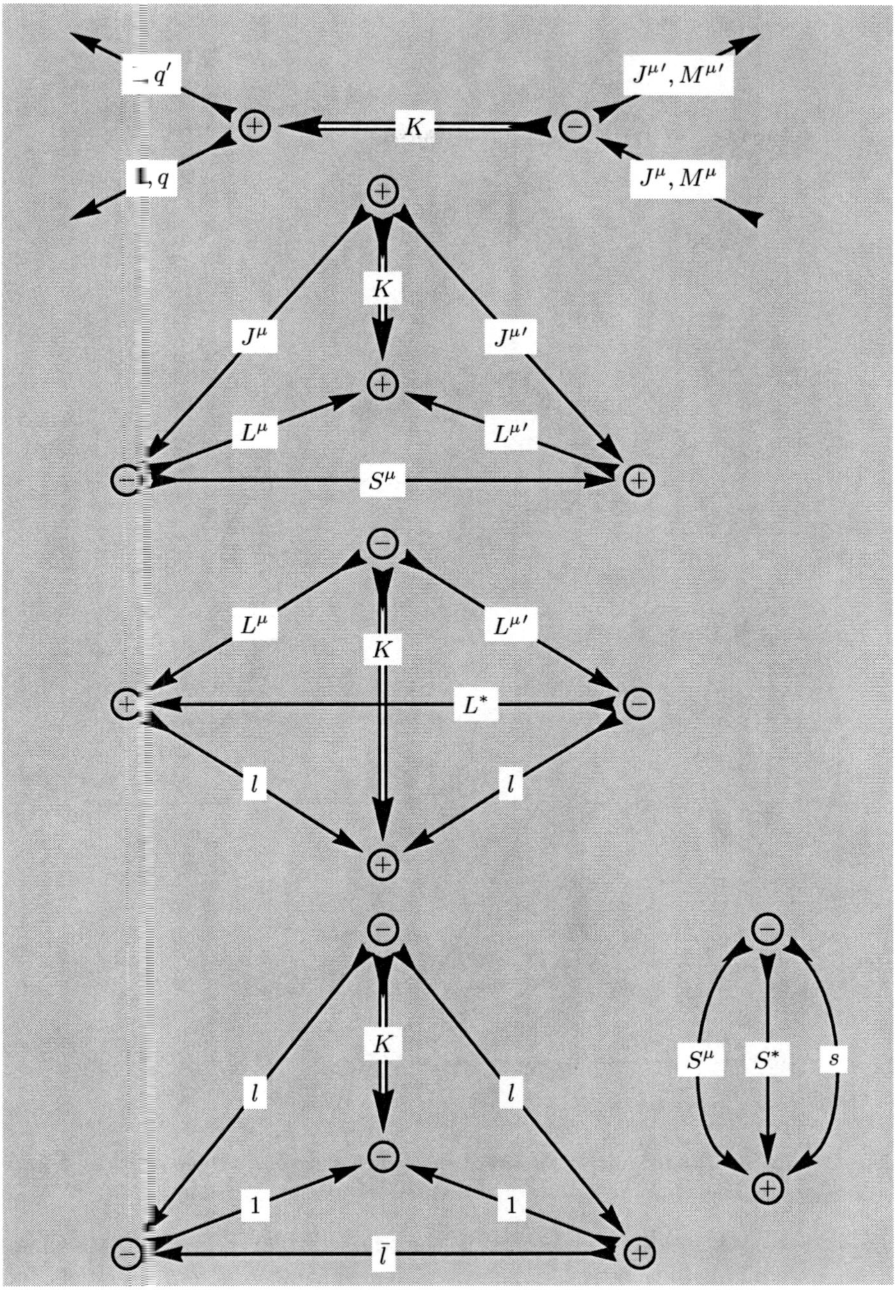

Fig. 10.41. The 9j-diagram has been separated into two 6j-symbols

Several line reversals and node sign changes are required to bring the 6j-symbols into their appropriate form.

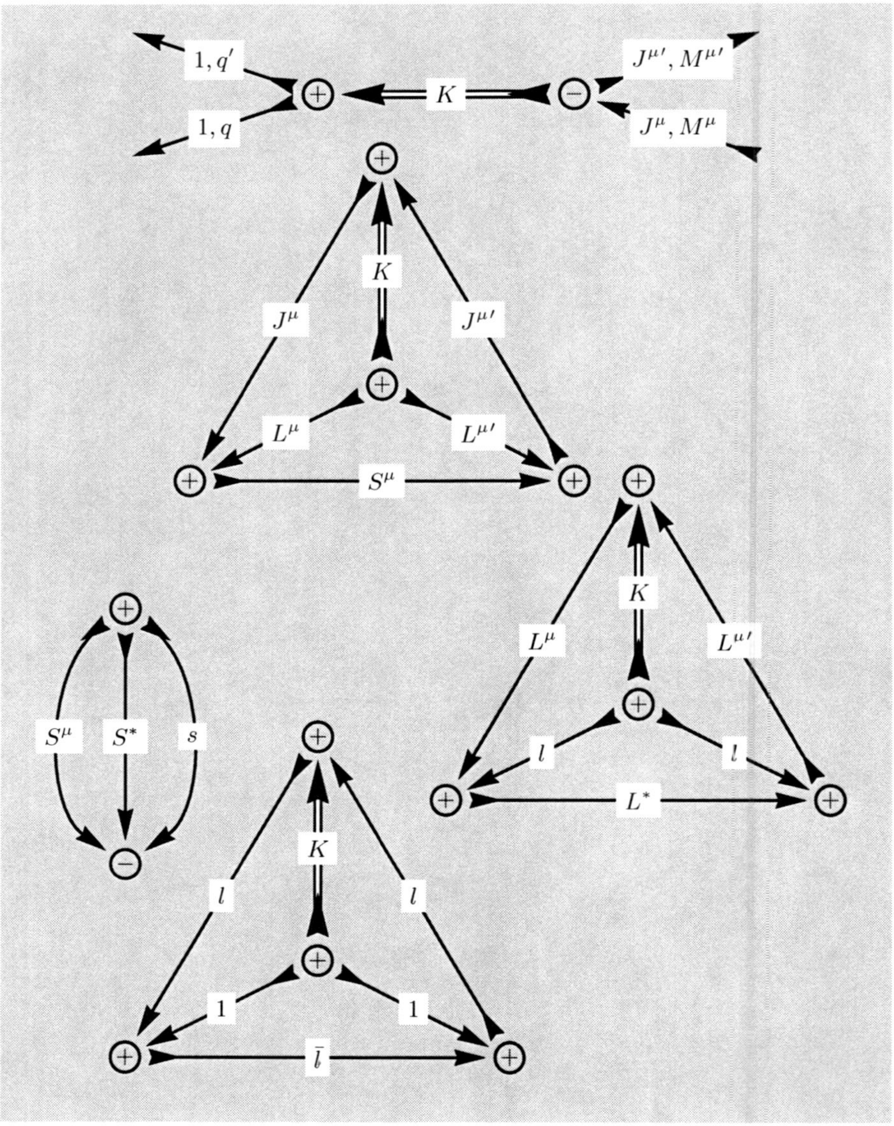

Fig. 10.42. All three $6j$-symbols have been transformed into their prescribed form which can be translated into the corresponding algebraic expression

This diagram is directly comparable to the result displayed in Fig. 10.33 where both $9j$-symbols have a zero angular momentum line. These $9j$-symbols can be transformed into the $6j$-symbols found in Fig. 10.42 here. This concludes our demonstration that both graphical treatments arrive at the same result, as given in (10.20).

References

1. L.C. Biedenharn, H. Van Dam *Quantum Theory of Angular Momentum*, (Academic Press, New York 1965)
2. E.U. Condon, G.H. Shortley *The Theory of Atomic Spectra*, (Cambridge University Press, Cambridge 1935)
3. G. Racah, *Theory of Complex Spectra*
   I: Phys. Rev. **61**, 186 (1942); II: Phys. Rev. **62**, 438 (1942); III: Phys. Rev. **63**, 367 (1943); IV: Phys. Rev. **76**, 1352 (1949)
4. A.R. Edmonds, *Angular Momentum in Quantum Mechanics*, 2nd edn. (Princeton University Press, Princeton NJ 1957)
5. M.E. Rose, *Elementary Theory of Angular Momentum*, (Wiley, New York 1957)
6. D.M. Brink, G.R. Satchler, *Angular Momentum*, 2nd edn. (Oxford University Press, London New York 1968)
7. B.R. Judd, *Operator Techniques in Atomic Spectroscopy*, (McGraw Hill, New York 1963)
8. A, de Shalit, I. Talmi *Nuclear Shell Theory*, (Academic Press, New York London 1963)
9. M. Rotenberg, R. Bivins, N. Metropolis, J.K. Wooten Jr., *The 3j- and 6j symbols*, (Technology Press, Cambridge Mass. 1959)
10. A.P. Yutsis, I.B. Levinson, V.V. Vanagas, *Mathematical Apparatus of the Theory of Angular Momentum*, (published for the National Science foundation, Washington DC, by the Israel Program for Scientific Translation Jerusalem 1962)
11. E. El Baz, B. Castel, *Graphical Methods of Spin Algebras in Atomic, Molecular, and Particle Physics*, (Marcel Dekker Inc., New York 1972)
12. D.A. Varshalovich, A.N. Moskalev, V.K. Khersonskii, *Quantum Theory of Angular Momentum*, (World Scientific, Singapore 1988)
13. D.M. Brink, G.R. Satchler, *Angular Momentum*, 3rd edn. (University Press, Oxford 1994)
14. J.S. Briggs, Rev. Mod. Phys. **43**(2), 189 (1971)
15. R.D. Cowan, *The Theory of Atomic Structure and Spectra*, (University of California Press, Berkeley 1981)
16. B.T. Thole, G. van der Laan, Phys. Rev. A **38**, 1943 (1988)
17. B.T. Thole, G. van der Laan, Phys. Rev. B **44**(22), 12424 (1991)

E. Balcar, S.W. Lovesey, *Introduction to the Graphical Theory of Angular Momentum*, STMP 234, 201–227, DOI 10.1007/978-3-642-03118-2_BM2,

18. G. van der Laan, B.T. Thole, Phys. Rev. B **48**(1), 210 (1993)
19. B.T. Thole, G. van der Laan, Phys. Rev. B **49**(14), 9613 (1994)
20. G. van der Laan, Int. J. of Modern Physics B **8**(6), 641 (1994)
21. B.T. Thole, G. van der Laan, M. Fabrizio, Phys. Rev. B **50**(16), 11466 (1994)
22. G. van der Laan, B.T. Thole, J. Phys.: Condens. Matter **7**, 9947 (1995)
23. P. Carra, M. Altarelli, Phys. Rev. Lett. **64**(11), 1286 (1990)
24. P. Carra, B.T. Thole, M. Altarelli, X. Wang, Phys. Rev. Lett. **70**, 694 (1993)
25. P. Carra, Jpn. J. Appl. Phys. **32**, 279 (1993)
26. P. Carra, H. König, B.T. Thole, M. Altarelli, Physica B **192**, 182 (1993)
27. Ch. Brouder, G. Brinkmann, J. Electron Spectroscopy **86**, 127 (1997)
28. J.-J. Labarthe, J. Phys. A: Math. Gen. **31**, 8689 (1998)
29. D. van Dyck, V. Fack, Comp. Phys. Comm. **151**, 354 (2003)
30. D. van Dyck, V. Fack, Discrete Math. **307**, 1506 (2007)
31. F.P. Temme, J. Phys. A: Math. Theor. **41**, 015210 (2008)
32. D.A. Varshalovich, private communication
33. J. Luo, G.T, Trammell, J.P. Hannon, Phys. Rev. Lett. **71**, 287 (1993)
34. S.W. Lovesey, J. Phys. Condens. Matter **9**, 7501 (1997)
35. S.W. Lovesey, E. Balcar, J. Phys. Condens. Matter **9**, 4237 (1997)
36. S.W. Lovesey, E. Balcar, J. Phys. Condens. Matter **9**, 10983 (1996)
37. S.W. Lovesey, K.S. Knight, E. Balcar, J. F. Rodríguez, Phys. Reports **411**, 233 (2005)
38. Ya.B. Zel′dovich JETP **6**, 1184 (1958)

A

Properties of Spherical Tensor Operators

Our exposition of graphical methods for coupling coefficients has been compiled in anticipation of its application in calculations directed at problems in the physical sciences. As a demonstration we have included a non-trivial example as Chap. 10. In this example matrix elements of spherical tensor operators play a central role, and the close association with coupling coefficients becomes obvious. In order to facilitate access and a deeper understanding of the calculation in Chap. 10 , we highlight here some definitions and properties of spherical tensor operators and their matrix elements.

While the basic concepts involved in calculations of matrix elements are firmly established, there is not, and never has been, consensus in the published literature about issues that are a matter of convention and choice. These issues must be taken seriously, however, for calculations can lead to error if such conventions are not consciously adopted and adhered to consistently. Notable examples are the order in the coupling of spin and orbital angular momentum (for the so-called SL-coupling, or Russell–Saunders coupling, we choose s-l here, as opposed to l-s coupling order), the phase of spherical harmonics (we follow Condon and Shortley [2]) and the definition of a Hermitian spherical tensor operator (we follow Racah [3]).

A.1 Spherical Components

The transition from Cartesian components to spherical components results from a different choice of (complex) base vectors for the representation of vectors or tensors. The main advantage of spherical components lies in the property described by the Wigner–Eckart Theorem. We will consider two vector operators $\boldsymbol{R}$ and $\boldsymbol{J}$, which differ in their symmetry properties, as introductory examples.

Spherical components of a (polar) vector $\boldsymbol{R} = (x, y, z)$ are taken to be $R_0 = z, R_{+1} = -(x + \mathrm{i}y)/\sqrt{2}, R_{-1} = (x - \mathrm{i}y)/\sqrt{2}$. Since x, y and z are real the spherical components are related by $(R_q)^* = (-1)^q R_{-q}$. We note that the angular part of $\boldsymbol{R}$ is proportional to a spherical harmonic $r_q = rC_q^{(1)}$, where $C_q^{(1)} = \sqrt{4\pi/3}Y_q^1$. The spherical components of the operator of the total angular momentum $\boldsymbol{J}$, which is an axial or pseudo- vector are completely analogous with $J_0 = J_z$ and $J_{\pm q} = \mp(J_x \pm \mathrm{i}J_y)/\sqrt{2}$.

A.2 Time Reversal Operator θ

Essential symmetry properties of tensor operators are their behaviour under inversion, time-reversal, rotation etc. The convention for the influence of the reversal of the direction of time $t \to -t$ is adopted from an analogy with the behaviour of eigen-functions, ψ, under time-reversal. These eigen-functions, being associated with solutions of a Schrödinger equation, may be simply replaced by their complex conjugates, i.e., $\psi(\boldsymbol{r}, t) \to \psi^*(\boldsymbol{r}, -t)$. In addition, we adopt the convention that the time-signature of a variable is the sign in the relation between time-reversal and complex conjugation. Since the two operations applied separately to spherical components R_q give identical results, the time-signature $= +1$, per definition.

Although the Cartesian components of $\boldsymbol{J}$ are Hermitian this is no longer true for the spherical components, in fact, $(J_q)^+ = (-1)^q J_{-q}$. With reversal in the direction of time these components change sign, of course. Let the time-reversal operator be denoted by θ; then $\theta J_x \theta^{-1} = -J_x$ with other components of $\boldsymbol{J}$ behaving in the same way. The operator θ includes the operation of complex conjugation and $\theta J_q \theta^{-1} = (-1)^{1+q} J_{-q}$ leads to $\{\theta J_q \theta^{-1}\}^+ = -J_q$.

The operator θ is antilinear and antiunitary by which we mean, first, $\theta(c\psi) = c^*(\theta\psi)$. For the second property a different notation is often used expressing a matrix element as a scalar product of two state vectors, $\langle\psi|B|\phi\rangle \equiv (\psi, B\phi)$, because the action of θ on a bra can be seen more clearly. Thus, secondly, an arbitrary operator B satisfies $(\theta\psi, \theta B\phi) = (\psi, B\phi)^*$ and from this property one arrives more or less directly at,

$$(\psi, B\phi) = (\theta\psi, \{\theta B\theta^{-1}\}\theta\phi)^* = (\theta\phi, \{\theta B\theta^{-1}\}^+\theta\psi) \tag{A.1}$$

where $\theta\psi$ and $\theta\phi$ change place in the last equality. Within the usual notation one would have,

$$\langle\psi|B|\phi\rangle = \langle\theta\psi|\theta B\theta^{-1}|\theta\phi\rangle^* = \langle\theta\phi|\{\theta B\theta^{-1}\}^+|\theta\psi\rangle\ , \tag{A.2}$$

and we set this aside for later use.

A.3 Wigner–Eckart Theorem

The Wigner–Eckart Theorem gives for a matrix element $\langle jm|B_Q^K|j'm'\rangle$ of a spherical tensor operator B_Q^K, with integer rank K and projection $Q(-K \leq Q \leq K)$,

$$\langle jm|B^K_Q\, j'm'\rangle = (-1)^{j-m}\begin{pmatrix} j & K & j' \\ -m & Q & m' \end{pmatrix}(j\|B^K\|j')\,. \tag{A.3}$$

The reduced matrix element $(j\|B^K\|j')$ can be complex and it is not Hermitian. Using in (A.3) our definition of a Hermitian spherical tensor operator, $(B^K_Q)^+ = (-1)^Q B^K_{-Q}$, we find $(j\|B^K\|j') = (-1)^{j'-j}(j'\|B^K\|j)^*$. The $3j$-symbol in (A.3) is invariant with respect to inversion of spatial co-ordinates, $(x,y,z,) \to (-x,-y,-z)$, and the parity of the matrix element on the left-hand side is the parity of the reduced matrix element. The parity operator commutes with the rotation operator, because rotations are generated by parity-even operators J_α. Thus states that differ only in projection have the same parity.

A tensor product of two operators, u^k_q and $v^{k'}_{q'}$, is

$$\{u^k \otimes v^{k'}\}^K_Q = \sum_{qq'}(kqk'q'|KQ)u^k_q v^{k'}_{q'}\,. \tag{A.4}$$

The Clebsch–Gordan coefficient in (A.4) is related to a $3j$-symbol, see (2.1).

If the two operators commute

$$\left[\{u^k \otimes v^{k'}\}^K_Q\right]^+ = (-1)^{k+k'+K+Q}\{u^k \otimes v^{k'}\}^K_{-Q}\,.$$

For non-commuting operators it is safer to use a symmetrized product,

$$\{u^k \otimes v^{k'} + v^{k'} \otimes u^k\}^K_Q = B^K_Q\,, \tag{A.5}$$

that automatically satisfies our convention $(B^K_Q)^+ = (-1)^Q B^K_{-Q}$.

The definition (A.4) follows the standard rule for coupling state vectors $|jm\rangle$ and $|j'm'\rangle$ to form $|JM\rangle$, namely,

$$|JM\rangle = \sum_{mm'} |jm\rangle|j'm'\rangle(jmj'm'|JM)\,. \tag{A.6}$$

A.4 Time Reversal and Parity

The time reversal operator applied to $|jm\rangle$ must reverse the sign of the projection, akin to the effect of an applied magnetic field. We choose the convention $\theta|jm\rangle = (-1)^{j-m}|j,-m\rangle$ for then $|JM\rangle$ in (A.6) behaves with respect to θ in the same way as the two state vectors from which it is formed.

The operator θ so defined is the product of an operator that takes the complex conjugate of a c-number and the operator $\exp(-i\pi J_y)$, which effects a rotation by π about the y-axis. Observe that $\theta^2|jm\rangle = (-1)^{2j}|jm\rangle = \pm|jm\rangle$ where the upper sign is for integer j and the lower sign is for half-integer j. When the Hamiltonian of the system commutes with θ and j is half-integer, we conclude that while $|jm\rangle$ and $\theta|jm\rangle$ must have the same energy they correspond to distinct states, i.e., there is two-fold degeneracy (also called Kramer's degeneracy). The degeneracy can be lifted by application of a time-odd field.

A number of interesting and useful results emerge from an interpretation of the closure relation for angular momentum states,

$$\sum_{jm} |jm\rangle\langle jm| = 1 \, . \tag{A.7}$$

Looking for $\langle jm|$ expressed in terms $|jm\rangle$ we treat the bra and ket as tensor operators with (A.7) a statement about the product of two tensors. The right-hand side of (A.7) is a scalar and, therefore, invariant under rotations. Such a quantity is proportional to a multipole of rank 0, a monopole. Our definition of a product of two tensors u^k and $v^{k'}$ is given in (A.4) and a monopole operator is obtained by taking $K = Q = 0$. Use of (2.8) yields $k' = k$, $q' = -q$, and,

$$\sum_q (-1)^{k-q} u_q^k v_{-q}^k = \sqrt{(2k+1)}\{u^k \otimes v^k\}_0^0 = \text{invariant} \, . \tag{A.8}$$

Comparison of (A.7) and (A.8) suggests that we adopt the rule $\langle jm| = (-1)^{j-m}|j,-m\rangle$.

In addition, we interpret the bra to mean the time-reversed ket. This interpretation is suggested, as we have seen in Sec. A.2, by the fact that solutions of the Schrödinger equation satisfy $\psi^*(\boldsymbol{r},t) = \psi(\boldsymbol{r},-t)$, and one equates $\psi^*(\boldsymbol{r},t)$ to a bra state.

We note that rotation of the ket $|jm\rangle$ by 180° about the y-axis generates $(-1)^{j-m}|j,-m\rangle$. Specifically, $\exp(-\mathrm{i}\pi J_y)|jm\rangle = (-1)^{j-m}|j,-m\rangle$ where J_y is the y-component of the angular momentum operator. This identity plays the central role in some discussions of time-reversed states in which the time-reversal operator, θ, is the product of $\exp(-\mathrm{i}\pi J_y)$ and the operator for complex conjugation. Thus $\theta(c|jm\rangle) = c^*(-1)^{j-m}|j,-m\rangle$ where c is a classical (c-) number. Above we have elected to use this definition of the operator for time reversal.

At the end of Chap. 3 we have indicated that the properties of the $3jm$-symbols, as well as the algebra of angular momentum states encourage the definition a metric tensor,

$$\begin{pmatrix} & j & \\ m & & m' \end{pmatrix} = (-1)^{j-m}\delta_{m,-m'} \, , \tag{A.9}$$

which leads to the relation

$$\begin{aligned} \sum_{m'} |jm'\rangle \begin{pmatrix} & j & \\ m & & m' \end{pmatrix} &= \sum_{m'} |jm'\rangle(-1)^{j-m}\delta_{m,-m'} \\ &= (-1)^{j-m}|j,-m\rangle = \langle jm| \, . \end{aligned} \tag{A.10}$$

Here we have applied the rule $\langle jm| = (-1)^{j-m}|j,-m\rangle$ and from (A.7) and (A.10) one has,

$$\sum_{m,m'} |jm\rangle|jm'\rangle \begin{pmatrix} & j & \\ m & & m' \end{pmatrix} = \text{invariant} \, . \tag{A.11}$$

However, in the main body of our text on the graphical method for coupling angular momenta we found no need to introduce the concept of metric tensor.

In the presence of spin, a relativistic quantity, application of time-reversal, θ, and parity, P_π, are conjugate operations that reverse signs of the four variables x, y, z, and t. Accordingly, the general form of our convention is $P_\pi\theta|jm\rangle = (-1)^{j-m}|j,-m\rangle$, or $\theta|jm\rangle = (-1)^{j-m}P_\pi^{-1}|j,-m\rangle$.

A.5 Reduced Matrix Elements

For an operator, B, with definite parity $P_\pi B P_\pi^{-1} = \sigma_\pi B$ where $\sigma_\pi = \pm 1$ according to whether B is parity-even ($\sigma_\pi = +1$) or parity-odd ($\sigma_\pi = -1$). Let B also have a definite time signature, σ_θ, and define σ_θ as we have through $\{\theta B\theta^{-1}\}^+ = \sigma_\theta B$. Using these definitions in the previously mentioned identity for a matrix element of B in terms of time-reversed states $\theta|jm\rangle$, we find the reduced-matrix element of B must satisfy,

$$(j\|B^K\|j') = (-1)^K\sigma_\pi\sigma_\theta(j\|B^K\|j')^* \ , \tag{A.12}$$

and thus,

$$(j'\|B^K\|j) = (-1)^{j'-j}(-1)^K\sigma_\pi\sigma_\theta(j\|B^K\|j') \ . \tag{A.13}$$

Note that $(-1)^{j-j'} = (-1)^{j'-j}$. By way of an application of this important condition we shall consider the reduced matrix element of the position variable.

For the variable R_q, $K = 1$ (dipole), $\sigma_\pi = -1$ (parity-odd or a polar vector) and $\sigma_\theta = +1$ (time-even) and (A.12) tells us that the reduced matrix element $(j\|R\|j')$ is purely real, which is a standard convention. The explicit result is,

$$(lj\|R\|l'j') = (-1)^{l'+j-1/2}[(2j+1)(2j'+1)]^{1/2}(l\|R\|l')\begin{Bmatrix} l & j & 1/2 \\ j' & l' & 1 \end{Bmatrix}, \tag{A.14}$$

with,

$$(l\|R\|l') = (-1)^{l-l'}(l'\|R\|l) = (-1)^l|R|[(2l+1)(2l'+1)]^{1/2}\begin{pmatrix} l & 1 & l' \\ 0 & 0 & 0 \end{pmatrix}. \tag{A.15}$$

Here we have expanded quantum labels to explicitly show orbital angular momentum, l, for the state $|jm\rangle$. It is evident from (A.15) that a non-zero reduced matrix element has $l + l'$ equal to an odd integer. Note that (A.14) and (A.15) demonstrate the relation we derived earlier between $(j\|B^K\|j')$ and $(j'\|B^K\|j)$. (The result (A.14) is correct for s-l coupling; the corresponding result for l-s coupling has a different phase factor in which l and j' replace l' and j in (A.14).)

Similarly the reduced matrix element of total angular momentum, J_q, is purely real, which is again a standard convention. For this variable, $K = 1$ (dipole), $\sigma_\pi = +1$ (parity-even or an axial vector) and $\sigma_\theta = -1$ (time-odd).

We have, using $\hbar = 1$,

$$(l\|\boldsymbol{L}\|l') = \delta_{l,l'}[l(l+1)(2l+1)]^{1/2} , \tag{A.16}$$

and,

$$(1/2\|\boldsymbol{S}\|1/2) = \sqrt{\frac{3}{2}} , \tag{A.17}$$

together with,

$$(j\|\boldsymbol{J}\|j') = \delta_{j,j'}[j(j+1)(2j+1)]^{1/2} . \tag{A.18}$$

Off-diagonal matrix elements of the magnetic moment $\boldsymbol{\mu} = \boldsymbol{L} + 2\boldsymbol{S} = \boldsymbol{J} + \boldsymbol{S}$ are, $j \neq j'$,

$$\langle jm|\boldsymbol{\mu}|j'm'\rangle = \langle jm|\boldsymbol{S}|j'm'\rangle = -\langle jm|\boldsymbol{L}|j'm'\rangle , \tag{A.19}$$

with reduced matrix elements,

$$(lj\|\boldsymbol{S}\|lj') = (-1)^{l+j'-1/2}[(2j+1)(2j'+1)]^{1/2}\sqrt{\frac{3}{2}}\begin{Bmatrix} 1/2 & j & l \\ j' & 1/2 & 1 \end{Bmatrix} , \tag{A.20}$$

and,

$$\begin{aligned}(lj\|\boldsymbol{L}\|l'j') \\ = (-1)^{l'+j-1/2}[(2j+1)(2j'+1)]^{1/2}(l\|\boldsymbol{L}\|l')\begin{Bmatrix} l & j & 1/2 \\ j' & l' & 1 \end{Bmatrix} .\end{aligned} \tag{A.21}$$

In (A.21) $l = l'$ because of the reduced matrix element of $\boldsymbol{L}$ given in (A.16).

A.6 Equivalent Operators

When tackling physics problems it is often useful to employ operators with particularly convenient mathematical properties, e.g. if they operate on their eigen-states. The ability to use such equivalent operators is one outcome of the Wigner–Eckart Theorem (A.3). Thus values of a time-even, parity-even quadrupole ($K = 2$) can be deduced from matrix elements of symmetrized products of J_α. (This is, however, only true for matrix elements between states $|JM\rangle$ for a fixed value of J as can be seen in (A.18)). In other cases a little more ingenuity is required in selection of such operators.

In nuclear physics, for example, there is interest in a spherical tensor of rank $K = 1$, termed anapole moment, which is time-odd and parity-odd. Following Zel'dovich [38], one may use the vector product of spin and position, $(\boldsymbol{S}\times\boldsymbol{R}) = -(\boldsymbol{R}\times\boldsymbol{S})$, for the anapole. Note from (A.12) that the corresponding reduced matrix element is purely imaginary. From an explicit calculation one obtains the result,

$$\begin{aligned}(lj\|(\boldsymbol{S}\times\boldsymbol{R})\|l'j') \\ = 3\mathrm{i}(-1)^{j+j'}[(2j+1)(2j'+1)]^{1/2}(l\|R\|l')\begin{Bmatrix} j & l & 1/2 \\ j' & l' & 1/2 \\ 1 & 1 & 1 \end{Bmatrix} ,\end{aligned} \tag{A.22}$$

which indeed is purely imaginary. Observe that (A.22) obeys

$$(j\|(\boldsymbol{S}\times\boldsymbol{R})\|j') = -(-1)^{j'-j}(j'\|(\boldsymbol{S}\times\boldsymbol{R})\|j)\ ,$$

since interchange of two rows in the $9j$-symbol creates a factor $(-1)^{l+l'+j+j'} = -(-1)^{l+l'+j-j'}$, and this finding complies with our general condition (A.13) for reduced matrix elements.

An alternative for the anapole operator is $(\boldsymbol{J}\times\boldsymbol{R}) - (\boldsymbol{R}\times\boldsymbol{J}) = \mathrm{i}[\boldsymbol{J}^2, \boldsymbol{R}]$. However, with $\boldsymbol{J}^2|jm\rangle = j(j+1)|jm\rangle$ this operator has no matrix element between states restricted to a single j-manifold and therefore the operator has limited value. In particular, it can not be used for a coherent discussion of the corresponding operator with $K = 0$ (monopole), because off-diagonal matrix elements of a scalar vanish. Thus one might consider constructing the anapole operator with the orbital angular momentum, $\boldsymbol{L}$, in place of $\boldsymbol{J}$, but closer examination reveals $\boldsymbol{L}\cdot\boldsymbol{R} = 0$ which rules $\boldsymbol{L}$ out of a discussion of the monopole. On reflection, a pellucid account of both the monople and anapole can be achieved with $\boldsymbol{R}$ and the magnetic moment $\boldsymbol{\mu} = \boldsymbol{L} + 2\boldsymbol{S} = \boldsymbol{J} + \boldsymbol{S}$, which has obvious physical bearing. As noted, the reduced matrix element of a monopole can only exist for $j = j'$ and the result is,

$$(lj\|(\boldsymbol{\mu}\cdot\boldsymbol{R})\|l'j) = (-1)^{l'+j-1/2}[6(2j+1)]^{1/2}(l\|R\|l')\begin{Bmatrix} l & 1/2 & j \\ 1/2 & l' & 1 \end{Bmatrix}\ . \tag{A.23}$$

The result (A.23) is purely real, of course, and it satisfies

$$(lj\|(\boldsymbol{\mu}\cdot\boldsymbol{R})\|l'j) = (l'j\|(\boldsymbol{\mu}\cdot\boldsymbol{R})\|lj)\ ,$$

as it must. Note $\boldsymbol{\mu}\cdot\boldsymbol{R} = \boldsymbol{R}\cdot\boldsymbol{\mu} = 2\boldsymbol{S}\cdot\boldsymbol{R}$ since $\boldsymbol{L}\cdot\boldsymbol{R} = 0$.

A.7 Many Electron Unit Tensors

In the remaining part of the Appendix, we look at the matrix element of a tensor operator for several electrons in an atomic shell. The shell has angular momentum l that may accommodate $2(2l+1)$ equivalent electrons, and all electrons have spin $s = 1/2$ and orbital angular momentum l. The task of calculating a matrix element is easy because tables exist for d- and f-electron states of so-called unit tensors, which are sums of fractional parentage coefficients, see e.g. [35].

If we create a tensor product of rank K from operators that work on the spin, z^a, and orbital, y^b, variables, following (A.4), the reduced matrix element of the tensor product is,

$$(s\|z^a\|s)(l\|y^b\|l)W^{(a,b)K}\ , \tag{A.24}$$

where $W^{(a,b)K}$ is the unit tensor. The reduced matrix element of rank $b = 0$ is $(l\|y^0\|l) = \sqrt{2l+1}$ and the same expression applies for $a = 0$, namely, $(s\|z^0\|s) = \sqrt{2}$. We give an application of the unit tensor before listing some of its properties.

Let us consider the spatial Fourier transform of the electron spin-density, which is a quantity that arises in diffraction by an ion. We work with a one-electron operator,

$$\mathrm{e}^{\mathrm{i}\boldsymbol{k}\cdot\boldsymbol{R}}\boldsymbol{s} = 4\pi \sum_{KQ} \mathrm{i}^K j_K(kR) Y_Q^K(\hat{\boldsymbol{k}})^* \boldsymbol{s} Y_Q^K(\hat{\boldsymbol{R}}) \,, \tag{A.25}$$

where the right-hand side is reached by a standard identity. The sum is over all positive integer values of K including 0, $-K \leq Q \leq K$, $\hat{\boldsymbol{k}}$ and $\hat{\boldsymbol{R}}$ are unit vectors and j_K is a spherical Bessel-function. If we label an equivalent electron by i the angular part of the required many-electron operator is,

$$f_{Q'}^{K'} = \sum_i (\boldsymbol{s}_q)_i Y_Q^K(\hat{\boldsymbol{R}}_i)(1qKQ|K'Q') \,. \tag{A.26}$$

Application of (A.24) gives for the reduced matrix-element of $f_{Q'}^{K'}$ the result,

$$(J\|f^{K'}\|J') = \sqrt{\frac{3}{2}}(l\|Y^K\|l)W^{(1,K)K'} \,, \tag{A.27}$$

which is used in,

$$\langle J|f_{Q'}^{K'}|J'\rangle = (-1)^{J-M}\begin{pmatrix} J & K' & J' \\ -M & Q' & M' \end{pmatrix}(J\|f^{K'}\|J') \,. \tag{A.28}$$

A value of (A.27) for one electron is quite readily derived and its value conforms to the unit tensor for one electron obtained from the general expression (A.33).

It should be evident in (A.24), (A.27) and (A.28) that we use an abbreviated notation in which the quantum numbers of the shell, including total spin, S, total angular momentum, L, and seniority are not displayed. The rank K of $W^{(a,b)K}$ satisfies the triangle relation $|a-b| \leq K \leq a+b$, while $|S-S'| \leq a \leq S+S'$, and $|L-L'| \leq b \leq L+L'$. These relations are manifest in the link between $W^{(a,b)K}$ and a second useful unit-tensor $W^{(a,b)}$, namely,

$$W^{(a,b)K} = \left\{\frac{(2J+1)(2K+1)(2J'+1)}{(2a+1)(2b+1)}\right\}^{1/2}\begin{Bmatrix} S & S' & a \\ L & L' & b \\ J & J' & K \end{Bmatrix} W^{(a,b)} \,. \tag{A.29}$$

The 9j-symbol with $S = S'$, $L = L'$, and $J = J'$ exists for even $a+b+K$.

The physical significance of $W^{(a,b)}$ is best seen through the equivalence of matrix elements in the J, M basis and the S, L basis. One passes from one basis to the other basis using,

$$(-1)^{J-M}\begin{pmatrix} J & K & J' \\ -M & Q & M' \end{pmatrix} W^{(a,b)K} \tag{A.30}$$

$$\Longrightarrow (-1)^{a+b+Q}\left\{\frac{2K+1}{(2a+1)(2b+1)}\right\}^{1/2} W^{(a,b)} \sum_{mn}\begin{pmatrix} a & K & b \\ -m & Q & -n \end{pmatrix}$$

$$\times(-1)^{S-M_S}\begin{pmatrix} S & a & S' \\ -M_S & m & M_S' \end{pmatrix}(-1)^{L-M_L}\begin{pmatrix} L & b & L' \\ -M_L & n & M_{L'}' \end{pmatrix} \,.$$

It is a simple matter to obtain the reduced matrix-element of $\boldsymbol{S}$ in terms of $W^{(1,0)}$ and the reduced matrix-element of $\boldsymbol{L}$ in terms of $W^{(0,1)}$ from (A.30).

A matrix element of the spin-orbit interaction in a shell is,

$$\langle JM|\sum_i (\boldsymbol{s}\cdot\boldsymbol{l})_i|J'M'\rangle$$

$$= -\frac{3}{\sqrt{2}}(l\|L\|l)W^{(1.1)0}(-1)^{J-M}\begin{pmatrix} J & 0 & J' \\ -M & 0 & M' \end{pmatrix} . \tag{A.31}$$

The $3j$-symbol is zero unless $J = J'$ and $M = M'$ in keeping with expectation for the matrix element of a scalar operator, of which (A.23) is a previous example.

For one electron,

$$W^{(a,b)} = \sqrt{[(2a+1)(2b+1)]} , \tag{A.32}$$

while,

$$W^{(a,b)K} = \sqrt{[(2J+1)(2K+1)(2J'+1)]}\begin{Bmatrix} 1/2 & 1/2 & a \\ l & l & b \\ J & J' & K \end{Bmatrix} . \tag{A.33}$$

For use of (A.33) in (A.27) take $a = 1$ and $b = K$.

B

List of Diagrams

B.1 $3jm$-Symbols

In each of the following figures a $3jm$- or Wigner-symbol, often with a phase factor, is opposed to two graphical equivalents. The diagrams consist of one node and three directed lines, each belonging to one of the angular momenta of the $3jm$-symbol. The node sign is negative if $a \to b \to c$ corresponds to a clock-wise cyclic order in the diagram and vice versa. The direction of the angular momentum lines goes into the node if the projection in the $3jm$-symbol is negative and out from the node for a positive projection.

$$(-1)^{a-\alpha}\begin{pmatrix} a & b & c \\ -\alpha & \beta & \gamma \end{pmatrix} =$$

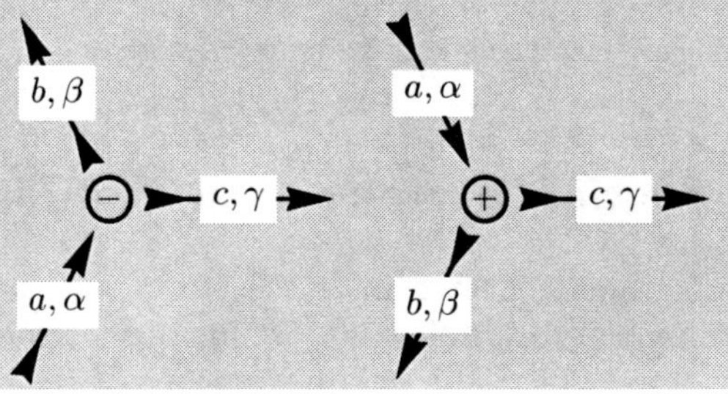

Fig. B.1. $3jm$-symbol I: one ingoing and two outgoing lines

$$(-1)^{b-\beta+c-\gamma}\begin{pmatrix} a & b & c \\ \alpha & -\beta & -\gamma \end{pmatrix} =$$

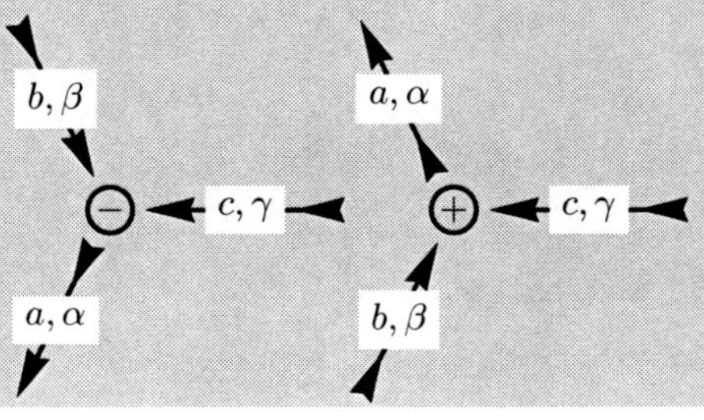

Fig. B.2. $3jm$-symbol II: one outgoing and two outgoing lines

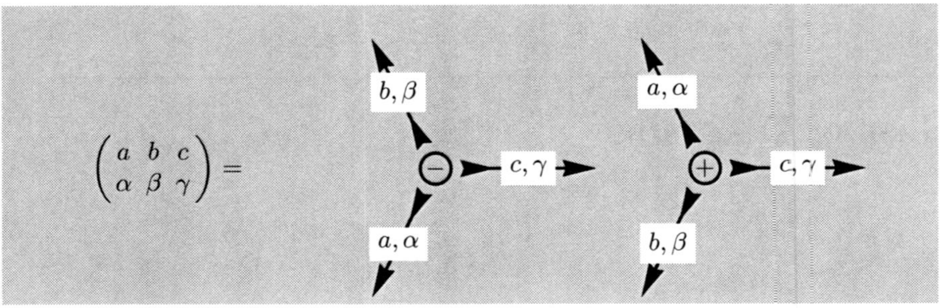

Fig. B.3. $3jm$-symbol III: three outgoing lines

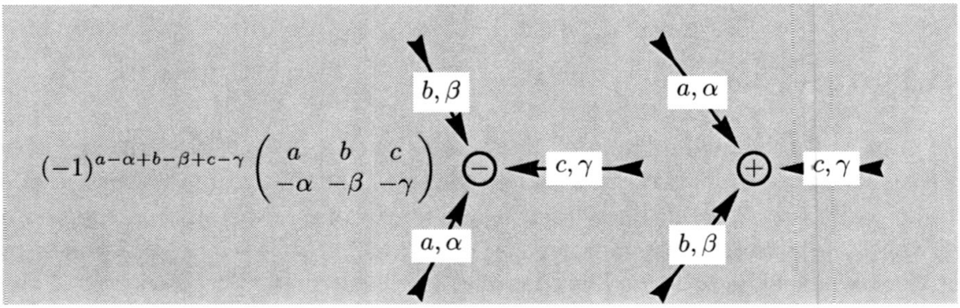

Fig. B.4. $3jm$-symbol IV: three ingoing lines

The $3jm$-symbols constitute the basic building block from which all other higher rank symbols can be constructed.

B.2 Special Diagrams with Two-Line Nodes

The diagrams collected in this list represent factors appearing frequently in diagrams as a result of separation or summation. Except for the first, the diagrams are, in fact, abbreviated versions of the diagrams with zero-lines given in the figures. Each figure shows a diagram that results from the following diagram by removing one node.

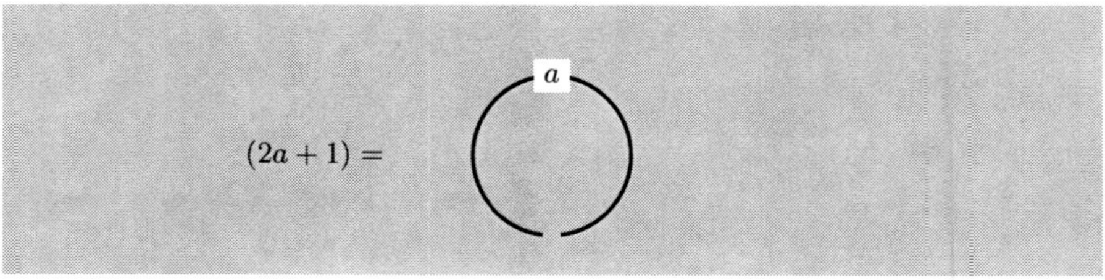

Fig. B.5. Nodeless and directionless diagram. The gap in the circle is a remnant of the removed node

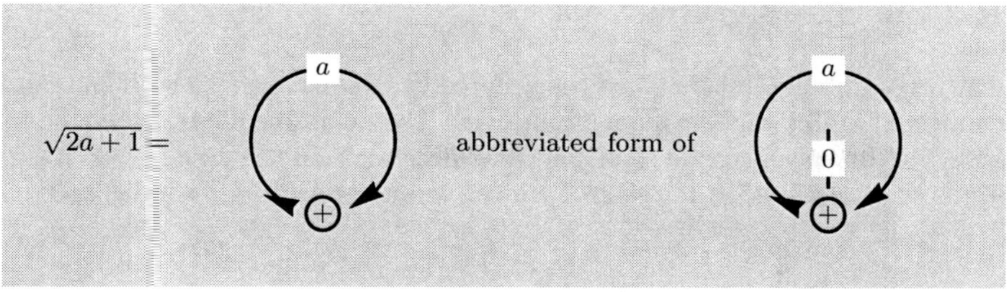

Fig. B.6. Diagram with a single two-line node

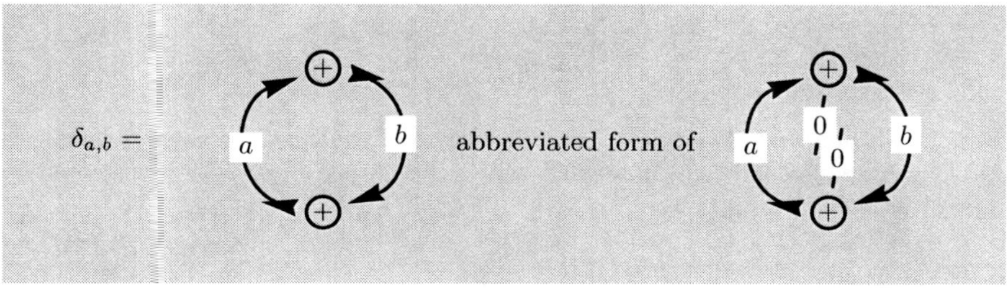

Fig. B.7. Diagram with two two-line nodes

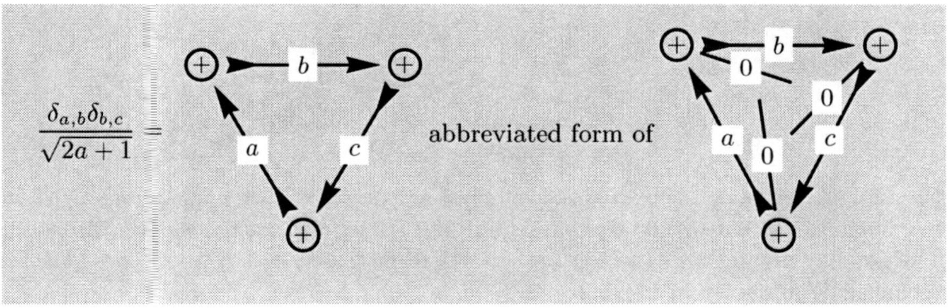

Fig. B.8. Triangular diagram with three two-line nodes

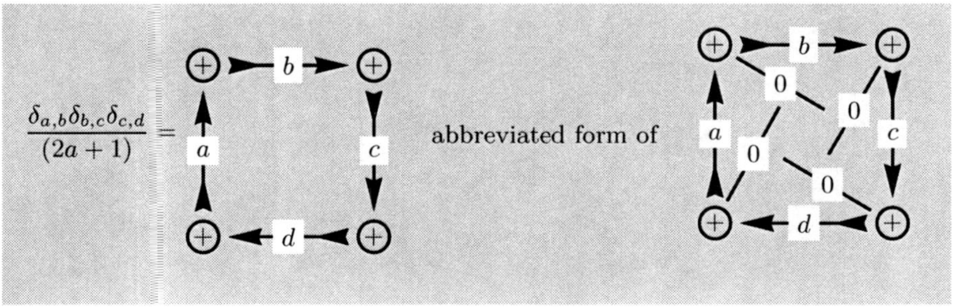

Fig. B.9. Quadrangular diagram with four two-line nodes

B.3 Specific $3nj$-Symbols

The $3nj$-symbols listed here are represented by closed diagrams with $2n$ nodes connected by $3n$ angular momentum lines. The definitive diagrams are characterized by specific node signs and line directions. In the case of $n = 4$ two kinds of symbols exist, for each of which two diagrams have been defined.

$$\{a\ b\ c\} = 1_{\Delta\{a,b,c\}}$$

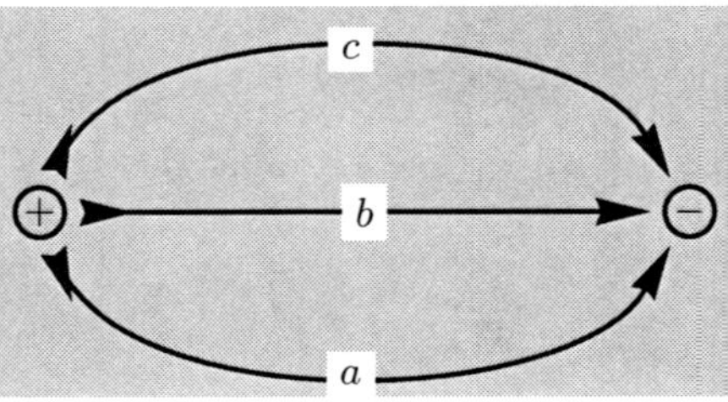

Fig. B.10. Simplest closed diagram with a value of 1 if the three angular momenta fulfil the triangle condition. In analogy with other $3nj$-symbols it is represented by $\{a\ b\ c\}$ as shown in Fig. 4.6

$$\begin{Bmatrix} a & b & c \\ d & e & f \end{Bmatrix} =$$

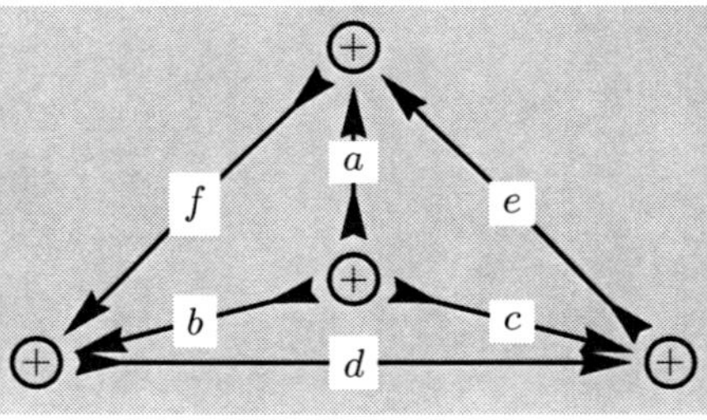

Fig. B.11. All node signs of the $6j$-symbol are positive. The angular momenta of the bottom row constitute the outside of the triangle in anti-clockwise direction. The angular moments of the top row go out from the central node to the opposite corner node as shown in Fig. 5.2

$$\begin{Bmatrix} a & b & c \\ d & e & f \\ k & l & m \end{Bmatrix} =$$

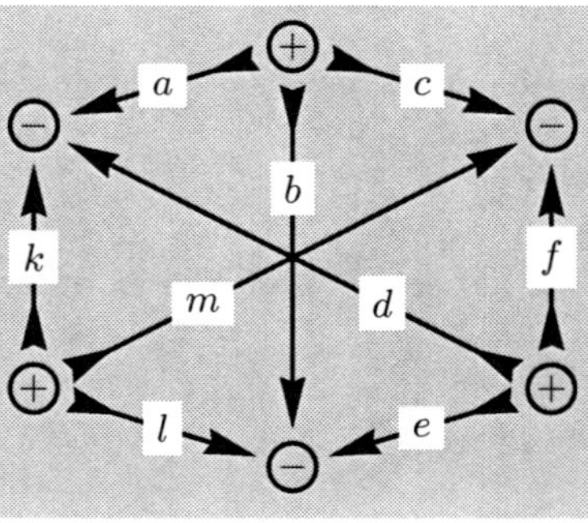

Fig. B.12. The diagram for the $9j$-symbol is a hexagon with alternative positive and negative nodes. Each of the positive nodes is attached to three angular momentum lines directed at the negative nodes as shown in Fig. 6.2

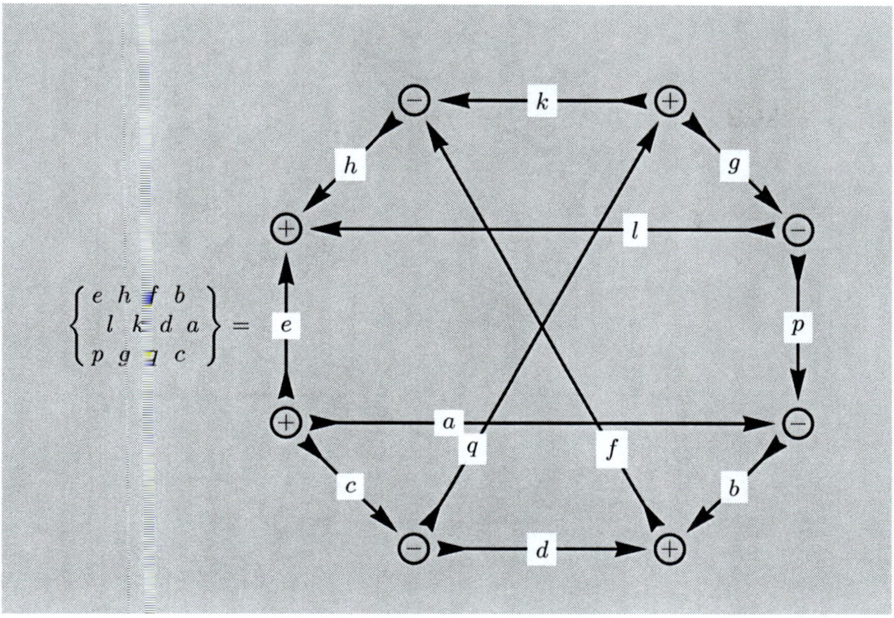

Fig. B.13. The diagram of the $12j$(I)-symbol (first kind) as shown in Fig. 8.59

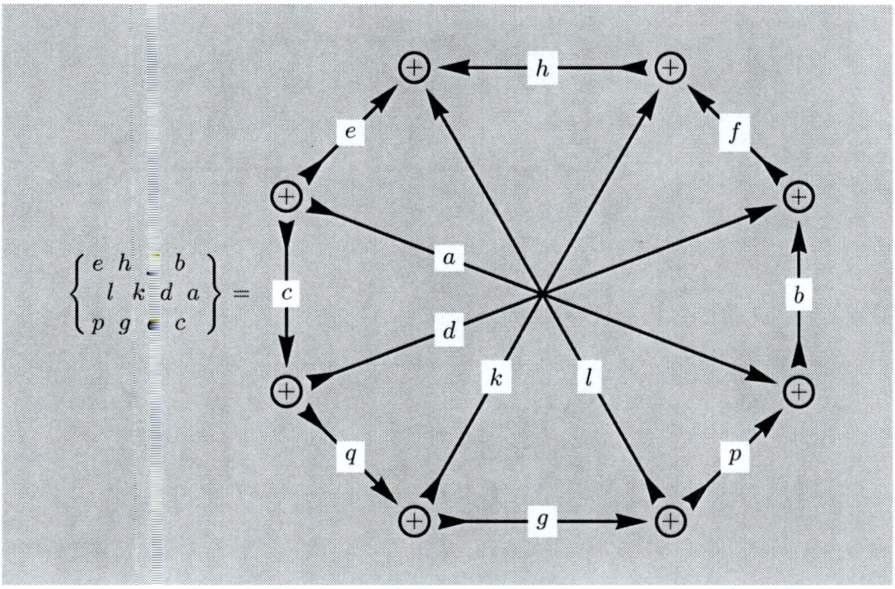

Fig. B.14. Alternative form of the diagram for the $12j$(I)-symbol (first kind) as shown in Fig. 8.62

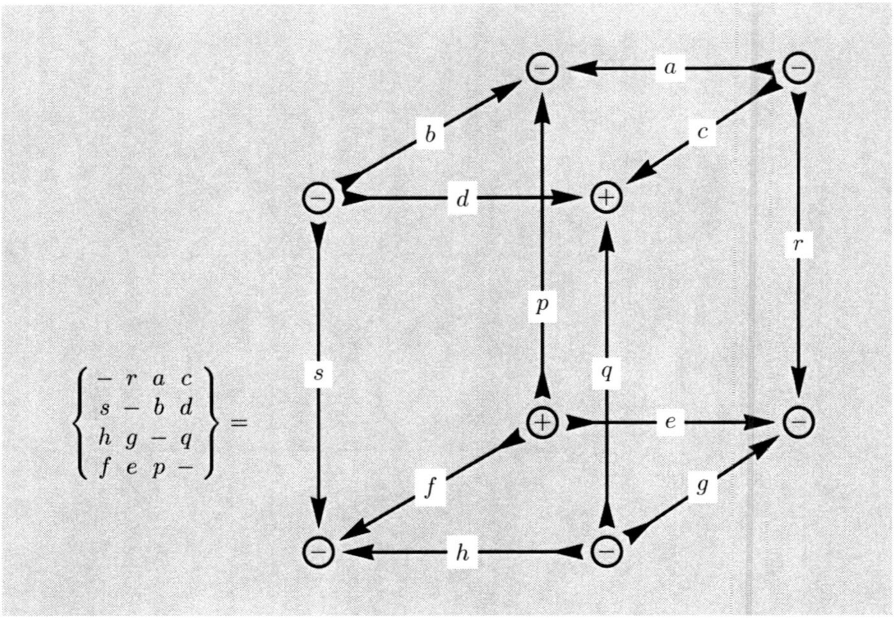

Fig. B.15. Diagram for the $12j(\mathrm{II})$-symbol (second kind) as shown in Fig. 8.45

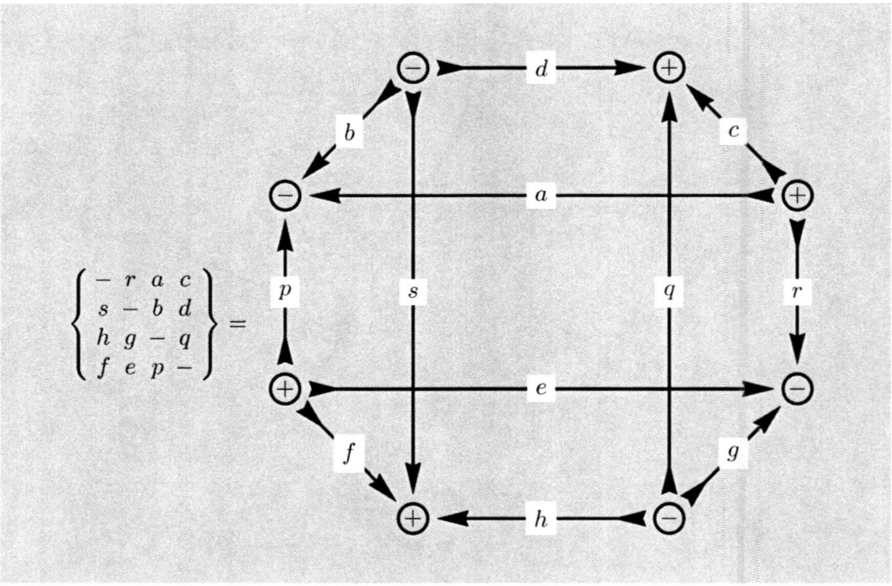

Fig. B.16. Alternative form of the diagram for the $12j(\mathrm{II})$-symbol (second kind) as shown in Fig. 8.47

B.4 Diagram Notation

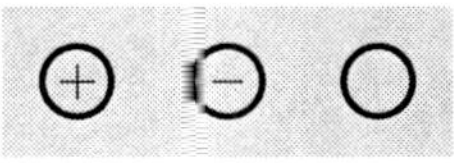

Signed nodes, unsigned node

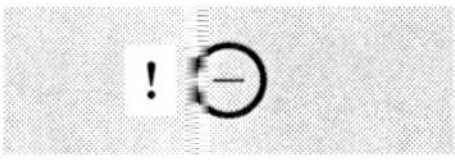

Change of node sign due to a change of the cyclic arrangement of lines is indicated by '!'

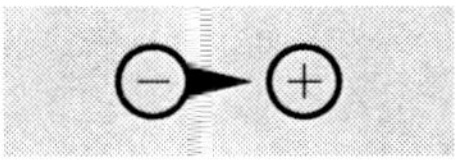

Direct change of a node sign that results in a contribution to the phase

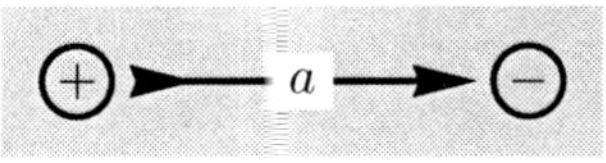

Directed node connection for angular momentum a with projection α. Summation on α is implied

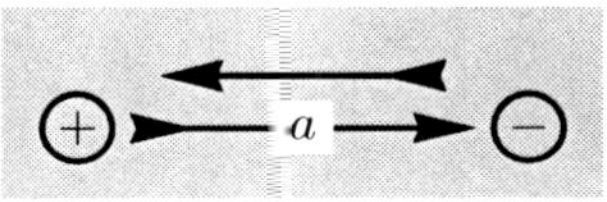

Directed node connection for angular momentum a with an antiparallel arrow indicating the intended change of direction

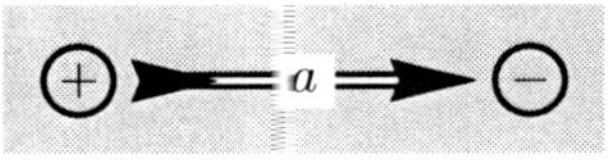

Double-line node connection indicating a sum over angular momentum a (and also α). Summation on an angular momentum a implies a factor $2a + 1$

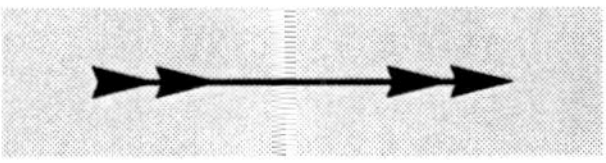

Change of line direction for an external line

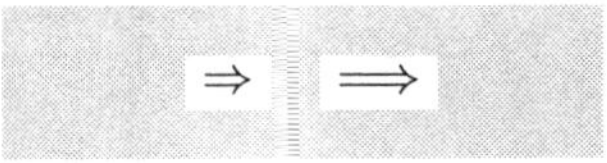

Indicator for the direction of development within one figure

C

List of Symbols

$j, j_1, j_2, j_3, \ldots$	angular momentum variables with integer and half-integer values
$m, m_1, m_2, m_3, \ldots$	projections or magnetic quantum numbers of angular momentum variables
$a, b, c, \ldots$	alternative notation of angular momentum variables
$\alpha, \beta, \gamma, \ldots$	alternative notation of projections or magnetic quantum numbers
J, M	total angular momentum and projection achieved by coupling
$(j_1 m_1 j_2 m_2 \vert JM)$	Clebsch–Gordan coefficient
$\begin{pmatrix} j_1 & j_2 & J \\ m_1 & m_2 & -M \end{pmatrix}$	$3jm$-symbol, Wigner-coefficient
$\langle j_1 m_1 \vert, \langle JM \vert, \vert j_1 m_1 \rangle, \vert JM \rangle$	bra- and ket-states with angular momentum quantum numbers
$\begin{pmatrix} & j & \\ m & & m' \end{pmatrix}$	metric tensor

$\triangle\{j_1, j_2, J\}$	triangle condition
δ_{j_1,j_2}, δ_{m_1,m_2}	Kronecker-δ symbols
$n!$	$= 2 \times 3 \ldots \times n$, factorial
$\|[(j_1, j_2)j_{12}, j_3]J\rangle$	ordered coupling-sequence
$\{j_1\ j_2\ j_3\}$	$3j$-symbol with value $1_{\triangle\{j_1,j_2,j_3\}}$
$\begin{Bmatrix} j_1 & j_2 & j_3 \\ j_4 & j_5 & j_6 \end{Bmatrix}$	$6j$-symbol
$\begin{Bmatrix} j_1 & j_2 & j_3 \\ j_4 & j_5 & j_6 \\ j_7 & j_8 & j_9 \end{Bmatrix}$	$9j$-symbol
T^K_Q, t^K_Q, u^a_α, v^b_β	spherical tensor operators of rank $K \ldots$ and projection $Q \ldots$, rank and projection take integer values
$\langle \alpha SLJM \| T^K_Q \| \alpha' S' L' J' M' \rangle$	many-electron matrix element
α, S, L, J, M	quantum numbers specifying the initial state with spin S, orbital momentum L and total angular momentum J, projection M; note that α takes, if necessary, in this context the role of an additional quantum label
α', S', L', J', M'	quantum numbers for the final state
α^η, S^η, L^η, J^η, M^η	quantum numbers for an intermediate state
α^*, S^*, L^*, J^*, M^*	quantum numbers for a valence state
s, $\bar{l}$, $\bar{j}$, $\overline{m}$	quantum numbers for a core state
$(\alpha SLJ \\| T^K \\| \alpha' S' L' J')$	reduced many-electron matrix element in the Wigner–Eckart Theorem

$W^{(a,b)K}$	unit many-electron tensor operator for coupled states
$W^{(a,b)}$	unit many-electron tensor operator for SL-states
$(s\|u^a\|s)$	reduced single-electron spin matrix element
$(s\|v^b\|l)$	reduced single-electron orbital matrix element
$(l^n\alpha SL\{\|l^{n-1}\alpha^* S^* L^*)$	fractional parentage coefficient
$f(\mu,\mu')$	scattering amplitude for initial and final equilibrium states μ and μ', respectively
E^μ,$E^{\mu'}$	energy of initial and final state
E^η,Γ_η	energy and width of the intermediate state
$\hbar\omega$	x-ray photon energy
$\boldsymbol{\varepsilon}\cdot\boldsymbol{r_i}$	scalar product of the polarization vector $\boldsymbol{\varepsilon}$ with the position vector of the i-th electron
X_Q^K	spherical tensor constructed by coupling the initial and final polarization
$\{u_\alpha^a\otimes v_\beta^b\}_Q^K$	tensor product of rank K, projection Q of u_α^a and v_β^b

Index

Lightning Source UK Ltd.
Milton Keynes UK
13 July 2010

156909UK00007B/111/P

9 783642 031175